CARBON NANOMATERIALS

SYNTHESIS, STRUCTURE, PROPERTIES AND APPLICATIONS

CARBON NANOMATERIALS

SYNTHESIS, STRUCTURE, PROPERTIES AND APPLICATIONS

RAKESH BEHARI MATHUR
BHANU PRATAP SINGH
SHAILAJA PANDE

CRC Press
Taylor & Francis Group
Boca Raton London New York

CRC Press is an imprint of the
Taylor & Francis Group, an **informa** business

Taylor & Francis
Taylor & Francis Group
6000 Broken Sound Parkway NW, Suite 300
Boca Raton, FL 33487-2742

Version Date: 20161109

ISBN 13: 978-1-4987-0210-2 (hbk)

Library of Congress Cataloging-in-Publication Data

Names: Mathur, Rakesh Behari, 1952- author. | Singh, Bhanu Pratap (Scientist in Physics), author. | Pande, Shailaja, author.
Title: Carbon nanomaterials : synthesis, structure, properties and applications / Rakesh Behari Mathur, Bhanu Pratap Singh, Shailaja Pande.
Description: Boca Raton, FL : CRC Press, Taylor & Francis Group, [2016]
Identifiers: LCCN 2016022639| ISBN 9781498702102 (hardback ; alk. paper) | ISBN 1498702104 (hardback ; alk. paper) | ISBN 9781498702119 (e-book) | ISBN 1498702112 (e-book)
Subjects: LCSH: Nanostructured materials. | Carbon. | Nanotubes.
Classification: LCC TA418.9.N35 M52135 2016 | DDC 620.1/15--dc23
LC record available at https://lccn.loc.gov/2016022639A

Visit the Taylor & Francis Web site at
http://www.taylorandfrancis.com

and the CRC Press Web site at
http://www.crcpress.com

Contents

Preface

Exploring the science of nanostructures has become, in just a few years, a new theme common to many disciplines in which scientists and engineers are beginning to manipulate matter at the atomic and molecular scales in order to obtain materials and systems with significantly improved properties. For decades, microstructures, which are thousands of times larger than nanostructures, have formed the basis for our current technologies, including ceramics, composites, fabrication of alloys, and microelectronics. Although microstructures are small on the scale of direct human experience, their physics largely remains as much the same as that of macroscopic systems.

Carbon is the unique element in the periodic table that has isomers from zero dimensions to three dimensions. It is interesting that sp2 hybridization leads to different nanoforms of carbon: a single-atom thick two-dimensional (2D) graphite called graphene, a planar local structure in the closed polyhedra (0D) of the fullerene family, and the one-dimensional (1D) cylinders called carbon nanotubes.

Carbon nanomaterials have a unique place in nanoscience owing to their exceptional thermal, electrical, chemical, and mechanical properties, and have found application in areas as diverse as superstrong composite materials, energy storage and conversion, supercapacitors, gas storage, smart sensors, targeted drug delivery, catalyst support, field-emission devices, quantum wires, paints, and nanoelectronics. The degree of competitiveness in sports has been remarkably impacted by nanotechnology making athletes safer, comfortable, and more agile than ever. Looking at the broad spectrum of possible applications there is no doubt that the 21st century belongs to these carbon nanomaterials. This is a field in which a huge amount of literature has been generated within recent years, and the number of publications is still increasing every year.

Of late there have been several books about these carbon nanomaterials, focusing mainly on an individual or selective carbon nanomaterial, and the subject was basically treated like any other nanomaterial. This book is the first attempt to cover all the existing carbon nanomaterials, namely, carbon nanofibers, vapor-grown carbon fibers (VGCF), graphite nanoribbons (GNRs), graphene quantum dots, and amorphous nanocarbons in addition to carbon nanotubes, fullerenes, and graphene in a single volume. Departing from the conventional pattern, an attempt has been made to create a platform for the reader to understand the fascinating material carbon in its various manifestations through historical developments, structure–property relations and so on at an introductory level and finally its transition from conventional bulk forms to nanocarbons. It is noteworthy that the field of carbon nanomaterials is still a gray area of research and therefore an attempt has been made to introduce the reader to the latest developments on the fundamentals of growth mechanism and synthesis techniques along with structure property relationships of individual nanomaterials leading to its application in modern-day technological breakthroughs. More than 500 references are cited in the book for the benefit of the readers who would like to have deeper insight into the subject.

The book will be very useful not only to experienced scientists and engineers who wish to broaden their knowledge of the wide-ranging carbon nanomaterials and willing to develop practical devices, but also to graduate and senior undergraduate students who look forward to making their mark on nanoscience and technology of carbon materials. It can also be useful for research and development personnel in industries, working in the area of conventional carbon materials, to upgrade the technology using carbon nanomaterials.

There are more than 10 different carbon nanomaterials addressed in this single book, which is its unique feature. Some of the results are the authors' own innovations and experience with these nanomaterials.

Chapter 1 introduces readers to the element carbon, its existence in various forms (crystalline and amorphous), and how carbon science and technology has grown over decades and centuries. The chapter assumes special significance in familiarizing students with the general background of the structure, property, and applications of various commonly used bulk forms of carbons and graphite, including carbon black, carbon fibers, pyrolytic carbon, synthetic carbon, and graphite. Hybridization in carbon plays a vital role in deciding the physical, chemical, and transport properties of carbons. In order to make the audience conversant with all these aspects before moving to the next chapters, it was thought essential to provide a brief background of the subject on the electronic and band structure of graphite to better correlate the structure and properties of nanoforms of carbon in subsequent chapters. A background of the two very important techniques—x-ray diffraction and Raman spectroscopy—given in this chapter to characterize carbons in different forms will prove to be very handy in understanding these aspects in the subsequent chapters. Special emphasis has been paid to explain the formation of pyrolytic carbon, which forms the basis of understanding the growth of carbon nanotubes, fullerenes, amorphous, and filamentous forms of carbons in the following chapters.

Chapter 2 discusses the history and growth mechanism of one of the most important allotropes of carbon in the form of hollow, closed cages of carbon atoms, the fullerenes. The most popular and stable molecule in the family of fullerenes is the highly symmetrical soccer-ball-shaped, 60-carbon atom buckminsterfullerene, or C60, which was first detected and discovered in 1985 at Rice University during laser vaporization experiments on graphite. Ever since fullerenes have become the subject of intense scientific research, their synthesis, structure, properties, and applications in materials science and nanotechnology are extensively studied. This chapter presents historical background and an overview of the Nobel Prize-winning discovery of fullerenes. It explores the methods of synthesis, the electronic and molecular structure, and the chemical and physical properties of fullerenes. The chapter also highlights the promising applications of fullerenes in the fields of electronics, materials science, and biological systems, with special emphasis on health and personal care.

Carbon nanotubes, with their extraordinary mechanical and unique electronic properties, have garnered much attention in the past decade or so with a broad range of potential applications, including nanoelectronics, composites, chemical sensors, biosensors, and nanoelectromechanical systems (NEMS). The scientific community, therefore, is more motivated than ever to move beyond basic properties and explore the real issues associated with carbon nanotube-based applications. Chapters 3 and 4 are entirely devoted to the synthesis and applications of single-walled and multiwalled carbon nanotubes. These nanotubes are made up of sp2-bonded carbon atoms that can be understood as rolled sheets of single-layer graphite or graphene with their diameter normally lying in the nanometer range while their length often exceed microns, thus making them 1D nanostructures. The chapters provide a brief history of single-walled and multiwalled carbon nanotubes followed by discussion of the growth mechanism, structure, and synthesis routes, together with their exceptional mechanical, electrical, and thermal properties. Different techniques to characterize these carbon nanotubes are discussed, including, scanning electron microscopy, transmission electron microscopy, Raman spectroscopy, ultraviolet-visible spectroscopy, and x-ray diffraction. Potential applications and the latest efforts by various research groups to use these tubes in energy storage, molecular electronics, composites, and biotechnology are also discussed. The chapters also provide comprehensive and up-to-date literature citations.

Chapter 5 introduces to the readers various filamentous forms of nanocarbons such as vapor-grown carbon fibers (VGCFs), graphite whiskers, carbon nanofibers, and nanocoils. A brief historical approach and the processing conditions for individual nanostructures is presented with respective applications. A detailed explanation has been provided for the experimental conditions and growth mechanism, morphology, and fracture behavior. Different processing parameters to register controlled structures are introduced. Also, the role of individual materials in particular applications is explained.

Chapter 6 deals with the planar monolayer of carbon atoms known as graphene, which is the basic building block of all other graphitic forms of carbon such as zero-dimensional fullerenes, one-dimensional carbon nanotubes, and three-dimensional graphite. Theoretically known to exist for several decades, graphene was isolated from graphite in 2004 by Konstantin Novoselov and Andre Geim. Much of the interest in graphene is due to its unique electronic properties. Graphene is a zero band gap semiconductor, wherein its charge carriers effectively behave like massless relativistic particles. Today graphene is considered as a star among the carbon nanomaterials with outstanding mechanical, electrical, thermal, and optical properties, and has paved the way for a variety of potential applications in many areas, such as transistors, transparent and flexible electronics, and nanocomposites. The chapter presents the history behind the discovery of graphene and explores the methods of synthesis, the electronic and molecular structure, the physical and chemical properties, and the applications of graphene especially in the field of electronics, optoelectronics, nanocomposites, and materials science. Chapter 6 also reviews one of the most important precursor materials for graphene—graphene oxide. Oxidation of graphite by insertion of strong acids and oxidizing agents within the interlayer space of graphite yields graphite oxide. Exfoliation of multilayered graphite oxide results in a few layers or individual layers of exfoliated sheets called graphene oxide. One of the most effective methods of graphene synthesis in terms of availability, simplicity, cost, yield, and scale-up advantages is the chemical reduction of graphene oxide.

Chapter 7 introduces one of the latest carbon nanomaterials: graphite nanoribbons (GNRs). Why and how GNRs can revolutionize the electronics industry in the 21st century is explained. A brief account behind the three different approaches to produce GNRs—bottom-up, lithography, and longitudinal unzipping of carbon nanotube (CNT) are introduced. Due to their outstanding properties, GNRs are the major candidate to become the silicon of the 21st century and open the era of carbon electronics. The science behind engineering the structure of GNRs for efficient field-effect transistors (FETs) is discussed at length. Narrowing the channel width in FETs, with a high heat dissipation rate and high current density play an important role together. The chapter also describes various other applications of GNRs such as in green energy generation and catalyst support.

An alternative, yet remarkable technique to introduce band gaps in graphene is to geometrically confine the size of graphene by producing nanometer-sized pieces of graphene, called graphene quantum dots. A well-confined shape, which exhibits pronounced quantum confinement and edge effects, leads to novel electrical and optical properties. Chapter 8 is primarily focused on the several synthesis routes of graphene quantum dots followed by different groups. The chapter also discusses the dependence of its optical behavior on its size and how its light-emitting powers have been put to good use as smart, luminescent materials with applications in biology as biological tracking agents, biomarkers, and smart sensors.

Chapter 9 introduces various amorphous nanoforms of carbon. We are aware of carbon black as an amorphous form of nanocarbon, which is fluffy and lightweight. However, carbon black itself can manifest into different forms, such as carbon black, thermal black, acetylene black, and lamp black. Carbon nanospheres and nano-onions are other forms of polycrystalline forms of carbon. Each form has a different synthesis route with a different structure property relationship. The chapter describes structural analysis of different forms of all these materials following Raman microscopy, electron microscopy, and x-ray diffraction. An effort has been made to explain how the properties of individual nanostructures have been exploited in different applications over a period of several decades.

It is the hope of the authors that the present book has brought together a large body of material in a very coherent fashion that will stimulate further advances in the fascinating field of science and technology of carbon nanomaterials, both in terms of fundamental aspects and engineering applications.

I would like to take this opportunity to pay my gratitude to CSIR-National Physical Laboratory (New Delhi, India), which provided me the platform to build my career on such a fascinating material, carbon. Special thanks to my colleagues with whom I have been associated for almost four decades and contributed to the advancement of carbon science and technology of carbon materials that has motivated me and my colleagues, Dr. Bhanu Pratap Singh and Dr. Shailaja Pande, to write this book. Above all I want to thank my wife, Kumkum, for her perseverance, support, and encouragement throughout the course of writing this book.

Finally, the authors would like to acknowledge the dynamic help and support from Ms. Aastha Sharma, senior acquisitions editor, CRC Press/Taylor & Francis Group, in completing this project on schedule.

Rakesh Behari Mathur

Authors

Dr. Rakesh Behari Mathur is scientist emeritus in the department of physics and the engineering of carbon, CSIR-National Physical Laboratory, New Delhi, India. He has been instrumental in developing and commercializing several carbon-based technologies, 160 research publications, and 12 patents during his more than 38 years of active research in the field of carbon science and technology. Dr. Mathur has worked in several reputed institutes and laboratories in Germany, France, the United Kingdom, the United States, Russia, and Australia, and has established active scientific collaborations with them.

Dr. Bhanu Pratap Singh is associated with the CSIR-National Physical Laboratory, New Delhi, India, as a senior scientist in physics and the engineering of carbon. He specializes in production and application of carbon nanotubes, graphene, and polymer composites for structural, electromagnetic interference (EMI) shielding, and energy applications, and has authored more than 80 publications.

Dr. Shailaja Pande earned her MSc (chemistry) and PhD (solvent extraction of coal) from the Indian Institute of Technology, Delhi. She was associated with the carbon group at CSIR-National Physical Laboratory, New Delhi, India, as a principal investigator under the Department of Science and Technology, Ministry of Science and Technology Women Scientists Scheme (WOS-A), and worked on the development of carbon nanotubes/polymer network composites for electromagnetic interference shielding and structural applications. She has written book chapters on polymer nanotubes and nanocomposites.

1

Introduction to Carbon and Carbon Nanomaterials

1.1 Introduction

Carbon is the fifteenth most abundant element in the earth's crust, and the fourth most abundant element in the universe by mass after hydrogen, helium, and oxygen. Carbon is also present as a major component in large masses of carbonate rocks, including limestone, dolomite, and marble. Coal is the largest commercial source of mineral carbon, accounting for nearly 8 billion tons or almost 80% of fossil carbon fuel. In its elemental form, carbon (C, atomic number 6) has a valency of 4 and is therefore placed in group IV of the periodic table along with Si, Ge, Sn, and Pb.

Carbon can exist in both crystalline and amorphous forms. Figure 1.1 shows the three crystalline allotropes of carbon, that is, with same chemical properties but with different physical forms. All carbon allotropes are solids under normal conditions, with graphite being the most thermodynamically stable form. In reality all the allotropes of amorphous carbons are made of microcrystals of graphite arranged in an irregular fashion. Diamond, graphite, and fullerenes are crystalline allotropes of carbon. The carbon atoms in diamond have a three-dimensional (3D) tetrahedral network of covalent bonds, which causes the electrons to be held tightly. Diamonds are therefore very hard and have high melting and boiling points. The structure is a closely packed structure and causes diamond to be denser than graphite. Since all its electrons end up in forming the covalent bonds, therefore it does not conduct electricity. In graphite only three of the four valence electrons of each carbon atom are used in bonding, leaving the fourth valence electron as free. Therefore graphite is a good conductor of electricity.

Diamond is transparent because it has no free electron to absorb radiations and make a transition in the optical region. Graphite on the other hand has one free electron, which can absorb all the radiations in the optical region and thus appear black. Figure 1.2 shows the two contrasting physical forms of naturally occurring allotropes of carbon, which are used for many industrial applications.

Recently fullerenes have been discovered as the third allotrope of carbon. They are molecular compounds made entirely of carbon atoms in the form of hollow spheres (C_{60}), ellipticals (C_{70}), and tubes. Spherical fullerenes are sometimes called buckyballs, and cylindrical fullerenes are called buckytubes or nanotubes. The fullerenes might have applications in synthetic cosmetics, pharmaceuticals, and organic photovoltaics.

Carbon that is deposited on the walls of the reactor during the destructive distillation of coal is called gas carbon and results into coal tar pitch. Carbon that is deposited on the walls of the distillation tower during the refining of crude petroleum is called petroleum coke. Both forms are used for making electrode in dry cells and are good conductors of electricity.

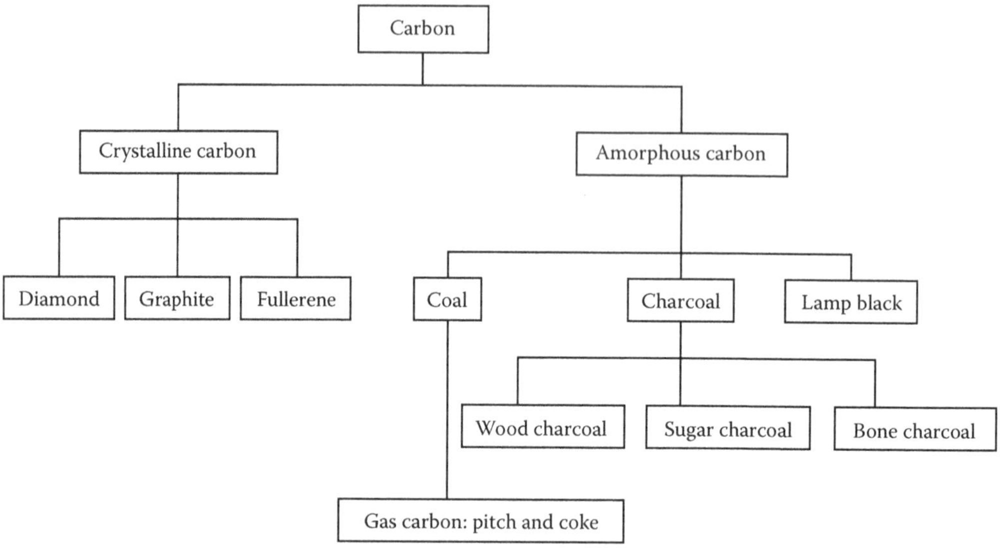

FIGURE 1.1
Various physical forms of carbon.

FIGURE 1.2
Physical forms of graphite and diamond crystals.

Wood charcoal is made by the destructive distillation of wood. It is a black, porous, brittle solid that is a good adsorbent, and good filtering and reducing agent.

Sugar charcoal is obtained by heating sugar in the absence of air. It is the purest form of amorphous carbon. It can be converted into activated charcoal to increase its adsorption capacity.

Animal charcoal is also called bone char. It is obtained by destructive distillation of bones. It contains about 10% to 12% of amorphous carbon.

Carbons have always been the friendliest materials to mankind. Charcoal from wood was the primitive artist's crude element for expression and communication as mural on cave walls. A candle burning with yellow flame would produce soot on an inverted plate and the black deposit known as soot (lamp black) when mixed with suitable base (water, oil) created pigments.

The advent of 20th century saw widespread use of graphite crystals as the writing points for several centuries and has now taken the form of lead in pencils. Printing formed the first major use of carbon in manufacturing technologies, many of which are still in use today.

Starting in 1940, controlling the processing parameters together with the variety of choices from carbonaceous materials, individually or in combination led to the development of advanced carbon products with contrasting properties. Examples include carbons that possess the highest thermal conductivity (carbon nanotubes, graphene, carbon–carbon composites), but at the same time are thermal insulators (carbon fiber felt/porous carbons), completely impervious to gases and fluids (glassy carbon with no open porosity) to highly porous carbons (activated carbon, surface area ~2000 m^2/g), very soft (graphite) and hardest (diamond), very low coefficient of friction (natural graphite, expanded graphite) to highly abrasive (hard carbons), low strength (monoliths) to strongest material (carbon fibers, carbon nanotubes, graphene). Because of such versatility, today carbon has entered into applications ranging from conventional carbon/graphite electrodes, brushes, clutch plates, mechanical seals and catalyst support, to highly advanced areas like carbon–carbon composites for space re-entry vehicles, aircraft brake discs, and carbon fiber composites for lightweight, high-strength structures.

The chapter is aimed to provide to the reader a brief idea about the basic structure of carbon material in the bulk form and how the discovery of the nanostructured forms of carbon has opened up use of carbon in newer applications. Since the majority of carbon materials are a result of sp^2 hybridization, the structure is explained on the basis of natural or single-crystal of graphite. This will form the basis of the understanding of electronic and band structure of nanoforms of carbon presented in the following chapters. X-ray diffraction and Raman spectroscopy are two important characterization tools that can give a lot of insight into the structure of various forms of carbon and are widely used by researchers to understand the novel and newer forms of nanocarbon. An introductory reference is also provided to x-ray diffraction and Raman spectroscopy used to characterize the structure of various forms of carbon and should help the reader in understanding the difference when we move from bulk to carbon nanomaterials.

1.1.1 Atomic Structure of Carbon

Carbon is a chemical element with the symbol C and atomic number 6 and represented by the electronic orbital structure illustrated in Figure 1.3. As a member of group IV of the periodic table, it is obtained in three naturally occurring isotopes, with ^{12}C (6 protons, 6 neutrons, nuclear spin I = 0) and ^{13}C (6 protons and 7 neutrons, nuclear spin I = 1/2) being stable, whereas ^{14}C (6 protons and 8 neutrons) is radioactive, decaying with a half-life of about 5700 years. The carbon-12 (^{12}C) isotope forms almost 99% of the carbon on earth, while carbon-13 (^{13}C) forms the remaining 1%. Though ^{14}C is only found in traces (~10^{-12} of all carbon atoms), due to its long half-life corresponding to reasonable life scale in human history, measurement on ^{14}C concentration in organic material (e.g., wood) allows one to date its biological history; this is called carbon dating.

A carbon atom with six electrons has the electron configuration $1s^2 2s^2 2p_x^1 2p_y^1$ in its ground state (lowest energy state), as shown in Figure 1.4. Note that there is a large energy difference between the 1s and 2s orbital and a very small difference in energy between 2s and 2p orbitals. The $1s^2$ orbital contains two strongly bound core electrons. Four more weakly bound electrons occupy the $2s^2 2p^2$ valence orbitals. In the crystalline phase, the

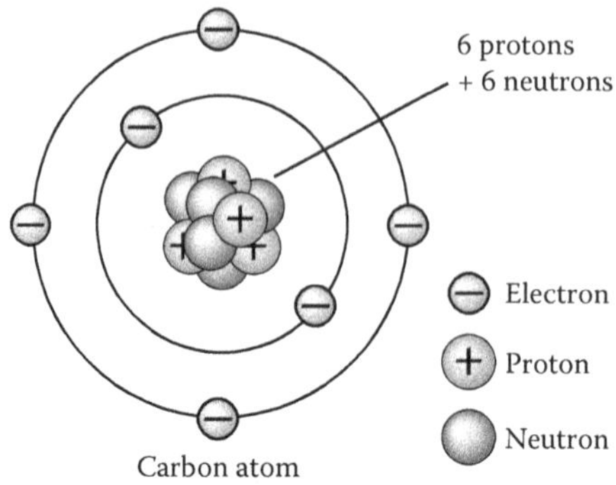

FIGURE 1.3
Orbital representation of carbon atom.

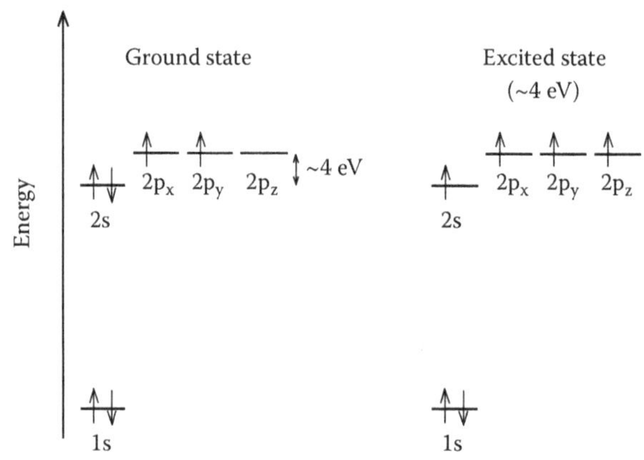

FIGURE 1.4
Electronic configuration for carbon in the ground state and in the excited state.

valence electrons give rise to 2s, $2p_x$, $2p_y$, and $2p_z$ orbitals, which are important in forming covalent bonds within carbon materials. Because the energy difference between the upper 2p energy levels and the lower 2s in carbon is small (~4 eV) compared to the binding energy of the chemical bonds, the electronic wave functions for these four electrons can readily mix with one another, thereby changing the occupation of the 2s and three 2p atomic orbitals, so as to enhance the binding energy of a C atom with its neighboring atoms. The general mixing of 2s and 2p atomic orbitals is called hybridization, whereas the mixing of a single 2s electron with one, two, or three 2p electrons is called sp^n hybridization with $n = 1, 2, 3$.

The vertical arrows in Figure 1.5 represent the direction of spin of an electron in the particular energy state. Notice that the sixth electron is placed in the $2p_y$ orbital rather than being paired with the fifth electron in the $2p_x$ orbital. This is the configuration resulting from **Hund's rule**, which states that the most stable arrangement of electrons in orbitals of equal energy is the one with the greatest number of parallel spins. The $2p_x$ and $2p_y$ electrons in the carbon atom are considered to have parallel spins, whereas if they both occupied the $2p_x$ orbital they would have to have opposite spins.

Carbon forms an excited state by promoting one of its 2s electrons into its empty $2p_z$ orbital and hybridized to the excited state. In the excited state configuration, carbon will

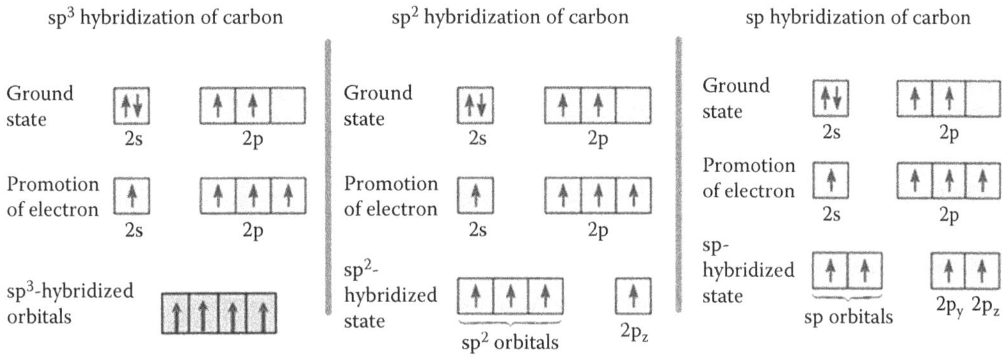

FIGURE 1.5
Hybridization of orbitals of a carbon atom.

be able to form four bonds of equal energy. Since both the 2s and the 2p subshells are half-filled, the excited state is relatively stable.

The reason why carbon is the only element that can make long chains is not only its lower atomic number or atomic mass but also the ease with which it can undergo different hybridization states due to its electronic structure.

1.1.2 Carbon Hybridization

Mixing of two or more atomic orbitals to form a new set of hybrid orbitals is called hybridization. The element carbon may form virtually an infinite number of compounds. Most plastics, pharmaceuticals, and food contain long chains of carbon atoms. This is largely due to the types of bonds carbon can form and the number of different elements it can join in bonding. Carbon may form single, double, and triple bonds, and the hybridization may lead to four single bonds (sp^3), two single and one double bond (sp^2), and one single bond and one triple bond (sp), as shown in Figure 1.5.

In sp^3 hybridization, one electron from the s orbital moves to the empty p orbital to give four hybridized orbitals of equivalent bonds (energy) of equal strength. In sp^2 hybridization, one p orbital does not combine with the other three orbitals and one electron is left, and the orbital looks like a lobe and in plane which is perpendicular to the other three hybridized orbitals. Together with other hybridized orbitals it makes a double bond, called the π-bond. In sp hybridization a similar process takes place, but instead two p orbitals stitch around to form a triple bond. sp^3 has the highest electron density, sp^2 lesser, and sp the least electron density. Every first bond made by a carbon atom is a σ bond, followed by π-bond. The σ bond is always formed by hybridized orbitals, whereas the π-bond is formed through leftover p orbital. A single bond can rotate, whereas a double bond cannot. Note that whereas all σ bonds undergo overlapping between the nucleus, the π-bonds overlap above and below the nucleus. A double bond is stronger than a single bond, but not double the strength; likewise a triple bond is also stronger than a double bond.

1.1.3 Diamond Structure

It is well known that carbon atoms can form chemical bonds by rearranging the configuration of the outer electrons with hybridization of 2s and 2p electrons. One possible hybridization scheme is found in diamond where four tetrahedral bonds are formed, each at an

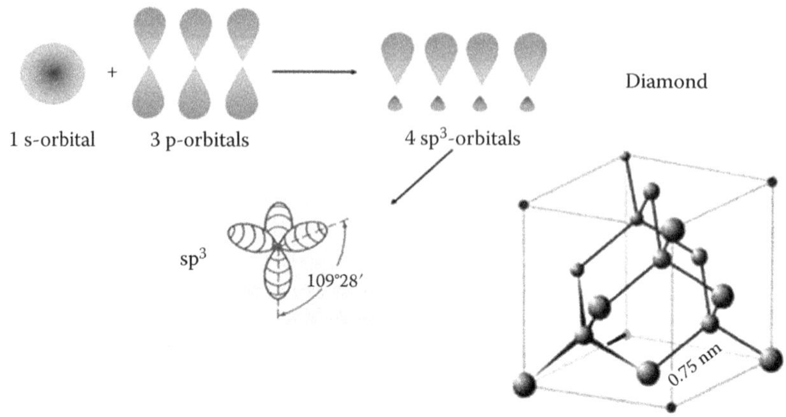

FIGURE 1.6
Illustration of the electron charge density for a carbon atom with sp^3 hybridization and crystal structure of diamond, showing a cubic cell.

angle of 109°28′. These very strong covalent bonds consist of hybridization of one s state and three p states (sp^3 bonding). The overall symmetry is cubic, as shown in Figure 1.6.

1.1.4 Graphite Structure

Another possible hybridization scheme for the bonding electrons occurs if the carbon atoms arrange themselves in hexagonal rings. Each layer is itself a giant molecule with very strong covalent bonds called graphene sheets. This is the case with graphite and aromatic molecules. Orbitals composed of one 2s and two 2p orbitals ($2p_x$ and $2p_y$ bonds) are formed. Each carbon atom in graphite is thus sp^2 hybridized. These form the strong covalent bond between carbon atoms in the plane and are known as σ bonds, as shown in Figure 1.7. The three sp^2 hybrid orbitals will arrange themselves in three-dimensional

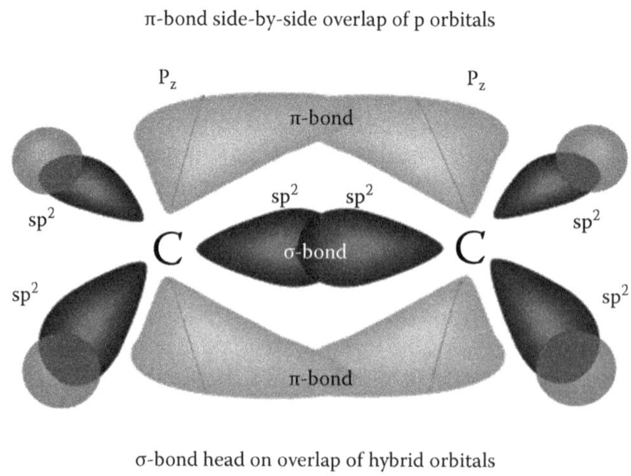

FIGURE 1.7
Schematic of the sp^2 hybridization of atomic orbitals in graphite with sigma and pi bond overlap in the graphite lattice.

space to get as far apart as possible. The geometry that achieves this is trigonal planar geometry, where the bond angle between the hybrid orbitals is 120°. The unmixed pure p_z orbital is perpendicular to this plane, called the π-orbital and provides the weak bonding between adjacent layers in the graphite structure. This weak interplanar bonding is sometimes referred to as van der Waals bonding.

The overlap of the π-orbitals on adjacent atoms accounts for the carbon–carbon double bond network in a given plane and is responsible for the high mobility of electrons in graphite. The p_z electron between the layers is delocalized (it is a free electron) and spreads uniformly over all carbon atoms and resonates between the valence bond structures. These π-electrons between the layers allow graphite to conduct electricity and heat. The layers can easily slide over each other making graphite

- Soft and slippery and an excellent lubricant (like oil)
- Account for the tendency of graphite materials to fracture along planes
- Cause the formation of intercalation or charge transfer compounds of graphite

1.2 Crystal Structure of Graphite

In the stable hexagonal layered structure, the carbon atoms are arranged in a honeycomb lattice with C–C bond separation of 0.142 nm, as shown in Figure 1.8 [1]. The ideal hexagonal graphite structure described earlier is composed of theoretically infinite basal planes with perfect ABAB stacking, with no defects whatsoever. The sheets align such that their two-dimensional hexagonal lattices are staggered in an ABAB pattern with lattice spacing between two identical sheets C_0 = 6.70 Å. The primed atoms in layer A on one graphene

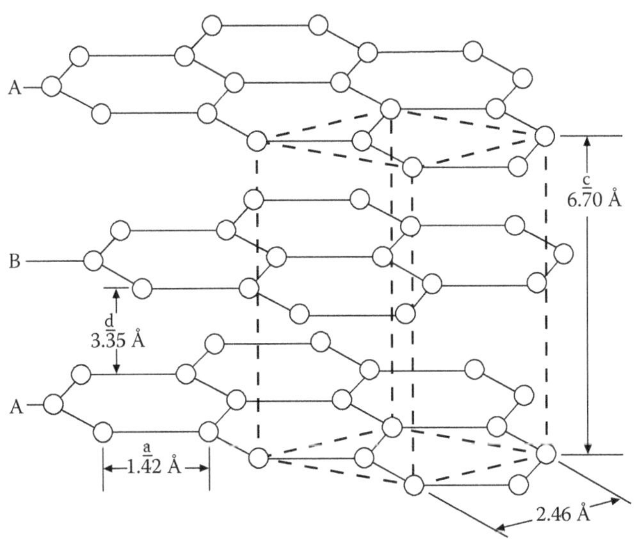

FIGURE 1.8
The hexagonal crystal structure of graphite. (From J. Bernal, *Proceedings of the Royal Society of London. Series A, Containing Papers of a Mathematical and Physical Character* 106, no. 740 (1924): 749–773.)

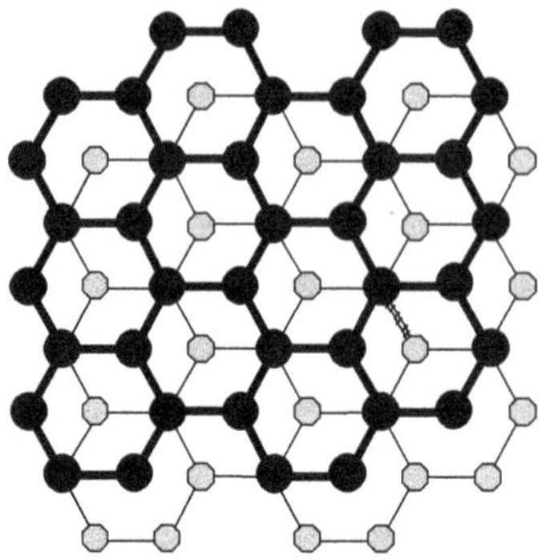

FIGURE 1.9
The crystal structure of graphite showing two adjacent hexagonal layers (top view).

layer are separated by half the orthogonal lattice spacing from the adjacent layer B. Figure 1.9 shows the positional relationship between two identical graphene planes A and B, and the graphite structure can be described as an alternate succession of these planes. The distance between adjacent layer planes (A and B) is 3.35 Å. Strong chemical bonding forces exist within the intraplanar or layer planes (150–170 kcal/g atom). However, the bonding energy between the planes is only about 2% of that within the plane (1.3–4 kcal/g atom). The dashed lines in Figure 1.8 form the three-dimensional unit cell of direct graphite lattice with an in-plane unit cell vector of length 2.46 Å.

Crystal density is 2.254 g/cc as compared to 3.53 g/cc for diamond with sp^3 hybridization. Such an ideal structure is, of course, never found, either in natural or synthetic graphite. By setting aside the interplanar interactions, the graphite may be treated to a first approximation as a two-dimensional lattice of a single graphene sheet.

The other form of graphite, called a metastable state, is sometimes present in a rhombohedral structure and has very similar physical properties, except the graphene layers stack slightly differently in an ABC ABC configuration, wherein layer C is displaced by the same amount with reference to layer B [1]. The hexagonal graphite may be either flat or buckled. The hexagonal form can be converted into rhombohedral form through mechanical treatment, and the rhombohedral form reverts to the hexagonal form when it is heated above 1300°C. Studies have shown that natural graphite may contain 15%–20% of the slightly less stable rhombohedral structure.

1.2.1 Two-Dimensional Lattice Structure: Unit Cell

As already mentioned in the last section, the carbon atoms in two-dimensional graphene condense in a honeycomb lattice. The honeycomb lattice is not a point lattice because two neighboring sites are not equivalent and atoms A and B have to move in different directions to reach the other nearest atom. As illustrated in Figure 1.10a one may view the honeycomb lattice as a triangular Bravais lattice with a two-atom basis (A and B) [2]. The distance between any two nearest carbon atoms is 0.142 nm, which is the average of the single (C–C) and double (C=C) covalent bonds, as in the case of benzene.

The unit cell, S_g, is spanned by the two vectors, $\mathbf{a}_1$ and $\mathbf{a}_2$, and contains two carbon atoms [2]. The basis vectors $(\mathbf{a}_1,\mathbf{a}_2)$ have the same length, $\mathbf{a} = \sqrt{3}\ a_{C\text{-}C} = 2.46$ Å, where $a_{C\text{-}C} = 1.42$ Å (interatomic distance) and form an angle of $\pi/3$. The area of the unit cell, S_g, is $A_g = 3a^2/2$.

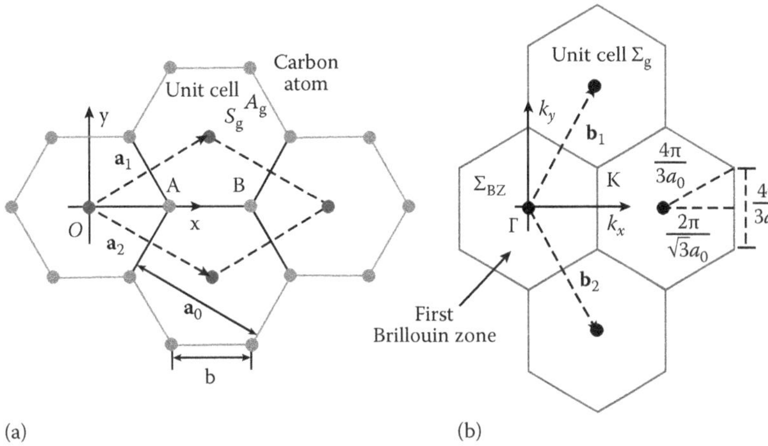

FIGURE 1.10
Honeycomb or direct lattice of graphite single layer, the structure of graphene. (a) Bravais lattice and (b) reciprocal lattice. (From A. Maffucci and G. Miano, *Applied Sciences* 4, no. 2 (2014): 305–317.)

The primitive unit cell can be considered as equilateral parallelogram with side a = 0.246 nm, where **a** is the lattice constant of graphene.

1.2.2 Reciprocal Lattice and Brillouin Zone

Every crystal structure has two lattices associated with it: the crystal lattice and the reciprocal lattice. An x-ray diffraction pattern of a crystal is a map or image of the reciprocal lattice of the crystal. Vectors in the direct lattice have the dimensions of length, whereas vectors in the reciprocal lattice have the dimension of the reciprocal of length (1/length). In fact the reciprocal lattice is the Fourier transform of the direct lattice so that every position in Fourier space may be described as a wave. It is much easier to describe the motion of an electron as a wave of wavelength λ and represent the lattice in terms of wavelength instead of lattice parameter "a." The advantage is that it is the motion of charge carriers, knowledge of which is essential for predicting the transport properties of materials. In a direct lattice the primitive cell contains all information of the physical and mechanical properties of the material. In a reciprocal lattice all information of wave motion can be understood in a Brillouin zone or Weigner Seitz cell. The length scale in BZ is smaller, i.e., reciprocal of "a." The motion of the electron is represented by the parameter $\mathbf{k} = 2\pi/\lambda$, known as wave vector. It gives the number of waves in length 2π. Further, each plane in the direct lattice is represented by a vector or a point in the reciprocal lattice. The d-spacings of the planes in the direct lattice is therefore related to the reciprocal lattice vector as $d_{hkl} = 1/|\mathbf{H}_{hkl}|$, where **H** is the reciprocal lattice vector.

In the reciprocal *k*-space depicted in Figure 1.10b, the 2D graphene lattice is characterized by the unit cell, spanned by the two vectors, $\mathbf{b}_1$ and $\mathbf{b}_2$, which have the same length $\mathbf{b}_1 = \mathbf{b}_2 = 4\pi\sqrt{3}a$ and form an angle of 2π/3. The area of the unit cell is given by $B_g = 8\pi^2\sqrt{3}\,a^2$.

The basis vectors of the direct space ($\mathbf{a}_1$,$\mathbf{a}_2$) and the basis vectors of the reciprocal space ($\mathbf{b}_1$,$\mathbf{b}_2$) are related by $\mathbf{a}i\cdot\mathbf{b}j = 2\pi\delta ij$ with $i,j = 1,2$ and the areas, A_g and B_g, are related by $A_gB_g = (2\pi)^2$.

The reciprocal lattice of graphene shown in Figure 1.10b is also a hexagonal lattice but rotated 90° with respect to the direct lattice. Physically, all sites of the reciprocal lattice represent equivalent wave vectors.

1.3 Electronic Band Structure of Graphite

1.3.1 Energy Dispersion Relation

The electrical and magnetic properties of graphite are conveniently explained on the basis of its band structure based on a two-dimensional model of the lattice since the spacings of the graphite layers is 0.335 nm, which is very large as compared to the separation to the individual carbon atoms in the hexagonal layers (0.142 nm). Therefore it appears reasonable to suppose that the electronic properties of two-dimensional graphite (neglecting interplanar effect) would be good approximation for the properties of the three-dimensional graphite. Moreover, the large anisotropies of the electrical conductivities and magnetic susceptibilities in and perpendicular to basal planes of graphite make the two-dimensional model appear as an adequate approximation.

The two-dimensional unit cell for graphite contains two carbon atoms, each carbon atom having four valence electrons. A tight binding model yields eight bands: four valence bands and four conduction bands. The two 1s electrons on each atoms are assumed to have negligible effect on the bonding because of their orbital localization. Of the four remaining valence electrons on each carbon atom, three sp^2 hybridized orbitals form the σ valence bands and σ* conduction bands. The remaining p_z orbitals of the carbon atoms, oriented perpendicular to the sheet (z-direction), form the π-valence band and π*-conduction band. Thus, considering the unit cell, there will be six σ-type energy bands and two π-type bands.

The four occupied bands are referred to as σ-valence bands and the remaining four are known as σ* conduction bands. Therefore the problem of the band structure of graphite sheet conveniently splits into a problem for the σ and σ* bands, and a problem for the π and π* bands. This approximation is apparently justified by the fact that the π-bands are more important than σ bands when considering electronic transitions near Fermi energy.

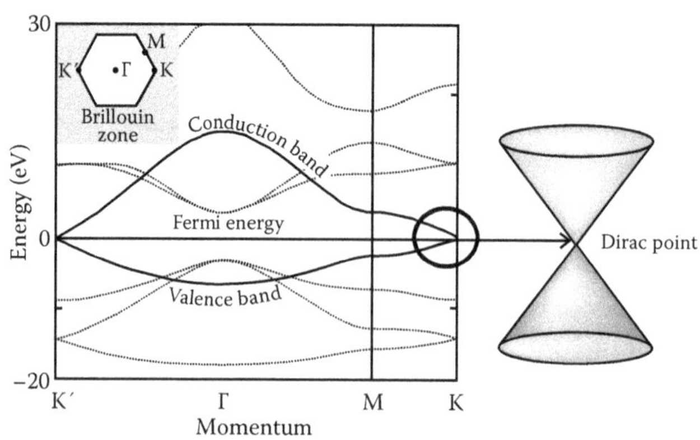

FIGURE 1.11
The Bloch band description of a single layer of graphene electronic structure; orbital energies depend on the momentum of charge carriers in the crystal Brillouin zone (inset left). The π and π* bands are decoupled from the σ and σ* bands and are closer to the Fermi energy. The bands form conical valleys that touch at two of the high symmetry points, conventionally labeled K and K′ in the Brillouin zone. Near these points the energy varies linearly with the magnitude of the momentum measured from BZ corners. (Reprinted with permission from Andrey K. Geim and Allan H. Macdonald, Graphene: Exploring carbon flatland, *Physics Today*, August, 36, Fig. No. Box-1. Copyright 2007, American Institute of Physics.)

This separation of the π and σ bands in the neighborhood of Fermi energy is shown in Figure 1.11 [3].

The resulting band structure of graphene is characterized with π and π* bands degenerate at the K points in the hexagonal Brillouin zone of the sheet [3]. This degenerate energy level corresponds to the Fermi energy, which shows that the graphene sheet is a zero gap semiconductor with a linear E versus K relationship (photon like spectrum) around the Fermi energy at the K point.

In the neighborhood of each Fermi point, the energy dispersion relation is linear and may be approximated as

$$E^{(\pm)} \approx \pm\hbar v_F|\mathbf{k} - \mathbf{k}_0| \tag{1.1}$$

where $\mathbf{k}_0$ is the wavenumber at a Fermi point, $v_F \approx 0.87 \times 10^6$ m/s is the Fermi velocity of the π-electrons, and h is the Planck constant. Such behavior of electrons in graphene makes it different and unique than any other form of carbon. In general, the band overlap increases with increasing number of layers. From 10 layers onwards the difference with the band overlap in bulk graphite (~40 meV) is less than 10%. There is also remarkable difference in the wave functions at the K point for the levels around the Fermi energy between a single graphene layer and system consisting of more graphene layers.

1.3.2 Density of States

The electronic density of states for graphite is illustrated in Figure 1.12. There is a large band gap between the σ orbitals forming bonding and antibonding bands [4]. However,

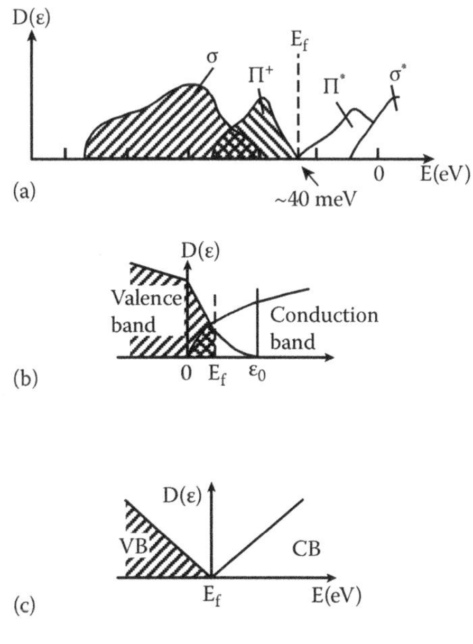

FIGURE 1.12
(a) Sketch of the density of states for graphite. Zero energy is taken at the vacuum level. (b) Density of states for 3D graphite near the region of band overlap. Zero energy is taken at the bottom of the conduction (pi*) band. (c) Density of states for a single sheet of carbon atoms (2D graphite) near the band touching region. (Adapted from I.L. Spain, Electronic transport: Properties of graphite, carbons, and related materials, *Chemistry and Physics of Carbon* 16 (1981): 119–304.)

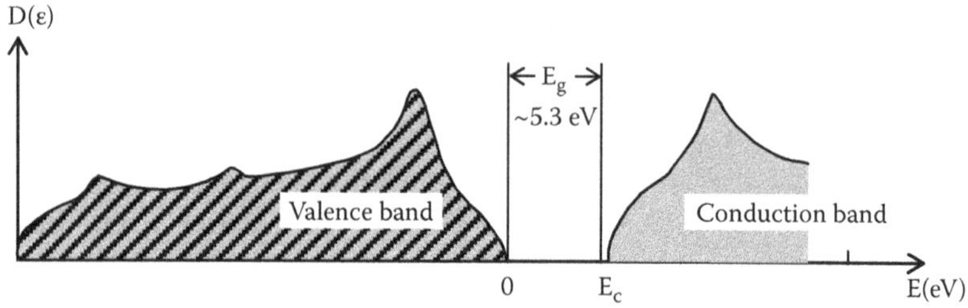

FIGURE 1.13
Theoretical sketch of the density of states of diamond. Zero energy is taken at the top of the valence or bonding band. (Adapted from I.L. Spain, Electronic transport: Properties of graphite, carbons, and related materials, *Chemistry and Physics of Carbon* 16 (1981): 119–304.)

the π-bands, which just touched at K point for two-dimensional graphite (a single layer of carbon atoms called graphene shown in Figure 1.11), overlap by ~40 meV for three-dimensional graphite, as illustrated in Figure 1.12a. This small overlap of bands makes the graphite semimetal. The density of states near the region of the overlap of the π-band (valence band) and π*-band (conduction band) is illustrated in Figure 1.12b. Assuming N number of atoms in the crystal, there are 4N bonding electrons, 3N of which are in σ band. If the π-bands just touched, as in the 2D graphite (Figure 1.12c) all of the N remaining electrons would lie in the valence band at zero Kelvin and the conduction band would be empty. However, in 3D graphite, few of the electrons ($\sim 10^{-4}$N) lie in the conduction band, leaving exactly the same number of empty states, called holes, in the valence band. Any deviation from the ideal graphite lattice can cause a change in the electronic structure to occur as the lattice becomes more defective resulting in change in the transport properties. Figure 1.12d shows the shift in Fermi level of boron doped graphite.

In case of diamond crystal lattice, the bonding states are separated from the antibonding states by approximately 5.3 eV, as shown in Figure 1.13. The pure diamond is therefore a very good electrical insulator.

1.4 Physical Properties of Graphite

Some of the important properties of the ideal graphite material, that is a material that most closely corresponds to an infinitely large graphite crystal which is closest to single crystal graphite, are summarized in Table 1.1. The properties discussed next are either calculated or based on the actual properties of graphite crystals closely approaching this ideal structure. As will be seen later in the text, a wide range of materials comes under the heading of carbon or graphite and these materials often have properties that are much different from those of the ideal graphite crystal. The material properties are sometimes confused with the nomenclature "carbon" or "graphite." Moreover, being a layered structure, properties of graphite are highly anisotropic, that is, the properties of the material may vary considerably when measured along the in-plane "ab" directions or within the plane and in the "c" direction, that is, perpendicular to the planes. Such anisotropy, especially in electrical and thermal properties, can often be very useful in several applications.

TABLE 1.1
Important Physical and Chemical Properties of Graphite

Property	Graphite
Crystalline form	Hexagonal
Lattice parameters	a = 0.246 nm, C_0 = 0.6708 nm
Color	Black
Density at 300 K, 1 atm	2.26 g/cc
Heat of fusion	46.84 kJ/mol
Heat of vaporization to monoatomic gas (estimated)	716.9 kJ/mol
Sublimation point at 1 atm (estimated)	4000 K
Pauling electronegativity	2.5
Atomic volume	5.315 cm^3/mol
Triple point (estimated)	4200 K
Boiling point (estimated)	4560 K
Moh's hardness	1-2

1.4.1 Electrical Properties of Graphite

In electrical conductors such as metals, the attraction between the outer electrons and the nucleus of the atom is weak; the outer electrons can move readily and, since an electric current is essentially a flow of electrons, metals are good conductors of electricity. In electrical insulators (or dielectrics), electrons are strongly bonded to the nucleus and are not free to move.

The atomic structure of graphite is such that the highest filled valence band overlaps the lowest empty conduction band by approximately 40 meV; a few holes and electrons are always available to carry current. In pure graphite the electron and hole densities are small and equal, and the carrier effective mass is low. The delocalized π-electrons can move readily along the basal planes under the influence of electric fields. Consequently, the electrical resistivity of graphite parallel to the basal planes (ab directions) is low and the material is a relatively good conductor of electricity. In the perpendicular direction (c direction), the spacing between planes is comparatively large, and there is no comparable mechanism for the electrons to move from one plane to another, that is, normal to the basal planes. As a result the electrical resistivity in that direction is high and the material is considered an electrical insulator. In some cases, it may be 10,000 times higher than in the ab directions. Some representative data of resistivity of different forms of graphite is summarized in Table 1.2 across the range of carbon materials and compared with some other commercially useful metals and alloys. On the one hand there are just about perfect crystals of graphite, and on the other there are nearly amorphous carbons like carbon black. Also, bonding between carbon atoms may differ.

Apart from its scientific interests, the electrical conductivity of carbon-based materials is of great importance because of its industrial applications. Brushes for electrical motors are made from polycrystalline graphite because of a combination of electrical and lubricating properties. High-temperature electrodes include furnace cores and electrodes for metallurgical applications. Carbon resistors are used as electrical components, and automobile ignition cables are often constructed with carbon cores to maximize suppression of radio frequency noise. The relationship between structural and electrical characteristics give a characteristic "fingerprint" by which different forms of carbon materials can be classified.

TABLE 1.2

Electrical Resistivity of Various Forms of Carbon and Graphite along with Important Metals and Alloys

Material	Electrical Resistivity Range (mΩ-cm)
Single crystal graphite	0.03–0.1
Pyrolytic graphite/HOPG	0.020 (ab), 40 (c)
Fine grained/High density (1.8g/cc) graphite	0.9
PAN based carbon fibers (T-300)	1.7
Pitch based carbon fibers (P-120)	0.226
Glassy carbon	4.2
Diamond	$>10^{15}$
Copper	0.0017
Gold	0.002214
Silver	0.0016
Stainless steel	0.069
Aluminum	0.00265
Silicon	6.40×10^7
Silicon carbide	10^5

1.4.2 Mechanical Properties of Graphite

As mentioned in Section 1.2, the bond between atoms within the basal plane of a graphite crystal is considerably stronger than the bond between the planes with an anisotropy ratio of approximately 75. This means that the strength in the ab directions is much higher than in the c direction. The three important elastic constants in the graphite crystal are given as

C_{11} = 1060 GPa (a direction)
C_{33} = 36.5 GPa (c direction)
C_{44} = 4.5 GPa (parallel to the planes or shear planes)

Young's modulus of elasticity of the crystal varies up to two orders of magnitude with the direction. The low value of C_{44} is the reason for poor interlaminar strength due to easy shear between basal planes.

It should be noted that such values are those of the ideal graphite crystal. The mechanical properties of "real" material, such as the molded or synthetic graphite materials and pyrolytic graphite, are considerably different since they are controlled by the dominant crystallite orientation, porosity, structural defects, and other factors. The result is large variations in the range of these properties, depending on the type of graphite and its manufacturing process.

The theoretical strength of single-crystal, defect-free graphite can be evaluated from the Orowan-Polanyi equation:

$$\sigma_\tau = (E_\gamma \gamma_a / a_d)^{1/2} \tag{1.2}$$

where σ_τ is the breaking stress, E_γ is Young's modulus (corresponding to elastic constant C_{11}), γ_a is the surface energy (i.e., the energy needed to separate two planes in the crystal per unit area; γ_a = 4.2 J/m^2 for the prismatic bonds [100 plane] in graphite), and a_d is the distance between the plane that is to be separated by the applied tensile stress.

TABLE 1.3

Mechanical Properties of Various Forms of Carbon and Graphite

Material	Tensile Strength	Young's Modulus
Single crystal graphite	~100 GPa	1060 GPa (a direction), 36.5 GPa (c direction)
Fine grained/high density (1.8 g/cc) graphite	40 MPa	12 GPa
Synthetic graphite	~10–30 MPa	6.0 GPa–10 GPa
Diamond	60 GPa	1220 GPa
PAN-based carbon fibers (high strength, type II)	3.53–6.37 GPa	230–295 GPa
PAN-based carbon fibers (high modulus, type I)	3.92 GPa	580 GPa
Pitch-based carbon fibers, P-120	2.20 GPa	820 GPa
Carbon–carbon composites	700 MPa	100 GPa
Stainless steel	520 MPa	190 GPa
Titanium	1000 MPa	110 GPa
Aluminum	70 MPa	69 GPa

However, for all practical applications the experimental strength lies well below the theoretical limit because of cracks and similar flaws in the carbon materials. Then Equation 1.2 is modified by Griffith's equation:

$$\sigma_\tau = (2E_\gamma\, \gamma_a/\pi L_f)^{1/2} \tag{1.3}$$

where L_f is the length of critical flaw that initiates rupture.

Mechanical properties of different forms of graphite including some important metals and alloys is compared in Table 1.3.

1.4.3 Thermal Properties of Graphite

Like electrical conductivity the acoustic and thermal properties of graphite are highly anisotropic, since phonons propagate quickly along the tightly bound planes but are slower to travel from one plane to another. The conductivity ratio in the a and c direction is around 200, much less than electrical conductivity because phonons also contribute to thermal conductivity (□). A large variation in thermal conductivity is found in different carbon forms and may vary from <50 W/mK for polycrystalline carbons or PAN-based carbon fibers to ~5000 W/mK for SWCNT (single-walled carbon nanotube) or graphene. A strong correlation is, however, found between □ and crystalline perfection of the specific carbon material. The thermal conductivity of the graphite is dominated first by the electronic contribution, as the temperature is increased from absolute zero, up to the conductivity peak temperature T_m □.

In practical applications of the thermal conductivity, it is often conventional to use the simplified Debye formula:

$$\square = 1/3\, C_\upsilon\, \nu\, l_\phi \tag{1. 4}$$

where C_υ is the specific heat at constant volume, ν is the average phonon group velocity (i.e., the velocity of sound), and l_ϕ is the phonon mean free path. In case of highly graphitic

TABLE 1.4
Thermal Conductivity of Various Forms of Carbon and Graphite Compared to Metals and Alloys

Material	Thermal Conductivity (W/mK)
Single-crystal graphite	~3000 W/mK in ab direction and 65–95 W/mK in c direction
Pyrolytic graphite/HOPG	1700 ± 100 W/(mK) ab direction and 8 ± 1 W/(mK) c direction
Fine grained/high density (1.8 g/cc) graphite	140
PAN-based carbon fibers (T-300)	10
Pitch-based carbon fibers (P-120)	640
Glassy carbon	5.8
Carbon–carbon composites	250–350
Diamond	2000
Copper	112–226
Gold	172
Silver	242
Stainless steel	16
Aluminum	226
Silicon	149
Silicon carbide	120

materials, the heat is carried by phonons and the electronic heat contribution is only negligible. The temperature dependence of the thermal conductivity of graphite obeys a $T^{2.3}$ law for temperatures above 30 K and reaches a maximum to about 3000 W/mK around 150 K and then decreases with a further increase in temperature due to the umklapp phonon–phonon scattering process.

Thermally conductive polymer composites offer new possibilities for replacing metal parts in several applications, including power electronics, electric motors, generators, and heat exchangers. Current interest to improve the thermal conductivity of polymers is focused on the selective addition of carbon nanofillers with high thermal conductivity. Unusually high thermal conductivity makes carbon nanotube (CNT) the best promising candidate material for thermally conductive composites. However, the thermal conductivities of polymer/CNT nanocomposites are relatively low compared with expectations from the intrinsic thermal conductivity of CNTs. The challenge primarily comes from the large interfacial thermal resistance between the CNT and the surrounding polymer matrix, which hinders the transfer of phonon dominating heat conduction in polymer and CNT. Carbon–carbon composites are one of the best lightweight (density ~1.8–1.9 g/cc) high temperature thermal conducting materials and can operate at temperatures > 2000°C. They are used in heat shields for space crafts, nuclear reactors, and aircraft brake discs. Their thermal conductivity values can be tailored close to copper by using specific carbon fillers like carbon fibers, VGCF, and CNT. A comparison of the thermal conductivity of the various forms of carbon vis-à-vis other important metals and alloys is presented in Table 1.4.

1.4.4 Chemical Properties of Graphite

Pure graphite is one of the most chemically inert materials. It is resistant to most acids, alkalies, and corrosive gases. However, impurities and imperfections are almost always present to some degree in both natural and artificial graphites and often have an important catalytic effect with a resulting increase in the chemical reactivity. The anisotropy of the graphite crystal is also reflected in its chemical behavior. The graphite crystal exhibits large differences in surface energy in the different crystallographic directions; these energies amount to ~5 J/m^2 in the prismatic plane but only about 0.11 J/m^2 in the basal plane. These differences account for the different rate of reaction, that is, slow at the basal plane and rapid at the edges or at the termination of the basal planes or at defect sites, such as dislocations, vacancies, and steps. Reaction with gases or vapors occurs preferentially at "active sites," that is, the edges of the basal planes of the crystal that are either zigzag or "arm chair." Consequently, graphite materials with large crystals and few defects have the best chemical resistance. The chemical reactivity is also appreciably affected by the degree of porosity, since high porosity leads to a large increase in surface area with resulting increase in reactivity. Obviously, high surface area materials such as activated carbon are far more reactive than dense, pore-free or closed-pore materials such as glassy carbon. Therefore there can be considerable differences in reactivity between one form of carbon and another.

Carbon reacts with oxygen at temperatures around 350°C to 400°C to form gaseous products like CO and CO_2, unlike other refractory materials like oxides and carbides of metals. Many carbides form a protective layer of oxide when reacted with oxygen and thus delays oxidation.

Graphite does not react with hydrogen below 1000°C–1500°C and forms CH_4. The reaction is, however, catalyzed and may occur at 500°C–600°C in presence of transition metals. Highly graphitized carbons like natural graphite or highly oriented pyrolytic graphite (HOPG) may form intercalation compounds with strong acids and alkali metals. These species penetrate between the graphene layers by charge transfer and have many useful properties and applications.

1.5 Nongraphitic Carbons

The relative inability of some carbons to graphitize is believed to be due to the presence of sp^3 cross-linking bonds between entangled graphitic layers, as shown in Figure 1.14a. At higher temperature the inability of the planes to grow and stack is due to the tangled arrangement of the planes. Nongraphitizable carbons are mechanically hard, and are sometimes referred to as hard carbons. These are sometimes also referred to as amorphous

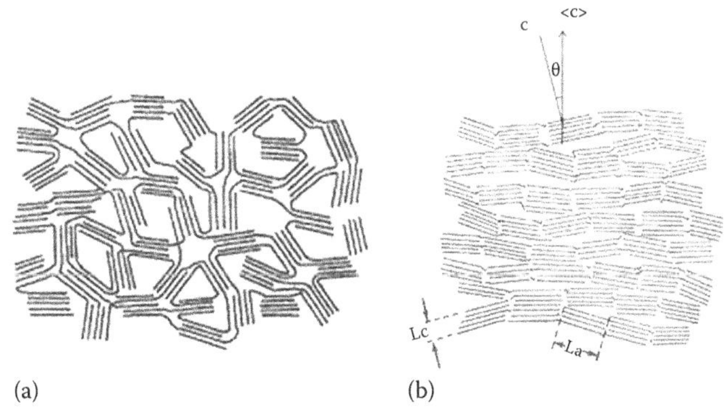

FIGURE 1.14
Schematic models for the structure of (a) nongraphitizing and (b) graphitizing carbons.

carbons. When examined by x-ray diffraction, these materials show only diffuse maxima at the normal scattering angles. This has been attributed to a random translation and rotation of the layers within the layer planes. This disorder is also referred to as turbostratic structure. It can be seen that vast differences do exist in graphitizable (preferentially parallel layer planes) microstructure with larger crystallite dimensions (see Figure 1.14b). The microstructure is directly related to raw materials and processing parameters.

1.5.1 Carbon Black

Carbon blacks are formed by the condensation of carbon from the gas phase into small spheres containing between ~10^5–10^9 atoms with no long-range order corresponding to the sphere diameter ~2–10 μm. However, aromatic layers are arranged to the surface only. Accordingly, it is difficult for the spheres to coalesce, and carbon black only partially graphitizes and is very close to a truly "amorphous carbon."

1.5.2 Coal

Nature provides an abundant supply of impure carbons in the form of coal. The carbonization process of the precursor material (rotted vegetation, peat) has taken place over a very long period of time deep inside the earth crust under conditions in which impurities cannot easily escape. They also fall into the category of nongraphitizing carbon. A wide variety of coals can be found and its quality is judged by its coal "ranking" depending upon its carbon and energy contents as well as ash contents. *Lignite,* or brown coal, is the lowest rank of coal (60%–75% carbon) and used almost exclusively as fuel for electric power generation. Lignite makes up almost half of our coal reserves. *Bituminous* coal is used primarily as fuel in steam-electric power generation, with substantial quantities used for manufacturing and to make coke. *Anthracite,* the highest rank of coal (>90% carbon), is a harder, glossy black coal, with virtually no moisture and volatile contents. It is used primarily for residential and commercial space heating because of its high-energy contents and smokeless fuel.

1.5.3 Chars

As mentioned earlier, not all aromatic hydrocarbons form coke. Some, such as phenanthrene and biphenyl, do not graphitize and are considered char formers. These compounds are branched aromatics (as opposed to the linear structure of the coke-former aromatics) with a preferred axis of growth. This characteristic prevents the formation of extensive graphitic planes and of a liquid mesophase. *Charcoal* is a traditional term used for chars obtained from wood and other organic related materials. The physical form of the parent material is retained, with highly developed pore structure. At the microscopic level, the basic structure is highly disordered with no detectable graphitic properties. Because of their large surface area and possibility to control the porosity, these are also used as activated carbons as filters and catalyst support.

1.5.4 Graphitizing Carbons

Most graphitizing carbons are mechanically soft and are sometimes referred to as soft carbons. These can be readily obtained from soft carbons, such as petroleum coke, by heating. When the heat treatment temperature is limited to about 1000°C, coke-type materials are formed. These carbonaceous materials have graphene layers of relatively small extent (about 1–4 nm), which are stacked in a roughly parallel fashion but with random rotations and translations between every pair of layers. This type of disorder is given the name "turbostratic disorder." As the carbons are heated (HTT, heat treatment temperature) from 1000°C to 2000°C, the lateral extent of the layers grows and the stacking becomes quite parallel as evidenced by a sharpening of the (002) Bragg reflection. However, at 2000°C there is still turbostratic disorder. Upon heating above 2000°C the turbostratic disorder is relieved in a more or less continuous way, the amount of remaining turbostratic disorder decreasing to zero monotonically after HTT > 3000°C.

Conceptually, graphite is a graphitic carbon with no or very little turbostratic disorder. In graphite carbon, the in-plane structure of graphene layers is almost the same as in graphite, except the lateral extent of the layers increases with heat treatment temperature.

1.5.5 Polycrystalline Graphite

The ideal hexagonal graphite structure described earlier is composed of theoretically infinite basal planes with perfect ABAB stacking, with no defects whatsoever. Such an ideal structure is, of course, never found, either in natural or synthetic graphite. Graphite materials, such as pyrolytic graphite, carbon fibers, carbon matrix composites (carbon–carbon), vitreous or glassy carbon, carbon black, chars, coke, and many other forms of carbon, are actually aggregates of graphite crystallites, in other words, polycrystalline graphites. These crystallites may vary considerably in size. For instance, the apparent crystallite size perpendicular to the layer planes (Lc) of some vitreous carbons may be as small as 1 to 2 nm, which is the length of a few atoms, or up to 100 nm found in highly ordered pyrolytic graphite (HOPG). Similarly, the length of crystallite (La) could range from 50 to 1000 nm, for highly disordered carbon to highly graphitic structure.

Figure 1.15 shows the network of twisted and folded carbon layers in a nongraphitic structure as a network of polycrystalline or glassy carbon material [5]. Interestingly this

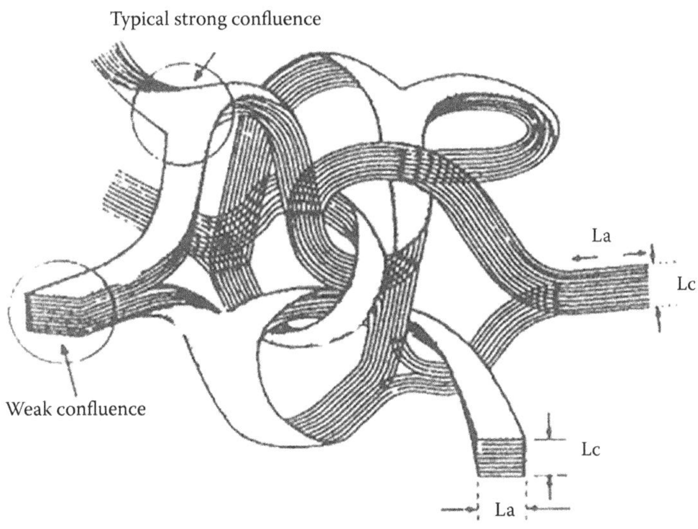

FIGURE 1.15
The ribbon model of polycrystalline or glassy carbon. (Reproduced with the permission from G. Jenkins, K. Kawamura, and L. Ban, Formation and structure of polymeric carbons, *Proceedings of the Royal Society of London A: Mathematical, Physical and Engineering Sciences,* The Royal Society, 1972, pp. 501–517.)

microstructural model of carbons introduces the concept of curvature of graphene layers observed in fullerenes and carbon nanotubes.

The layer planes may or may not be perfectly parallel to each other, depending on whether the material is graphitic or nongraphitic carbon. The aggregates of crystallites also have widely different sizes and properties. Some, such as soot, are extremely small and contain only a few small crystallites. In such cases, the properties are mostly related to the surface area. Other aggregates may be relatively large and free of defects and essentially parallel to each other, in which case the structure and its properties closely match those of the ideal graphite crystal. Such large aggregates are often found in pyrolytic graphite. In other aggregates, the crystallites have an essentially random orientation. This occurs in turbostratic (i.e., showing no evidence of three-dimensional order) or amorphous carbon. In such cases, the bulk properties are essentially isotropic. Thus in each graphitic material, the size, shape, and degree of imperfection of the basic crystallite, the general orientation of these crystallites, as well as the bulk characteristics such as porosity and amount of impurities, may vary considerably from one material to another. As a result, the properties of these various materials may show considerable differences.

These carbons can be further subdivided into graphitizing and nongraphitizing carbons. In general these carbons are classified as soft carbons and hard carbons respectively. As a guideline, soft carbons are derived from the liquid phase like coal tar or petroleum pitch, and hard carbons are derived from the solid phase. Other forms of carbon, like carbon black or pyrolytic carbon, are obtained from the gas phase.

1.6 Synthetic Carbon and Graphite

The range of materials included under this heading consists of products of pyrolysis of organic compounds at temperatures above 700°C and molded graphite. These carbons are largely made up of substantially plane condensed aromatic layers, stacked roughly parallel but may differ from natural crystalline graphite in the following important aspects:

- Interplanar spacing between d_{002} basal planes and mutual orientation of the layer planes
- Dimension of the layer planes (La) and stack height (Lc), that is, the number (N) of basal planes in the stack
- Density
- Missing carbon atoms in the layer planes
- Extrinsic atoms within the layer planes
- Defects in the layer planes, identified with the total pore size and pore volume between the stacks

These structural differences lead to considerable departure in the properties from those of single-crystal graphite in terms of electronic properties, transport properties, surface properties, and chemical reactivity. This, however, offers the opportunity to tailor the properties of carbon materials to suit the particular application.

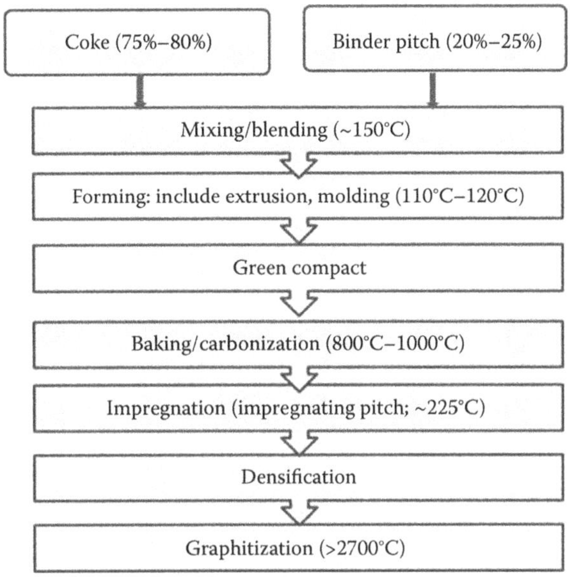

FIGURE 1.16
The major processing steps in the production of conventional synthetic graphite.

1.6.1 Synthetic Graphite: Processing, Properties, and Applications

The carbon and graphite artifacts prepared by various processes (e.g., molding, extrusion, chemical vapor deposition) can be defined as synthetic (or artificial) graphitic products. The compaction process is carried out on a mixture of carbon filler and organic binder, which is subsequently carbonized and graphitized generally up to a temperature of 1000°C and >2500°C, respectively, in a vacuum or inert atmosphere. The basic process was invented by E.G. Acheson, who produced the first molded graphite in 1896. The original applications of molded graphite were electrodes for electric arc furnaces and cinema arc carbons for movie projectors. Many improvements have been made over the years, and the applications have increased considerably in scope. Molded graphite is found in almost every corner of the industrial world and forms the base of the traditional graphite industry.

The major processing steps in the processing of a conventional polygranular graphite consists of two phases—a filler material and a binder phase—and the process is summarized schematically in Figure 1.16.

1.6.2 Filler

Carbons heat-treated to a condition where they are relatively pure are often referred to as cokes. The predominant filler material is petroleum coke made by the delayed coking process or pitch coke derived from coal tar.

Petroleum coke is the filler of choice in most applications. It is a porous by-product of the petroleum industry and an almost-pure solid carbon at room temperature. It is produced by destructive distillation without the addition of hydrogen, either by a continuous process (fluid coking) or, more commonly, by a batch process (delayed coking). The product of the delayed coker is green coke, which has seen a maximum temperature of about 450°C and contains a high percent of volatile materials. This green coke is then calcined (thermally processed), that is, heated to about 1300°C in an inert atmosphere to produce the calcined coke prior to being crushed and blended. This is one of the most important feedstocks of graphite anode.

Another high-end filler is termed "needle coke." A by-product of petroleum refining, this type of coke can be prepared when the precursor pitch has a low concentration of insoluble particles, so that only a few mesophase spheres are nucleated. A needle coke initially breaks up into needles when mechanically ground, then into very small platelets (~100 μ dia) like other coke.

1.6.3 Binder

The most common binder is coal tar pitch, which is a hard, brittle, and glassy material. It is a by-product of metallurgical-coke production and is obtained by the distillation or heat treatment of coal tar. Large-scale carbonization of hard coal to produce blast furnace coke is performed at temperatures between 1000°C and 1200°C and residence time of 14 to 20 hr. Beside coke as the main product (75 wt.%, relative to feed coal), coke oven gas, water, benzene, ammonia, and crude tar are obtained. The pitch is removed from the tar distillation plant in liquid form or after solidification after cooling with water. The tar yield accounts for approximately 3–4 wt.% relative to feed coal. From 35 to 60 kg of pitch are produced from every metric ton of coal.

Like a polymer or plastic, it has a specific softening point (SP) depending on its molecular weight. The as-produced pitch has a softening point around 70°C. However, for some applications, pitches with higher SP are needed; so-called hard pitch, having a SP >150°C can be produced from soft pitch by air blowing or thermal treatment under pressure and subsequent removal of low boiling point constituents.

Coal tar pitch has wide molecular weight distribution ranging from approximately 200 to 3000 amu. Approximately 95% of the carbon occurring in coal tar pitch is present in the form of aromatic compounds.

Both coal tar pitch and petroleum pitch are complex mixtures of numerous organic compounds, predominantly polycyclic aromatic hydrocarbons (PAHs). In addition to PAHs coal tar pitch also contains hetroaromatic compounds, whereas petroleum pitch is relatively rich in alkyl-substituted PAHs.

The binder helps in plasticizing the filler coke particles so that they can be formed into the desired shape. Commonly used forming processes include extrusion, molding, and isostatic pressing. The binder phase is carbonized during the subsequent baking operation (~1000°C). Mostly, engineering graphite is reimpregnated with impregnating grade pitch (lower SP) to densify the carbon artifact followed by rebaking. A useful increase in density and strength are obtained with up to 3–6 impregnation cycles.

The final stage of the manufacturing process is graphitization (2500°C–3000°C) during which carbon atoms in the baked material migrate to form the thermodynamically more stable graphite lattice.

The crystals in a manufactured, polygranular graphite are less than perfect with approximately one layer plane in every six constituting a stacking fault.

1.6.4 Mesophase Pitch

Mesophase, a term originally suggested by Friedel, is the anisotropic liquid crystals formed in the transformation of an isotropic aromatic parent liquid pitch to an anisotropic solid coke mosaic texture. It is formed by thermal treatment of pitch at temperatures between 400°C and 500°C for 1 to 12 hours. It was Brooks and Taylor [6] and Mochida et al. [7] who

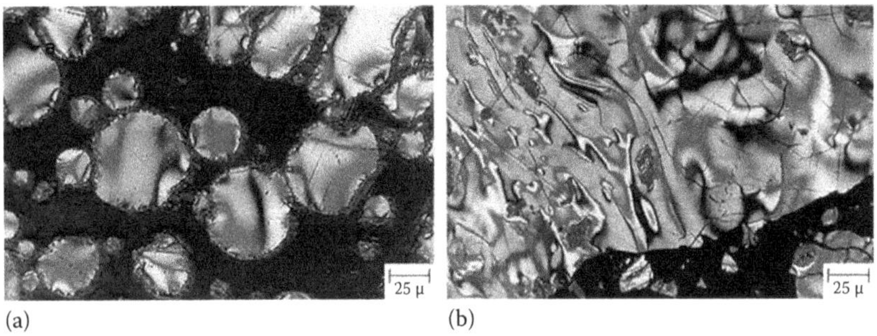

FIGURE 1.17
(a) Isolated mesophase spheres in isotropic pitch matrix. (b) Transform of mesophares into pitch coke after heat treatment to show flow texture.

proposed an explanation of growth of anisotropic features in cokes in terms of liquid crystals shown in Figure 1.17.

In the early stages of formation, mesophase appears as symmetrical spheres (Figure 1.17a) in which planar molecules are arranged in equatorial configuration. The mesophase spheroids range from about 2 μ to 100 μ. The normal or natural primary quinoline insoluble (QI) concentrate (free carbons or particles) at the interface between mesophase spheroids prevents its further growth. However, removal of QI contents allows condensation to take place, and large polyaromatic molecules are formed with a molecular weight averaging 1000. These polycyclic, spherulitic liquid crystals gradually increase in size to build up sufficient mutual van der Waals attraction to start promoting their alignment and form pre-cokes or "green cokes." The green cokes still retain from 6% to 20% volatile matter at 600°C. These volatiles are gradually removed when the temperature is raised. The coke so formed shows optical birefringence characteristic of disk-like, nematic liquid crystals. A typical flow texture of cokes is shown in Figure 1.17b.

The mesophase pitch besides being an important binder material acts as a precursor to high modulus carbon fibers, mesocarbon microbeads, carbon dots, and high-density carbon–carbon composites.

1.6.5 Mixing and Blending

A blend of the proper particle sizes of the petroleum coke and crushed granular pitch is blended in a mixer. A blend of about 75 wt.% of coke and 25 wt.% of pitch is generally used. The mixing is carried out at about 150°C before it is fed to die molds or extrusion press to make green compacts of desired size and shape. The binder coal tar pitch as obtained from coal tar refining is not a single phase but contains solid dispersed matter, obtained from gas phase pyrolysis of vapors. These are carbon black particles forming long chains of agglomerates of 100 particles or 1 micron diameter carbon particles of agglomerates of 10. These primary QI contents (~3%–10%) have influence on rheological properties of pitch and also on the quality of the baked product. For high-quality graphite products QI-free (<1% QI) pitch is often used as binder.

1.6.6 Carbonization

Carbonization is basically a heating cycle. The general carbonization mechanism of polyaromatic hydrocarbons is relatively simple, since it proceeds by the rupturing of the carbon–hydrogen bonds and the removal of the hydrogen. Some of these hydrocarbons first go through an intermediate liquid or plastic stage that occurs at temperatures above approximately 400°C–450°C.

The green molded artifact is heated slowly in a reducing or inert environment, over a range of temperatures that varies with the nature of the particular binder and may extend to 1250°C–1300°C. The organic material is decomposed into a carbon residue, and volatile compounds diffuse out to the atmosphere. The process is complex and several reactions may take place at the same time such as dehydrogenation, condensation, and isomerization. The binder pitch also converts into coke and the carbon content of the residue is a function of the nature of the precursor and the pyrolysis temperature. It usually exceeds 90 wt.% at 900°C and almost 99 wt.% at 1300°C.

Most of the synthetic graphite production is related to graphite electrodes. These electrodes of a particular diameter (1–2 m) and length (8–10 m) are then carbonized at a very slow heating rate up to a temperature of >2500°C in electrically heated furnaces. The baking may take from 20 to 60 days. The electrodes are removed, machined, and packed. After the aluminum industry, the second largest user of baked carbon and graphite is the steel industry.

1.6.7 Graphitization

Graphitization occurs in a series of steps that begin as the increasing temperature passes the carbonization temperature, that is, ~1300°C. Hydrogen, sulfur, and other elements, which might still be present after carbonization, are gradually removed, and, as the temperature reaches 2000°C, essentially none remains. The weight loss during graphitization is also attributed to the removal of interlayer chemical species, mostly interstitial carbon.

Over a fairly large range of HTT the interlayer spacing d_{002} is nearly equal to 0.344 nm, which is characteristic of well-developed layer planes that are stacked parallel to each other but are rotated at random with respect to each other, about the c-axis of the crystallite. This random rotational arrangement of planes is termed turbostratic. However, materials with turbostratic crystallites are not graphite but carbon.

Above 1800°C, the conversion from a turbostratic structure to a graphitic structure (shown in Figure 1.18) begins slowly at first then more rapidly as the temperature passes 2200°C. The gradual graphitization of the structure is readily confirmed by x-ray diffraction (Figure 1.18). If HTT exceeds a certain limit, for example, 10–15 minutes of treatment above 2500°C, the following transformation takes place in the structure: (1) removal of most defects within each graphite layer plane as well as between the planes, (2) gradual shifting and growth of the crystallites, (3) removal of cross-linking bonds, (4) beginning of the planes to stack into regular ABAB sequence of single-crystal graphite, and (5) shifting of carbon rings or single atoms to fill vacancies and eliminate dislocations. The crystallite size (La) increases from 5 nm, which is a typical size for turbostratic crystallites, to approximately 100 nm or more [8]. At the same time, the interlayer spacing (d_{002}) is reduced from 0.344 nm to a minimum of 0.335 nm, which is the spacing of the graphite crystal. The process is called graphitization.

The carbonization and graphitization processes are kinetic in nature, so the length of time of application at HTT is also important. Most graphitizable materials (cokes) require a temperature of 3000°C in order to reach full graphitization with a minimum value (i.e., 0.3354 nm) of the interlayer spacing. At 3000°C, full graphitization is usually obtained within 2 to 3 hours. Lower temperatures require considerably more time.

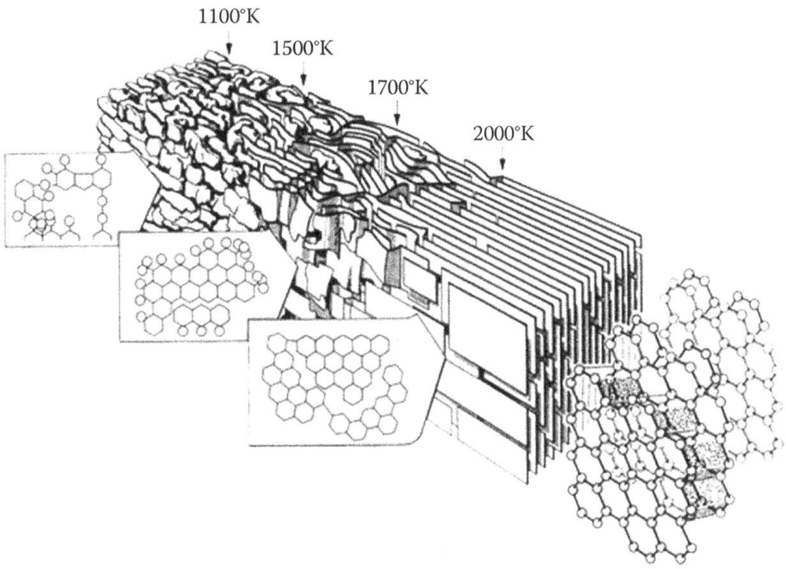

FIGURE 1.18
Schematic representation of the structural changes during the pyrolysis of carbonaceous polymers. A heat treatment temperature beyond 2500°C results in growth of structure from amorphous to polycrystalline to highly oriented parallel graphite planes with long-range order. (From H. Marsh and G. Griffith. New process and New Applications. *Proc. Int. Symp. on CARBON*, Kagaku Gijitsu-sha, Toyohashi, Tokyo, Japan, p. 81, 1982.)

Many types of organic precursors do not graphitize when heat-treated. Even at 3000°C the layer size, La, may not exceed 5 nm. All thermosetting polymers fall into this category, forming disordered carbons with glass-like properties. Normally these materials are porous because of the large quantity of gases evolved, much of it trapped in the pores. This class of carbons is referred to as nongraphitizing, and is sometimes characterized as glassy or vitreous.

Other forms of synthetic carbons are produced from gaseous phase and are termed pyrolytic carbons. These carbons produced in the range of 1000°C to 1200°C by decomposition of hydrocarbons show the development of faceting upon HTT >3000°C.

The world demand for the synthetic carbons is illustrated in Figure 1.19. The largest consumption of graphite electrodes is in the steel and aluminum industries, whereas several grades of carbon fibers find application in lightweight, high-strength composites for the aerospace sector.

1.6.8 Structural Analysis of Carbons by X-Ray Diffraction

X-ray diffraction, the most important of the scattering techniques, has been used as a dominant tool to study the structure of all types of carbon materials. For example, the line width of the d_{002} reflections give the c-axis crystallite size Lc. The a-spacing or the in-plane crystallite size is provided by study of the (100) diffraction line and its half width at half maximum intensity. The observation of the (101) and (112) profiles provide information on the interplanar correlation, since turbostratic graphite does not show well-defined (101) and (112) diffraction lines, but rather shows "Bragg rods."

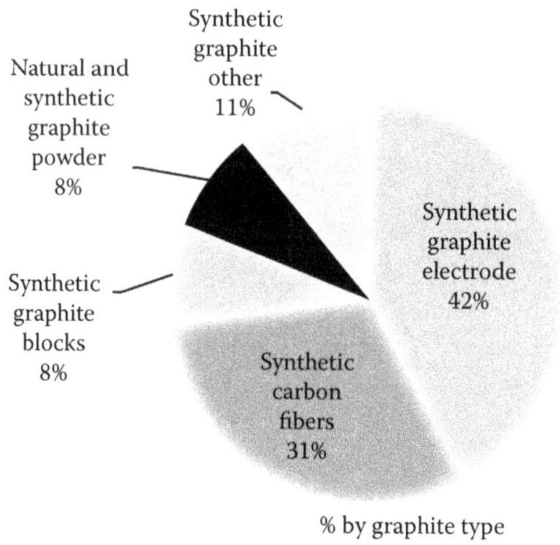

FIGURE 1.19
World graphite market. (From Ashbury Carbons.)

Figure 1.20 shows x-ray powder diffraction photographs showing increase in size of graphite crystallites in cokes derived from pitch heat-treated at successively increasing temperature. It is observed that as the crystallite size increases with an increase in HTT, the corresponding diffraction rings become sharper, and many new lines corresponding to *hkl* reflections appear as the graphitization approaches natural graphite.

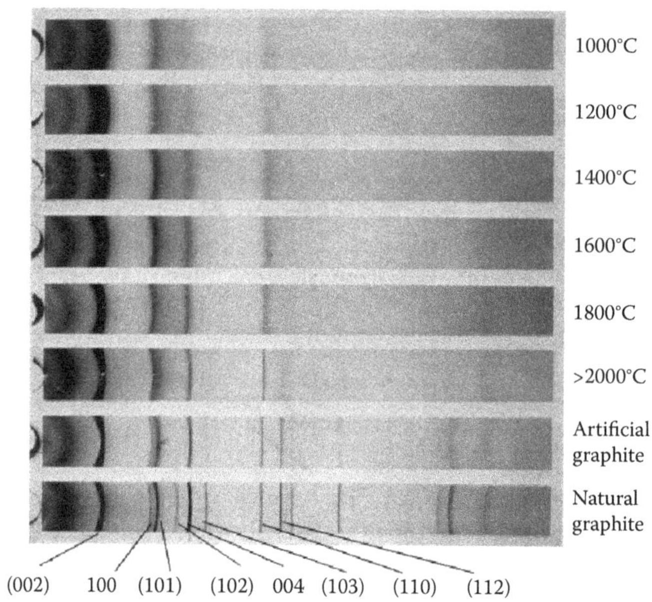

FIGURE 1.20
X-ray diffraction photographs of coke heat-treated to different temperatures. (Courtesy of H.P. Rooksby, Hirst Research Centre, GEC, Wembley.)

Looking at the pattern, the following inferences can be drawn:

- The existence of a line or band near (002) suggests existence of parallel layers.
- Sharpening of this line and appearance of lines (00l) with higher index, i.e., (004) and (006), suggest growth of crystal.
- Appearance of lines in *hkl* positions suggests emergence of three-dimensional ordering

The individual reflection in the powder diffraction photograph in Figure 1.20 is labeled with reference to x-ray powder diffraction records in Figure 1.21. Figure 1.21 shows the x-ray diffraction record of the highly graphitized commercial sample [9].

Intensity of the strongest (002) peak is taken as 100% for comparison. Appearance of the (112) peak at $2\theta = 82°46'$ suggests the three-dimensional ordering of the graphitic structure.

The FWHM (B) and peak angles of the (002) and (100) peaks are used in a Scherrer formula (Equation 1.5) to determine average crystallite thickness and average graphene sheet diameter, respectively. Here t is the crystalline or graphene sheet dimension, λ is the wavelength of the x-ray source, θ is the scattering angle (in radians), and κ is a constant:

$$t = \frac{\kappa\lambda}{(B\cos\theta)} \tag{1.5}$$

Since there is a large difference in the structure of different carbons and some of the carbons do not have three-dimensional crystalline state, two different forms of Scherrer formula are used. For the dimension of turbostratic crystallites perpendicular to the graphene sheets (Lc), (002) data is used and κ equals 0.9. For the dimension in graphene sheet planes (La), (100) data is used and κ equals 1.84.

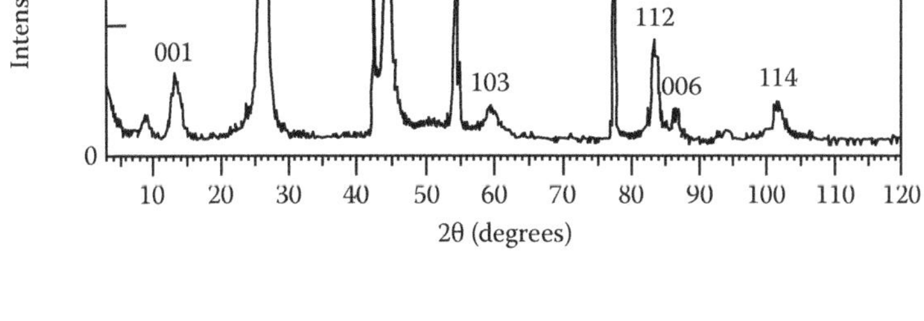

FIGURE 1.21
X-ray diffraction pattern of heat-treated commercial graphite sample. X-radiation; λ CuKα = 1.5437Å. (From M.S. Seehra, Interim report, West Virginia University, Morgantown, 1992. With permission.)

1.7 Pyrolytic Carbon and Graphite

The chemical vapor deposition (CVD) of carbon materials is not new. Its first practical use was developed in the 1880s in the production of incandescent lamps to improve the strength of filaments by carbon deposition and a patent was issued over a hundred years ago, covering the basis of the CVD of carbon. Pyrolytic graphite is the only graphitic material that can be produced effectively as a coating. The coating can be made sufficiently thick and can be removed from the substrate as a free-standing object. Pyrolytic graphite is a key element in the technology of coating specialty molded graphites, carbon foam, carbon fibers, metals, and ceramics. It plays a major role in processing of carbon–carbon components, such as high temperature crucibles, reentry heat shields, rocket nozzles, and carbon–carbon brake discs.

1.7.1 Thermodynamics and Kinetics

The CVD of pyrolytic graphite can be optimized by experimentation. The carbon source (hydrocarbon gas), the method of activating the decomposition reaction (e.g., thermal, arc discharge or plasma, laser), and the deposition variables (e.g., temperature, pressure, gas flow) can be changed until a satisfactory deposit is achieved. However, this empirical approach may be too cumbersome and, for more accurate results, it is sometimes combined with a theoretical analysis or modeling. Basically the CVD process is governed by two factors: (1) thermodynamics, that is, the driving force that indicates the direction the reaction is going to proceed; and (2) kinetics, which defines the transport process and is a rate-determining step, that is, how fast it is going. Chemical thermodynamics is concerned with the transfer of energy from one chemical system to another in accordance with the first and second laws of thermodynamics. In the case of CVD, this transfer occurs when the gaseous compounds, introduced in the deposition chamber, react to form the carbon deposit and removal of by-products gases. The C/H ratio of the gas mixture (e.g., CH_4 and H_2) entering the reaction chamber is an important factor to control the nature of the deposition [10–12].

1.7.2 Isotropic and Laminar Deposits

The isotropic carbon has little graphitic characteristic and essentially no optical activity. It is composed of very fine grains without observable orientation and for this reason, it is known as isotropic carbon rather than isotropic graphite. It is often obtained in fluidized-bed deposition, possibly due to continuous surface regeneration by the mechanical rubbing action of the bed. Generally, isotropic deposits are obtained at higher temperatures (>1300°C), and laminar and columnar deposits are obtained at lower temperatures.

It is reported that higher C/H ratios (1/4) favor laminar or columnar deposition, and lower ratios (1/14) favor isotropic deposition.

1.7.3 Graphitization of Columnar and Laminar Deposits

The columnar and laminar deposits just described generally have a turbostratic structure in the as-deposited condition, with a large interlayer spacing (~0.344 nm) as revealed by x-ray diffraction. The material graphitizes readily when heat-treated at 2500°C for 1–2 hr. The heat treatment at 2500°C causes the reordering of the structure. The various crystallite imperfections, such as vacancies, stacking faults, dislocations, and rotational disorders, tend to heal and disappear; the crystallite size La and Lc increases; and the (002) diffraction peak narrows considerably and becomes close to the position of the ideal graphite line. The interlayer spacing (d_{002}) decreases to approach that of the ideal graphite crystal (0.3354 nm).

When columnar or laminar pyrolytic graphites are annealed above 2700°C, usually under a pressure of several atmospheres, further ordering and stress relieving of the structure occur within each plane and between planes. The material is known as highly oriented pyrolytic graphite (HOPG). It is soft and structurally close to the ideal graphite crystal with an angular spread of the c-axes of crystallites of less than one degree.

Unlike columnar and laminar pyrolytic deposits, isotropic carbon does not graphitize readily and is, in this respect, similar to glassy or vitreous carbon. The d-spacing rarely decreases below 0.344 nm. The crystallite size (Lc) also remains small.

1.8 Nano Forms of Carbon

Though carbon has been known for thousands of years, during the last three decades three new forms of carbons have been discovered and the inventors of two of them (C_{60} and graphene) have won the Nobel Prize. Some of the novel nanostructures are illustrated in Figure 1.22. The classical example of closed shell stabilization of carbon nanostructures is the formation of C_{60} molecules and other fullerenes by electric arc evaporation of graphite discovered by Smalley, Kroto, and Curl in the year 1985. Subsequently the team won the Nobel Prize in chemistry in the year 1996. The diameter of a single C_{60} fullerene molecule is 1.4 nm. Here stabilization is achieved by the formation of closed shell structures that obviate the need for surface heteroatoms to stabilize the dangling bonds, as in the case of bulk crystals of diamond and graphite.

Other forms of carbon are carbon nanotubes or elongated fullerenes, either closed at both ends or only at one end. Although various carbon cages were studied, it was only in 1991 when Iijima observed for the first time tubular carbon structures. The nanotubes consisted of up to several tens of graphite shells, so-called multiwalled carbon nanotubes

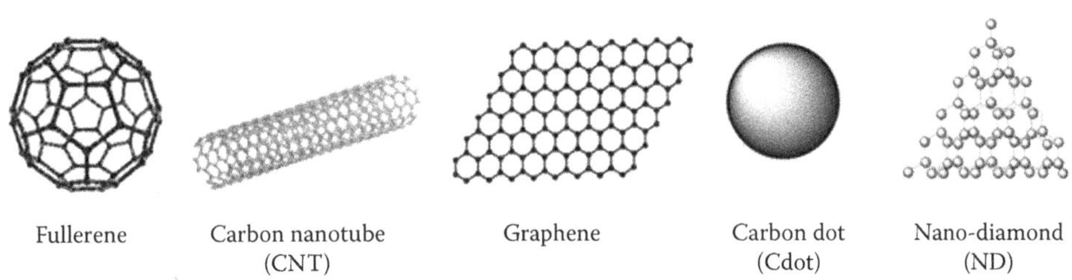

FIGURE 1.22
Nanoforms of carbon, from left, o-D fullerene, 1-D carbon nanotube, 2-D graphene, carbon dot, and nanodiamond.

(MWCNTs) with adjacent shell separation of ~0.34 nm, diameters of ~1 nm, and a large aspect ratio. Two years later Iijima and Ichihashi and Bethune et al. synthesized single-walled carbon nanotubes (SWCNTs).

Nanocarbons, which have been obtained by controlling the structure in nanometer scale as well as the bonding nature of carbon atoms, can provide the highly functional advanced performances, which are difficult to obtain from conventional carbons.

Fullerenes and carbon nanotubes and graphene are unique in the larger family of nanocarbon-based materials as interrelated prototypes for zero-dimensional quantum dots (fullerenes), one-dimensional quantum wires (CNT), and two-dimensional quantum wells (graphene) (see Figure 1.22).

Likewise carbon nanotubes, which are capped at each end by half of fullerene (Semi spherical sheet), have aroused great interest in the research community because of their exotic electrical and mechanical properties. The unique properties of fullerenes and carbon nanotubes described in the following chapters are expected to be of interest for practical applications.

Early during his pioneering work, Bacon in 1960 synthesized graphite whiskers that he described as scrolls, using essentially the same conditions as for the synthesis of carbon nanotubes except for the use of helium pressures higher by an order of magnitude to synthesize the scrolls. It is believed that the cross-sectional morphology of multiwalled nanotubes and carbon whisker scrolls is different.

Vapor-grown carbon fibers, which were accidently found inside furnaces containing hydrocarbon gases and CO for melting metals, are typically several nanometers in diameter, several microns in length with a tubular microstructure. The formation of these filaments at relatively low temperatures <1000°C, in presence of metal catalysts can be a source of cheap high modulus carbon fibers for reinforcement in composites.

Carbon nano-onions (CNOs), which consist of concentric graphitic shells, represent another new allotropic nanophase of carbon materials. CNOs have already been shown to offer a variety of potential applications such as solid lubrication, electromagnetic shielding, fuel cells, heterogeneous catalysis, gas and energy storage, and electro-optical devices owing to their outstanding chemical and physical properties. According to a recent study, CNOs can also be used to produce ultrahigh-power micrometer-sized supercapacitors due to their accessible external surface area for ion adsorption.

Direct current (DC) arc discharge between graphite electrodes is one of the versatile processes to produce a variety of carbon nanomaterials besides fullerenes and carbon nanotubes. These include different forms of carbon nanostructures like carbon soot, carbon nanoshells, metal encapsulated carbon shells, sea urchins, bucky onions, helical coils, spinning cones, and nanohorns.

Formation of carbon nanoshells during the arc discharge process demonstrates that it is possible to encapsulate metal nanoparticles inside the shells for potential applications in magnetic nanoparticles as data storage, ferrofluids, confinement of radioactive waste, and targeted drug delivery nanocapsules. A host of publications on this aspect are available.

1.8.1 Quantum Confinement in Carbon Nanomaterials

In bulk semiconductors the wave vector **k** can take a very large number of values in general. Therefore, available density of states is very large in the directions of k_x, k_y, and k_z. However, there is quantization of energy levels if one of the dimensions become in the range of the de Broglie wavelength of thermal electrons, the energy levels transform from continuum to discrete, and the wavelength λ_B is given by

$$\lambda_B = \frac{h}{p} = \frac{h}{\sqrt{2m^* E}} \tag{1.6}$$

where h is Planck's constant, p is the electron momentum, m^* is the electron effective mass, and E is the energy. The wavelength λ_B is typically on the order of tens of nanometers for carriers with typical thermal energy. The spatial confinement of the carriers in lower dimensional semiconductors leads to dramatically different energy–momentum relations in the directions of confinement, which results in a completely new density of states, compared to the bulk case, as depicted in Figure 1.23. The density of states is no longer continuous or quasi-continuous (Figure 1.23a) but becomes quantized, and the energy–momentum relations dramatically transform into subbands. Therefore, if one of the dimensions of a semiconductor is comparable to, or less than λ_B, the electrons will be confined to two dimensions and behave as charge carriers in a quantum well (QW) structure.

In case of graphene, like in thin sheet, the confinement of electron is in z-direction and the allowed values of k become highly discretized. As shown in Figure 1.23b, for certain values of kz the density of states remain constant and it follows in steps (conduction staircase).

In case of carbon nanotubes these can be treated as quantum wire (QWR), which is a one-dimensional confined structure in the sense that the carrier movement is free in one dimension (i.e., along the length of the tube) and confined in two others (i.e., along the diameter of the tube) (x-y graphene plane in an unzipped nanotube). Figure 1.23c shows the density of states for such structures. These nanotubes are characterized by a number of singularities where each peak corresponds to a single quantum subband.

Finally, a quantum dot in the present case, is a nanostructure in which the electron and hole movement is confined in all three dimensions. Quantum dots are thus tiny clusters of semiconductor material having all three dimensions of only a few nanometers. In the case of quantum dots, the charge carriers occupy only a restricted set of energy levels rather like the electrons in an atom and the valence and conduction bands have sharp energy

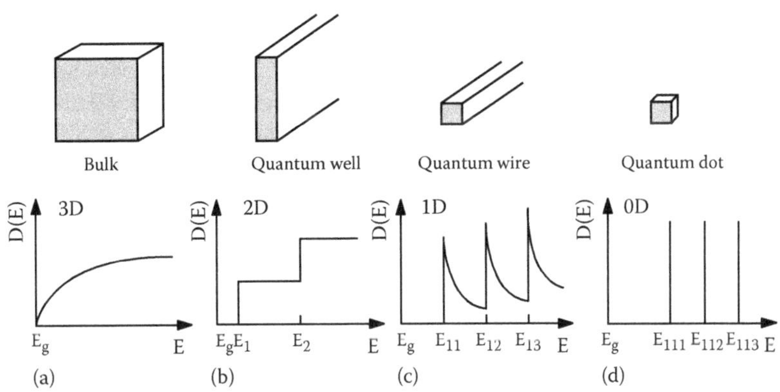

FIGURE 1.23
Schematic diagram illustrating the representation of the electronic density of states in conduction band depending on dimensionality. (a) 3D graphite, (b) graphene, (c) carbon nanotubes, and (d) carbon nanodot.

levels as shown in Figure 1.23d. For this reason, quantum dots are sometimes referred to as "artificial atoms." It is important to stress, however, that quantum dots actually contain hundreds of thousands of atoms.

The confinement of electrons to lower dimensional structures results in completely new optoelectronic properties, compared to bulk graphite. Importantly it is basically the density of states that decides the performance of the device.

1.8.2 Raman Spectroscopy of Nanocarbons

Raman spectroscopy is the inelastic scattering of light by matter. Being highly sensitive to the physical and chemical properties of the materials, it has become one of the most important tools for nanoscience and technology. Raman spectra are very sensitive to changes that break the translational symmetry. For this reason Raman spectroscopy is most sensitive to highly symmetric covalent bonds with little or no natural dipole moment. The carbon–carbon bonds that make up these materials fit this criterion perfectly and as a result Raman spectroscopy is highly sensitive to these materials and able to provide a wealth of information about their structure.

Historically the spectroscopy played an important role in the study and characterization of graphite materials, being largely used over the last four decades to characterize pyrolytic graphite, carbon fibers, fullerenes, glassy carbons, nanographite ribbons, and graphene. For sp^2 nanocarbons such as graphene and carbon nanotubes, Raman spectroscopy can give information about crystallite size, clustering of the sp phase, the presence of sp^2–sp^3 hybridization and other crystal disorders, number of graphene layers, nanotube diameter, chirality, and so on. It can also account for the introduction of chemical impurities, doping defects, and finally metallic versus conducting behavior.

Every band in the Raman spectrum corresponds directly to a vibrational frequency of a bond within the molecule. The vibrational frequency and hence the position of the Raman band is very sensitive to the orientation of the bonds and the weight of the atoms at either end of the bond.

In the case of diamond, where the material consists of highly uniform C–C bonds in a tetrahedral crystal structure, the Raman spectrum is very simple (Figure 1.24) [13]. It consists only of a single band because all of the bonds in the crystal are of the same orientation and strength, resulting in a single vibrational frequency.

The Raman spectrum of crystalline graphite is marked by the presence of strong peaks due to sp^2-bonded carbon in planar sheets in which the bond energy of sp^2 is higher than the sp^3 bond in diamond. The two prominent peaks are centered at 1580 and 2700 cm^{-1} and known as G and G′ bands, respectively, where the "G" label refers to graphite.

The G band is a first-order Raman allowing feature originating from the zone center (phonon wave vector $q = 0$), corresponding to in-plane E_2g_2 optic modes. The G′ band (~2700 cm^{-1}) is a second-order (two phonons) feature that allows exploration of $q = 0$ phonons.

In the Raman spectra obtained with samples with small crystallite size La ~0.5 micrometer, that is smaller than the wavelength of light, the presence of an additional dispersive peak centered at approximately 1350 cm^{-1} is observed when using lower excitation at 2.1 eV (414 cm^{-1}). The feature is assigned to the breathing of the carbon hexagons that become Raman active at the border of the crystallite areas owing to the loss of translation symmetry.

The frequency of this feature is about half of the second-order G′ frequency and is related to the phonon of the first-order process of G′ peak.

Since the peak at ~1350 cm^{-1} is observed in the presence of defects in an otherwise perfect infinite graphite structure, it has been named as the D band (D for defect or disorder).

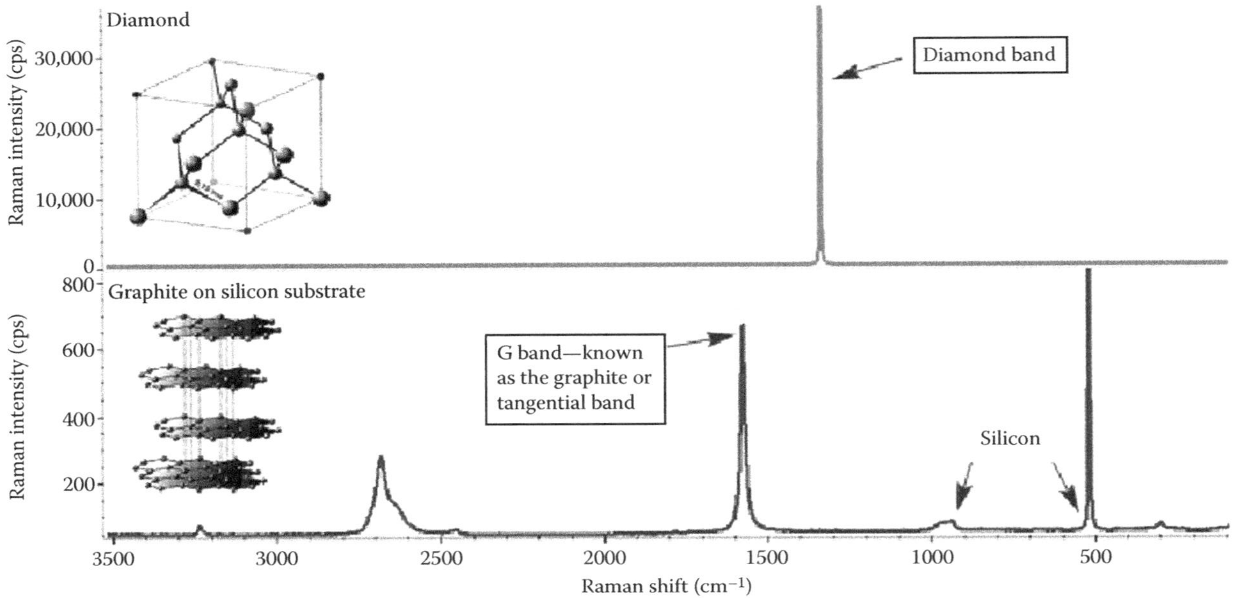

FIGURE 1.24
Raman spectra of diamond and of a sample of graphite on a silicon substrate. (From J. Hodkiewicz, *Thermo Scientific Application* Note 51901 (2010).)

This disorder refers to the crystallite boundary and the ratio I_D/I_G provides a parameter that can be used for quantifying disorder.

Another disorder-induced band centered at 1620 cm^{-1} is usually obtained in the Raman spectra of disordered graphitic materials, although with smaller intensity than the D band. This feature has been named the D′ band and also depends on the crystallite size La and $E_{laser.}$

Fullerenes are essentially hollow carbon shells of various size. The most well known of these is a 60-carbon atom unit called buckminsterfullerene or C_{60}. There are many other fullerenes with a few to many hundred carbon atoms. Figure 1.25 compares the spectra of C_{60} and C_{70} [13]. The main feature of the C_{60} spectrum is a relatively sharp line at around 1462 cm^{-1}, known as the pentagonal pinch mode. This tells us that C_{60} is composed of sp^2-bonded carbon. The sharpness of the band also tells us that the bonds are for the most part very uniform in nature. In contrast the spectrum of C_{70} is littered with numerous bands. These are due to reduction in molecular symmetry, which results in more Raman bands being active.

Graphene is the fundamental building block of many important carbon materials including graphite. Graphite consists of stacks of sp^2-bonded planar graphene sheets. When comparing Raman spectra of graphene and graphite, at first glance the spectra look very similar. However, there are some signature differences as we can see from Figure 1.26 [13]. The most obvious is the difference in the band at 2700 cm^{-1}, which is known as the G′ band. It is much more intense than the G band in graphene compared to graphite. The G′ band is also referred to as the 2D band; both 2D and G′ are accepted names for this band. Upon a closer look at shape and the position of the G′ band in these two materials one can see that both the shape of the band and the position are different.

The peak shift in graphite is a result of interaction between the stacked graphene layers, which has a tendency to shift the bands to higher frequency. The G′ band in a single layer graphene spectrum fits to a single band, whereas curve fitting reveals several underlying bands in the graphite spectrum. These bands are a result of the different interlayer interactions that occur at different depths within the graphene.

When the bond lengths and angles of graphene are modified by strain, caused by the interaction with a substrate or with other graphene layers or due to external perturbations, the hexagonal symmetry of graphene is broken. The G band is therefore highly sensitive to strain effects in sp^2 nanocarbons and can be used to probe any modification to the flat geometric structure of graphene, such as the strain induced by external forces.

Curvature effects in carbon nanotubes give rise to multiple peaks in the G band spectrum for a SWCNT. Figure 1.27 shows the Raman spectrum of SWCNT in which we can see well-defined G and G′ bands as they are in graphene and graphite [13]. We also see a prominent band around 1350 cm^{-1} from a hybridized vibrational mode associated with graphene edges and it indicates the presence of some disorder to the graphene structure. The band is often referred to as the disorder or D band, and its intensity relative to that of the G band is often used as a measure of the quality of carbon nanotubes.

There is another series of bands appearing at the low frequency end of the spectrum known as radial breathing mode or RBM bands. The RBM bands are unique to SWCNTs as the name suggests, corresponding to the expansion and contraction of the tubes. The frequency of these bands can be correlated to the diameter of the SWCNTs and they can provide important information on their aggregation state.

MWCNTs have very similar spectra to those of SWCNTs, as is apparent from Figure 1.27. The primary differences are the lack of RBM modes in MWCNTs and a much more prominent D band. The RBM modes are not present because the outer tubes restrict the

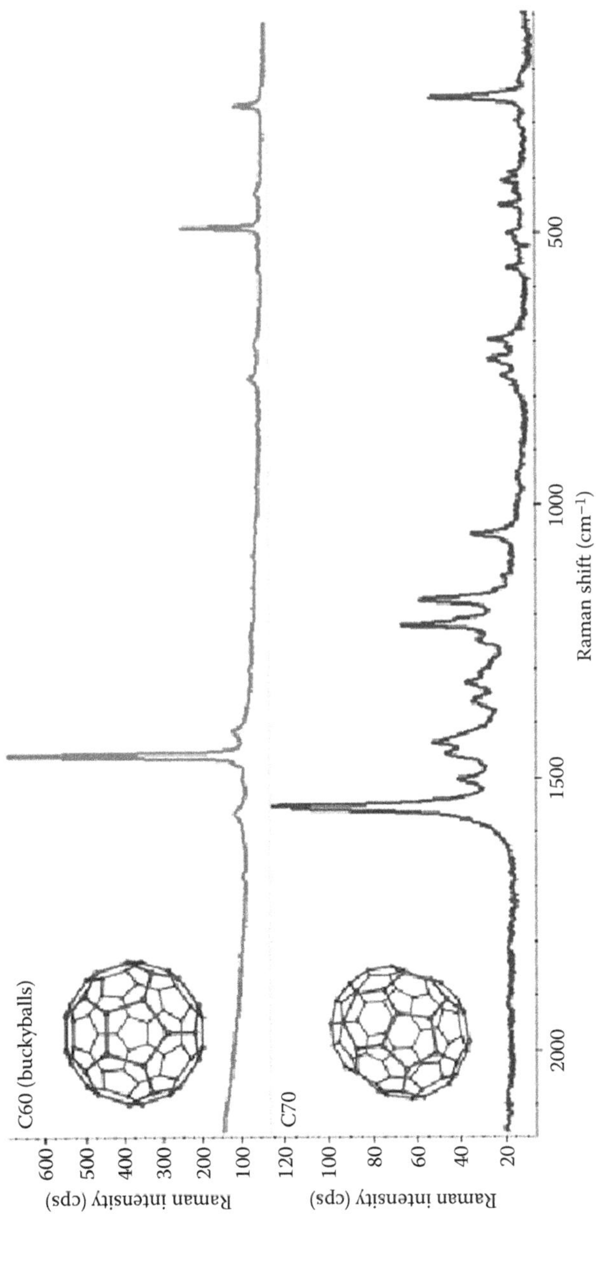

FIGURE 1.25
Raman spectra of two fullerenes C60 and C70. (From J. Hodkiewicz, *Thermo Scientific Application* Note 51901 (2010).)

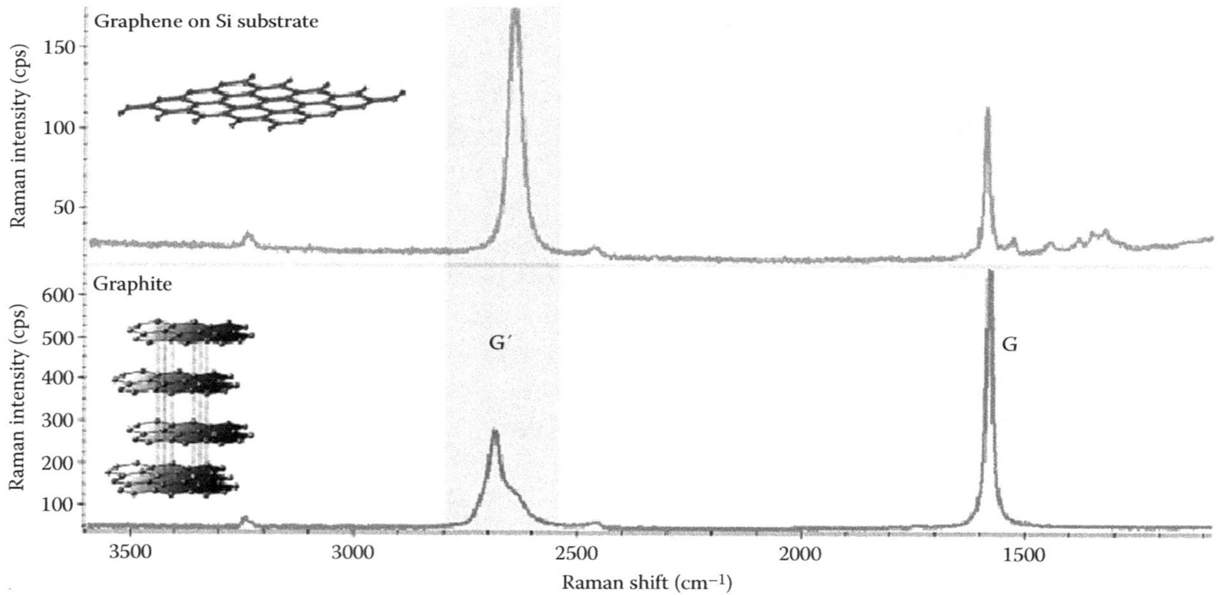

FIGURE 1.26
Comparison of Raman spectra of graphene and graphite. (From J. Hodkiewicz, *Thermo Scientific Application* Note 51901 (2010).)

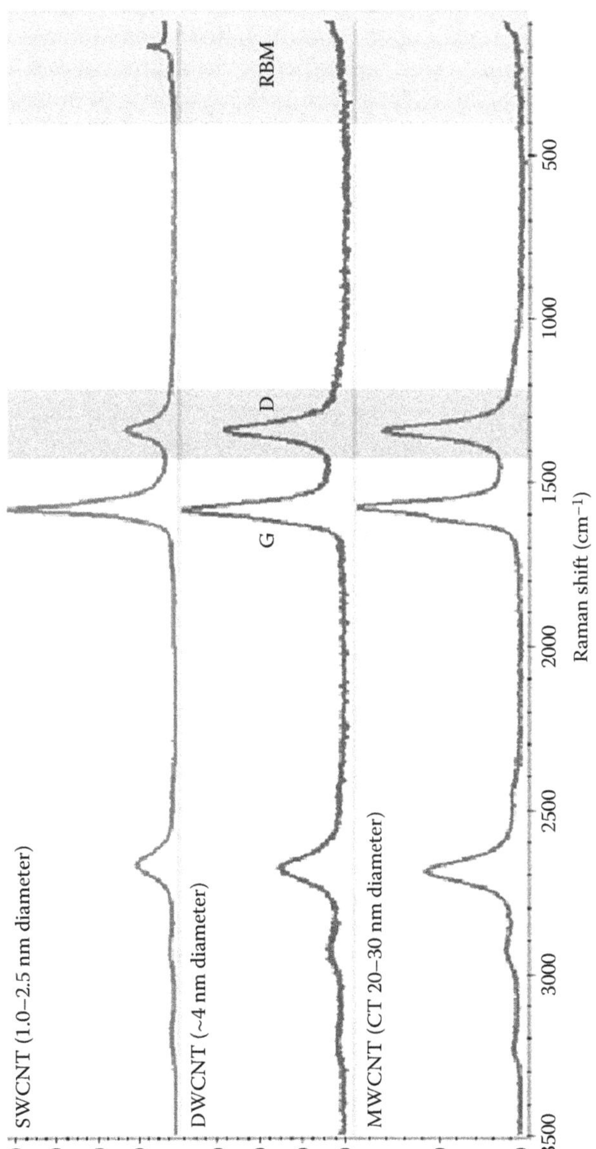

FIGURE 1.27
Comparison of Raman spectra of SWCNTs, DWCNTs, and MWCNTs. (From J. Hodkiewicz, *Thermo Scientific Application* Note 51901 (2010).)

breathing mode. The more prominent D band in MWCNTs is to be expected to a certain extent given the multilayer configuration and indicates more disorder in the structure. Notice that the RBM mode completely disappears as we go to a double wall tube and that the D band G′ bands get proportionately larger as we add layers to the walls of the tubes.

Nanostructured materials have gained tremendous industrial and academic interest in the last two decades due to their potential applications in super strong composites, field emitters, paints, Li battery, fuel cell, super capacitors, hydrogen storage, antibacterial and drug delivery systems, quantum wires, nanoelectronics, smart sensors, catalyst support, and fundamental and basic research. However, if this promise is to be realized nothing less than a complete revolution in the synthetic strategies available for materials production may be necessary. This new area of materials science, therefore, presents one of the greatest technical challenge of the 21st century.

References

1. J. Bernal. The structure of graphite. *Proceedings of the Royal Society of London. Series A, Containing Papers of a Mathematical and Physical Character* 106, no. 740 (1924): 749–773.
2. A. Maffucci and G. Miano. Electrical properties of graphene for interconnect applications. *Applied Sciences* 4, no. 2 (2014): 305–317.
3. I.E. Abbott. Graphene: Exploring carbon flatland. *Physics Today* 60, no. 8 (2007): 35.
4. I.L. Spain. Electronic transport: Properties of graphite, carbons, and related materials. *Chemistry and Physics of Carbon* 16 (1981): 119–304.
5. G. Jenkins, K. Kawamura, and L. Ban. Formation and structure of polymeric carbons. *Proceedings of the Royal Society of London A: Mathematical, Physical and Engineering Sciences,* The Royal Society, 1972, pp. 501–517.
6. J. Brooks and G. Taylor. The formation of graphitizing carbons from the liquid phase. *Carbon* 3, no. 2 (1965): 185, IN9, 187–186–IN 18–193.
7. I. Mochida, Y. Korai, C.-H. Ku, F. Watanabe, and Y. Sakai. Chemistry of synthesis, structure, preparation and application of aromatic-derived mesophase pitch. *Carbon* 38, no. 2 (2000): 305–328.
8. H. Marsh and G. Griffith. New process and new applications. *Proc. Int. Symp. on CARBON,* Kagaku Gijitsu-sha, Toyohashi, Tokyo, Japan (1982), p. 81.
9. M.S. Seehra. Interim Report. West Virginia University, Morgantown, 1992.
10. R. Baker. Catalytic growth of carbon filaments. *Carbon* 27, no. 3 (1989): 315–323.
11. N.M. Rodriguez. A review of catalytically grown carbon nanofibers. *Journal of Materials Research* 8, no. 12 (1993): 3233–3250.
12. G.G. Tibbetts. Why are carbon filaments tubular? *Journal of Crystal Growth* 66, no. 3 (1984): 632–638.
13. J. Hodkiewicz. Characterizing carbon materials with Raman spectroscopy. *Thermo Scientific Application* Note 51901 (2010).

Further Reading

Following is a partial list of some of the important literature available dealing with the broad subject of carbon.

Burchell, T.D. ed. *Carbon Materials for Advanced Technologies*. Elsevier Science Ltd., Oxford, 1999.

Donnet, J.B. and Andries Voet. *Carbon Black*. Marcel Dekker, New York, 1976.

Donnet, J.B., T.K. Wang, S. Rebouillat, and J.C.M. Peng, eds. *Carbon Fibers*, 3rd ed. Marcel Dekker, 1998.

Dresselhaus, M.S., G.S. Dresselhaus, K. Sugihara, I.L. Spain, and H.A. Goldberg. *Graphite Fibers and Filaments*. Springer-Verlag, New York, 1988.

Figueiredo, J.L., C. Bernardo, R.T.K. Baker, and K.J. Huttinger, eds. *Carbon Fibers, Filaments and Composites*. Kluwer Academic, The Netherlands, 1989.

Fischbach, D.B. and A. Pacault. The kinetics and mechanism of graphitization, in *Chemistry & Physics of Carbon*, vol. 7. Marcel Dekker.

Fitzer, E. and L.M. Manocha. *Carbon Reinforcement and Carbon/Carbon Composites*. Springer-Verlag, New York, 1998.

Jenkins, G.M. and K. Kawamura. *Polymeric Carbons*. Cambridge Press, Cambridge, 1976.

Kelly, B.T. and R. Taylor. The thermal properties of graphite, in *Chemistry & Physics of Carbon*, vol. 10. Marcel Dekker.

Mantell, C.L. *Carbon and Graphite Handbook*. Interscience, New York, 1968.

Marsh, H., E.A. Heintz, and F. Rodriguez-Reinoso, eds. *Introduction to Carbon Technologies*. University of Alicante, Secretariado de Publications, Spain, 1997.

Ruland, W. X-ray diffraction studies of carbon and graphite, in *Chemistry & Physics of Carbon*, vol. 4. Marcel Dekker.

Savage, G. *Carbon-Carbon Composites*. Chapman & Hall, 1993.

Spain, I.L. Electronic transport properties of graphite, carbons and related materials, in *Chemistry & Physics of Carbon*, vol. 16. Marcel Dekker.

Zabel, H. and Solin, S.A. eds. *Graphite Intercalation Compounds*, vols. I and II. Springer 1990 and 1992.

2

Fullerenes

2.1 Introduction

Fullerenes are a class of carbon allotropes in the form of hollow, closed cages of carbon atoms. Harold W. Kroto, Robert F. Curl, Richard E. Smalley, and their team discovered fullerenes serendipitously in 1985 at Rice University during a series of laser vaporization experiments on graphite in which they were probing by mass spectrometry the carbon clusters formed in the atmospheres of giant red stars [1–3]. The prominent all-carbon structure formed during those experiments on graphite consisted of 60 carbon atoms or C_{60} [1,2]. Architect Buckminster Fuller's 1960s geodesic domes provided a clue to the structure of C_{60}, which was proposed to have a soccer ball-like structure [1,4]. In honor of Fuller the C_{60} molecule was named buckminsterfullerene and the new class of all-carbon caged molecules was named fullerenes [1,2,5]. For their discovery of C_{60}, Kroto, Curl, and Smalley were awarded the Nobel Prize in Chemistry in 1996.

In 1990, Krätschmer and colleagues reported a method to generate fullerenes in a workable amount from a carbon soot deposit prepared by resistive heating of graphite [6]. The fullerenes were isolated from the soot by extracting in an organic solvent [4,7,8]. Mass and infrared (IR) spectra of the extracted material and x-ray analysis of the crystalline material obtained from the deposit extract provided the first proof of the C_{60} structure as proposed earlier by Kroto et al. [2]. An independent study of similarly arc-processed carbon by Taylor et al. showed that chromatographically separated soluble extracts of C_{60} and C_{70} separated into magenta and red fractions, respectively, and their C-13 NMR analyses gave definitive evidence of the proposed structures of C_{60} and C_{70} [9]. The method reported by Krätschmer et al. revolutionized research in fullerenes because a simple way was now available for producing high yields of C_{60} and higher fullerenes that could be isolated, solubilized, chromatographically separated, and characterized [2,8]. Prior to the discovery of fullerenes there were two well-known forms of pure and ordered carbon: diamond and graphite, which are the network solids [4,10]. Fullerenes, on the other hand, were a new class of discrete molecules [4]. The most stable and abundant representative of the fullerene family is the buckyball containing 60 carbon atoms [2,4,10,11]. Most of the studies on fullerenes reported in this chapter are therefore focused on C_{60}. Since the discovery of fullerenes the field of carbon nanomaterials has grown with the subsequent discoveries of carbon nanotubes [12] (considered as the fourth allotrope of carbon) and graphene [13]. The discovery of fullerenes paved the way for nanoscience and nanotechnology of carbon materials and the field has been growing ever since.

2.2 Discovery of Fullerenes

The discovery of fullerenes has a fascinating history. It is amazing how a molecule even before its discovery was hypothesized to exist and its structure was proposed even before it was isolated. Prior to the actual discovery of fullerenes in 1985, the possibility of hollow all-carbon cage molecules had been suggested by D.E.H Jones in 1966 [14]. He suggested that graphene sheet could curl up to make hollow molecules [15]. Eiji Osawa of Japan was perhaps the first to propose the idea of a buckyball in 1970 [2]. He suggested that the corannulene molecule was a subset of a soccer ball and hypothesized that a full body shape could exist. In 1973 Huckel calculations had been performed on C_{60} and by 1980 Davidson had determined and published the molecular orbital diagram of C_{60} [2].

During the 1980s there was much discussion on the presence of linear carbon chains in the interstellar space based on the radio waves detected from space [10]. Harold Kroto from Sussex University had theorized that such linear chains in interstellar space may arise from chemical reactions in the outer atmosphere of carbon-rich red giant stars [4]. Carbon clusters were known to exist in the atmospheres of carbon-rich red giant stars [16]. In 1985 Kroto got interested in understanding the great abundance of the linear carbon chains in interstellar space using a laser vaporization cluster beam apparatus developed by Richard Smalley and his team at Rice University [4,17], since the laser vaporization technique seemed to offer a way to simulate the stellar nucleation conditions if graphite was vaporized. Kroto collaborated with Smalley and Curl and together with graduate students James Heath, Sean O'Brien, and Yuan Liu simulated the conditions of their experiments as in the atmospheres of carbon-rich red giant stars by laser vaporizing graphite in a pulsed jet of helium [1,3,4]. The carbon clusters produced were probed by a mass spectrometer [2]. They found that along with small linear

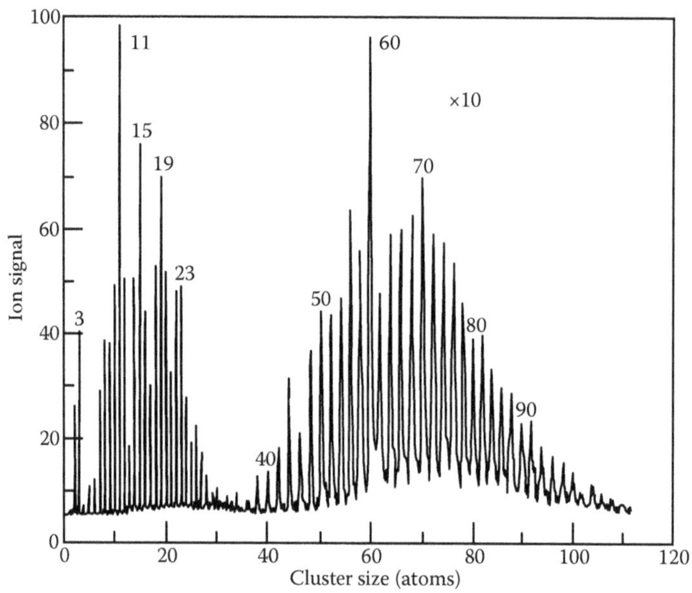

FIGURE 2.1
Time-of-flight mass spectrum of carbon clusters produced by laser vaporization of graphite. (Reprinted with permission from E.A. Rohlfing, D.M. Cox, and A. Kaldor, *The Journal of Chemical Physics* 81, no. 7: 3322–3330. Copyright 1984, American Institute of Physics.)

carbon chain distribution of clusters there was a distribution of large even-numbered carbon clusters with the 60 carbon atom cluster being more abundant than any other even-numbered cluster between 50 and 70 atoms. It is interesting that in 1984 a research group at Exxon had studied carbon clusters produced by laser vaporization technique by using a copy of the cluster beam apparatus developed at Rice University [4,18]. While they reported detection of linear carbon chains containing both odd- and even-number carbon atoms (1–30 carbon atoms) and large clusters containing only an even number of carbon atoms (30–190) (Figure 2.1) [18], they missed the abundance of the two even-numbered members, C_{60} and C_{70} clusters, which were later identified as fullerenes by the research group at Rice [1,4] (Figure 2.2).

By varying the vaporization conditions Smalley and his team were able to identify the optimum clustering conditions under which the peak corresponding to a mass of 720 (C_{60} peak) was the dominant cluster signal [1,2,4]. The C_{60} peak was invariably followed by a peak corresponding to a mass of 840 (C_{70} peak) and the C_{60}/C_{70} ratio was 5/1 in general [2,4]. The challenge was to model a 60-carbon atom molecule of unusual and remarkable stability. Smalley and his team based their model on the mathematical truth Fuller

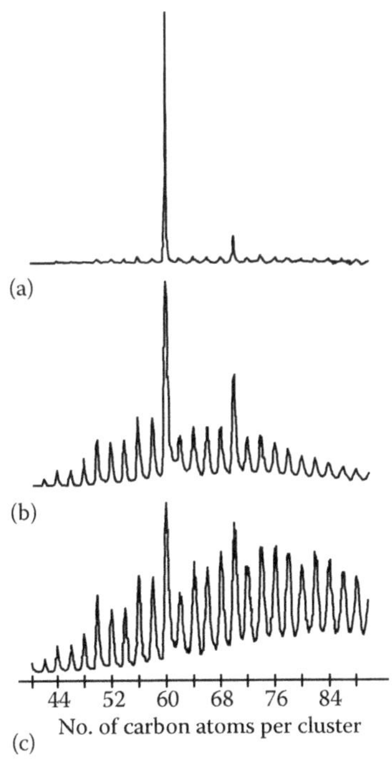

FIGURE 2.2
Time-of-flight mass spectra of carbon clusters prepared by laser vaporization of graphite and cooled in a supersonic beam. The three spectra shown differ in the extent of helium collisions occurring in the supersonic nozzle. In (c), the effective helium density over the graphite target was less than 10 torr and the observed cluster distribution of even-numbered clusters with 38–120 atoms here is believed to be due simply to pieces of the graphite sheet ejected in the primary vaporization process. The C_{60} peak was largest but not dominant. The spectrum in (b) was obtained when maximum helium density roughly 760 torr was present over the graphite target at the time of laser vaporization. The enhancement of C_{60} and C_{70} is believed to be due to gas-phase reactions at these higher clustering conditions. The spectrum in (a) was obtained by maximizing these cluster thermalization and cluster-cluster reactions by using an 'integration cup' shown in Figure 2.3 to increase the time between vaporization and expansion. Under these optimum clustering conditions, C_{60} is the dominant species followed by C_{70}. (Reprinted by permission from Macmillan Publishers Ltd., *Nature*, H.W. Kroto, J.R., Heath, S.C. O'Brien, R.F. Curl, and R.E. Smalley, *Nature* 318, no. 6042: 162–163, copyright 1985.)

employed in his domes that a sheet of hexagons can be made to curl by using twelve pentagons [1,5,16,19]. A perfect sphere could be formed with 60 carbon atoms, which was the model of a soccer ball [1,19]. Smalley and his team could explain the stability of C_{60} to further nucleation by a molecular structure with the perfect spherical symmetry of a soccer ball on the basis of geodesic structural concepts [1,4].

The all-carbon molecule C_{60} was proposed to have a truncated icosahedral cage structure or a truncated icosahedron (I_h symmetry) shape—a polygon with 60 vertices and 32 faces, 12 of which are pentagons and 20 are hexagons [1,20]. The C_{60} molecule was named after Buckminster Fuller, the designer of geodesic domes, and was called buckminsterfullerene (buckyball) [1]. The structure proposed for C_{70} had 25 hexagons and 12 pentagons, similar to the shape of a rugby ball [1,4]. The hollow-cage structural explanation or geodesic dome shapes could explain the stability of all the even-numbered carbon clusters greater than about 32 atoms (although less stable than C_{60} or C_{70}) [4]. In honor of Fuller, the new class of molecules was named fullerenes [4]. For their discovery of $C_{60,}$ Harold W. Kroto, Robert F. Curl, and Richard E. Smalley were awarded the Nobel Prize in Chemistry in 1996. Buckminsterfullerene has been designated as the official state molecule of Texas, home to Rice University, where this scientific breakthrough was achieved [16].

2.3 Explosion in Fullerenes Research

Discovery of C_{60} and because of its stability when formed under the most violent conditions led Kroto et al. to believe their assumption that C_{60} may be widely distributed in the universe [1]. However, though the buckminsterfullerene had been discovered, it had not been isolated to confirm its existence. The laser vaporization technique produced not bulk but only microscopic quantities of fullerenes enough for detection [1,17]. A major breakthrough in fullerene research came in 1990 when physicists W. Krätschmer and D.R. Huffman, during their studies of interstellar dust, which they assumed to consist mainly of particles of carbon, were attempting to produce interstellar-like small particles of carbon by the vaporization of graphite in a helium atmosphere, for the first time produced isolable quantities of C_{60} by resistive heating of two graphite rods in a helium atmosphere and extracting the carbon condensate so formed using an organic solvent [4,6,8,21]. These studies were actually carried out between 1983 and 1990 and started with the discovery of unexplained ultraviolet (UV) absorptions in the soot samples they had produced and ended with the extraction of fullerenes in crystalline form from these samples [4,21]. They had observed peculiar features (named camel hump) in the UV spectrum of the soot deposits in their initial studies on absorption features of carbon soot, which at that time could not be identified and were actually a signature of fullerene production [4]. Following the discovery of fullerenes, Krätschmer et al. started studying IR emissions from giant red stars and found that when helium pressure was just right, the IR spectrum of the soot produced by graphite vaporization showed four sharp, characteristic strong absorption bands as predicted by theory for buckminsterfullerene [2,6]. When they prepared their soot sample using the heavier isotope of carbon C^{13}, the four IR absorption bands were displaced by the expected amount [10]. They used benzene to solubilize the black carbon soot and obtained a wine-red solution or a brown liquid, depending on the concentration and thus, the first concentrated solution of fullerenes ever seen was obtained [4,8]. Tiny crystals were obtained upon evaporating the solvent [8]. The crystals redissolved in the solvent. The crystals could also be sublimed onto a substrate at 400°C in a vacuum or an inert atmosphere to obtain a crystalline material, which they called fullerite [8]. The x-ray analyses of the crystalline material showed the material to consist of 7 Å diameter spheroidal molecules confirming the presence of C_{60} (or C_{70}) [2,8]. The mass spectrometry data also showed evidence of C_{60} and C_{70} [21]. The IR bands of a coating of solid C_{60} on silicon substrate were at the same positions as for the graphitic soot sample before purification with the four most intense lines at 1429, 1183, 577, and 528 cm^{-1} [8]. These results were the first confirmation of the C_{60} fullerene structure proposal [2,4,8]. In a separate and parallel study, Taylor et al. performed mass spectral analyses of benzene soluble material extracted from arc-processed graphite [9]. The red soluble extract was processed by Soxhlet extraction by treating carbon deposit with benzene [2]. While the extract showed unequivocal evidence of C_{60} (720 mass peak) and C_{70} (840 mass peak), the mass spectra also showed the presence of other even-numbered carbon species in the extract such as C_{58} and C_{68} [2,3]. The chromatographically separated soluble magenta extract gave a single ^{13}C NMR line corresponding to C_{60}, which proved that all 60-carbon atoms are equal as expected for the truncated icosahedral structure of buckminsterfullerene [9]. The ^{13}C NMR of red fraction confirmed the structure of C_{70}, which was proposed to have D_{5h} symmetry [2,9].

The reports that isolable quantities of fullerenes could be synthesized and the confirmation of the identity of fullerenes by spectroscopic, crystallographic, and control experiments using ^{13}C-enriched material generated a huge amount of interest in fullerenes [6,8–10,22–24]. By 1991, it was relatively easy to produce fullerenes inspired by the techniques of Donald Huffman, Wolfgang Krätschmer, and Konstantinos Fostiropoulos [2,4,7,25,26]. Later on Smalley et al. developed an electric arc discharge technique to vaporize graphite electrodes instead of a resistive heating method to obtain carbon soot containing fullerenes [3,4,26,27]. Higher fullerenes such as C_{76}, C_{78}, C_{84}, C_{90}, and C_{94} and giant fullerenes were synthesized and studied by these techniques [24,26,28,29].

C_{60} and other fullerenes have also been noticed in normal candle soot, as had also been predicted by Kroto et al. [4,5,30]. The flame on the wick provides sufficient conditions for fullerene formation processes to occur. In 1992, fullerenes were found in a family of carbon-rich metamorphic rock known as shungites in Karelia, Russia [17,31].

In 2010, fullerenes (C_{60}) were discovered in a cloud of cosmic dust surrounding a distant star 6500 light years away [32]. Using NASA's Spitzer infrared telescope, the scientists spotted the molecules with unmistakable IR signature [32]. Sir Harry Kroto commented: "This most exciting breakthrough provides convincing evidence that the buckyball has, as I long suspected, existed since time immemorial in the dark recesses of our galaxy" [32].

2.4 Methods of Synthesis of Fullerenes

The most important fullerene synthesis methods are the laser vaporization technique, resistive heating, and electric arc discharge technique. These are high-temperature methods (up to 4000°C) that involve vaporization of graphite to produce fullerenes [7].

2.4.1 Laser Vaporization Method

The laser vaporization method was developed by Smalley's team at Rice University [1]. In this method, a pulsed laser beam using a Nd:YAG laser is made to strike against a rotating graphite target under a helium atmosphere [1,4,20]. In the original method used

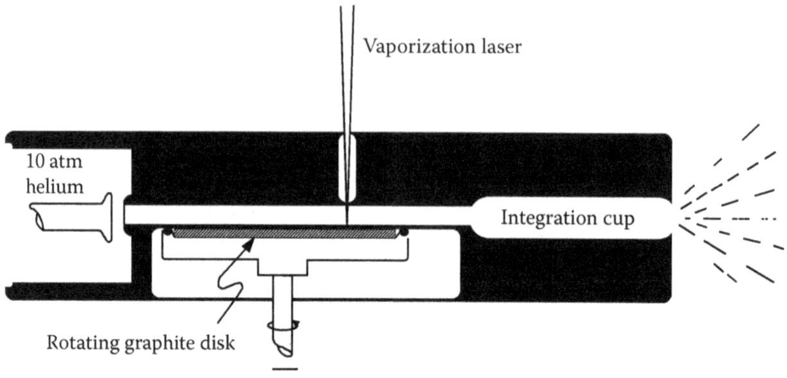

FIGURE 2.3
Schematic diagram of laser vaporization apparatus used to generate carbon-cluster beams that are probed downstream with a time-of-flight mass spectrometer. (Reprinted by permission from Macmillan Publishers Ltd., *Nature*, H.W. Kroto, J.R., Heath, S.C. O'Brien, R.F. Curl, and R.E. Smalley, 318, no. 6042 (1985): 162–163, copyright (1985).)

by Smalley et al. [1] on laser ablation, graphite was ablated in an inert gas buffer at room temperature, which resulted in microscopic quantities of fullerenes [2,4]. The dense carbon plasma created during laser vaporization cooled too quickly for growing carbon clusters to anneal into stable fullerenes. Smalley et al. modified the method, and graphite ablation was carried out in a high-temperature furnace that increases the yield of fullerenes since the laser-vaporized carbon plume expands farther while it is still hot enough to anneal [4]. In this method the graphite target was kept in an oven at 1200°C while passing helium over it slowly [4,23]. The presence of high-density helium in the vaporization zone provides the necessary thermal conditions (annealing) for slow condensation and clustering. Free expansion of the cluster-laden gas through a supersonic nozzle cools the cluster into a supersonic beam that is analyzed by a mass spectrometer. By using an integration cup (Figure 2.3) [1], the time between vaporization and expansion increases and results in cluster equilibration to the most stable carbon species, that is, fullerene C_{60} [1,2].

2.4.2 Resistive Heating of Graphite Electrodes

The landmark experiment of Krätschmer et al. led to an explosion of research on fullerenes [6,8,21,28]. In this method carbon soot is produced through resistive heating of graphite electrodes having a very small contact surface [26]. The tip of the graphite rod acting as an anode is sharp and is in contact with a flat graphite rod acting as a cathode. An alternating current (AC) applied between the two graphite electrodes in an atmosphere of helium evaporates the sharp anode by resistive heating. The contact heats up to 2500°C–3000°C and the electrode tips start glowing [33]. Smoke is generated, which rises up and is deposited as the carbon soot (fullerenes) on the cool walls of a glass bell jar reactor that is filled with inert quenching gas helium at about 150 torr. Helium gas pressure is thoroughly controlled to ensure the optimal yield of fullerenes [31]. The soot-containing fullerenes is collected from the walls of the reactor. Extraction and purification of the carbon soot yields fullerenes in about 15% [31,33] to almost 25%–30% yield [24,26]. An improvement in the yield of extracted carbon material as well as optimized separation procedures, enriched samples of higher fullerenes, which are otherwise present in trace amounts in raw extract, can be collected in milligram quantities [24,26]. The role of buffer gas is to cool the plasma by collisions with the carbon vapor. Therefore, gas has to be inert to avoid reactions with the smaller carbon clusters or atoms formed initially [33]. The yield reduces drastically by using nitrogen, the nitrogen atoms formed in the hot zone of the reactor react with carbon fragments. The use of helium as the buffer gas results in the highest yield of fullerenes. Also, electrodes with a diameter of about 3 mm are employed, one of them with a pointed end to provide the smallest possible contact area. This ensures evaporation of graphite from its tip. If thicker graphite electrodes are used, the glowing area expands over the whole electrode, thus diminishing evaporation of graphite from its tip [31].

2.4.3 Electric Arc Discharge: Arc Heating or Arc Vaporization of Graphite Electrodes

Electric arc discharge, a method of bulk synthesis of fullerenes, involves passing a high current (direct current, DC) between two pointed or sharp graphite electrodes, one electrode acting as the anode and the other as the cathode and to strike an arc between the electrodes kept at a specific distance [4,26,31,33]. The power is thus dissipated in the arc rather than heating the entire electrode [24,33]. The electrodes are housed within a vacuum chamber filled with an inert gas such as helium at a pressure of about 150 to 250 torr. Arc discharge between the ends of the graphite rods produces plasma. The temperature (~4000 K) in the resulting plasma between the two graphite electrodes is sufficient to vaporize the anode. Under optimized conditions of current, voltage, and helium pressure the vaporizing carbon atoms condense to form fullerenes in the carbon soot, which is collected from the walls of the chamber. The yield of dissolvable fullerenes is typically between 10% and 20% of the vaporized carbon [4,31,33]. The inert gas helium helps to avoid any reactions with the carbon species formed during vaporization while holding the fullerene precursors long enough near the arc to grow and anneal effectively and to cool the plasma by collisions with the carbon vapor. The carbon soot containing fullerenes is then subjected to a series of extraction steps to isolate and purify C_{60}, C_{70}, and larger fullerenes. This method allows efficient evaporation of carbon with thicker graphite electrodes up to 6 mm graphite rods [33]. However, with larger dimensions the yield of fullerenes decreases almost linearly with increasing diameter of the rods due to the sensitivity of the fullerenes to the intense UV radiation originating from the central portion of the arc plasma due to which the initially formed fullerenes react with other carbon clusters present leading to a nonvaporizable insoluble carbon product [31,33].

2.4.4 Other Methods of Fullerene Synthesis

Several other methods such as hydrocarbon pyrolysis, hydrocarbon combustion, electron beam irradiation of graphene, and solar radiation have also been used to synthesize C_{60} and other fullerenes. Electrodes made of coal have also been used in the DC electric arc discharge method to produce fullerenes [31]. Thermal processes like pyrolysis or combustion provide building blocks for fullerene growth by using hydrocarbons as starting materials.

2.4.4.1 Pyrolysis

Pyrolysis or thermal cracking of hydrocarbons such as polycyclic aromatic hydrocarbons (PAHs) as carbon precursors allows synthesis to be carried out at lower temperatures since hydrocarbons decompose at a lower temperature than graphite and PAHs already contain structural elements of the fullerene cage [7,31]. Pyrolysis of PAHs such as naphthalene has been used to synthesize fullerenes C_{60} and C_{70} at 1000°C in an inert gas [34,35]. Dehydrogenative coupling reactions lead to fullerene formation [27]. Laser pyrolysis of hydrocarbons has also been used to synthesize fullerenes containing carbon soot [27,36].

2.4.4.2 Combustion of Hydrocarbons

Combustion of hydrocarbons to synthesize fullerenes was reported by Howard et al. [23,30]. Existence of fullerenes in sooty flames was first suggested by Smalley's group [4]. The combustion system used in the flame synthesis method of fullerene synthesis consists of a premixed laminar flame stabilized on a water-cooled burner [30]. Premixed laminar flames of fuel (benzene or acetylene) and oxygen (either pure or mixed with argon, helium, or nitrogen) are produced under different sets of flame conditions such as pressure; carbon-to-oxygen ratio; type and concentration of diluent gas argon, helium, or nitrogen; and various gas velocities at the burner [30]. Benzene is the most common source of carbon and the concentration of oxygen and temperature of combustion are other important parameters [31]. Each flame is maintained for 1 to 3 h, during which a sample of condensed compounds and soot is withdrawn from the flame, extracted with toluene and analyzed by mass spectrometry [10,30]. The combustion technique has the advantage that the properties of the combustion flame may be varied and used to control the size distribution of fullerene clusters [23]. Combustion of benzene–oxygen mixtures in laminar flow flames produces C_{60} and C_{70} in varying ratios depending on the temperature, pressure, and carbon–oxygen ratio [7,23,30]. Sooty flames of benzene/oxygen/argon at pressures of around 40 torr contain C_{60}, C_{70}, and higher fullerenes such as C_{76}, C_{78}, C_{80}, C_{84}, C_{86}, C_{88}, C_{90}, C_{92}, C_{94}, C_{96}, C_{98}, C_{100}, C_{102}, C_{108}, and the abundance of higher fullerenes is higher than that of arc discharge [7]. By choosing suitable reaction parameters an increased portion of C_{70} up to 80% in the fullerene soot can be obtained by combustion of hydrocarbons [31].

2.4.4.3 Fullerene Synthesis Using Coal

Another method of fullerene synthesis includes using coals [37,38]. The idea behind using coals as starting materials is to make the process economical. Coals can be carbonized and graphitized to obtain carbon electrodes for the DC electric arc discharge process. Studies have been carried out using different types of Australian and Chinese coals from brown coals, bituminous coals, and semianthracite to anthracite coals [37,38]. Depending upon the carbon content and ash content of the starting coal the yields of C_{60} and other fullerenes vary.

2.4.4.4 Electron Beam Irradiation of Graphene under Transmission Electron Microscopy

Electron beam irradiation is a process of fullerene formation from graphene sheet. Under transmission electron microscopy (TEM), when graphene is exposed to electron beam, the carbon atoms in graphene rearrange to form fullerenes [7,39]. The high energy of the electron beam is transferred to the carbon atoms of the graphene causing fragmentation of large sheets of graphene into smaller flake-like structures. The final product of these transformations is often a perfect fullerene molecule. The transformations are readily observed in TEM. This top-down mechanism of fullerene formation has been explained in four steps: (1) loss of carbon atoms at the edge of graphene, leading to (2) the formation of pentagons, which (3) triggers curving of graphene into a bowl-shaped structure and which (4) subsequently zips up its open edges to form a closed fullerene structure [39].

2.4.4.5 Solar Generators

Solar furnaces are used to produce fullerenes by using parabolic mirrors for collecting sunlight and focusing the solar radiation on a graphite target housed in a pyrex tube, and enclosed by a tungsten preheater in an inert gas atmosphere such as argon [27,34,40,41]. High temperatures around 3000°C are achieved, which vaporize graphite to form carbon clusters that anneal to form fullerenes [7,27]. In this arrangement, although the sunlight is used to evaporate graphite, the exposure of generated fullerenes to radiation is far less extensive than with resistive heating or arc vaporization techniques [27]. The condensing carbon vapor quickly moves from the intensive sunlight, cools, and deposits on the upper walls of the tube.

2.5 Purification, Isolation, and Characterization of Fullerenes

2.5.1 Purification of Fullerene Soot

The fullerene soot produced is a mixture of mainly C_{60} and C_{70} with small amounts of C_{76}, C_{78}, C_{82}, C_{84}, C_{88}, C_{90}, C_{92}, C_{94}, C_{96}, and higher fullerenes [10,26,31,42]. The fullerene mixture is soluble in nonpolar solvents such as benzene, toluene, and carbon disulfide [8,23,34,42]. The first step in the purification process is the extraction of soot enriched in C_{60} and C_{70} in a solvent such as toluene using the Soxhlet apparatus [23,26,31,43,44]. Alternatively, the toluene-soot mixture can be sonicated at room temperature or refluxed and then filtered to remove insoluble carbon products from toluene solution of fullerenes [23]. Soxhlet is preferred over reflux due to recycling of the hot solvent [24]. The yield of red soluble extract in benzene increases from 14% by refluxing in benzene for 24 h to up to 26% by Soxhlet extraction in benzene for 24 h [24]. The yield of fullerene extract in a solvent also depends on the Soxhlet extraction time and the yield of soluble fullerenes increases as extraction time increases from 4 h to 24 h [24]. The fullerene extract still contains hydrocarbon impurities such as linear hydrocarbons [23]. To remove the hydrocarbon impurities from the extract, the soluble fullerene product is evaporated to dryness using a rotary evaporator [23,24]. The resulting black solid is washed with diethyl ether to solubilize and selectively remove the hydrocarbons. Solid that remains after washing contains different fullerenes.

Parker et al. [24] performed extensive studies on the extraction of carbon soot in various solvents. Mass spectrum of the benzene extract obtained from fullerene soot showed existence of larger clusters up to C_{266} in small amounts. They devised a sequential extraction scheme for fullerene soot and successively extracted the solid (remaining fullerene soot) left behind in the Soxhlet thimble with high boiling solvents following a series of benzene-pyridine-1,2,3,5-tetramethylbenzene (TMB) to increase the yield of soluble fullerenes and extract higher fullerenes in these solvents [24]. The total yield of soluble materials consisting of fullerenes C_{60} to C_{466} is 44% for this series of extractions. The yield of soluble extract increases with increase in boiling point of the solvent. Use of high boiling polar aprotic solvents like N-methyl-2-pyrrolidone (NMP) resulted in extracting higher fullerenes and showed evidence of giant fullerenes [24]. Greater solvent penetration into the soot matrix results in higher extraction yields in NMP. The NMP extract is black in color. They showed that the colors of raw soot extracts are different in different solvents, an indication that different molecular weight ranges are extracted in different solvents. The raw soot extract colors are pink in pentane and hexane, orange in heptane, red in toluene and benzene, greenish-brown in TMB, and black in NMP. Their elaborative studies on extraction of carbon soot showed that up to 94% of the soot is fullerenes [24].

2.5.2 Isolation of Fullerenes

Fullerenes obtained after extraction of raw soot are chromatographically separated into pure C_{60}, pure C_{70}, and higher fullerenes by column/liquid chromatography using alumina columns as the stationary phase or activated charcoal supported on silica gel as the stationary phase and toluene as the mobile phase or using HPLC [23,26,31,42]. In solution, each of the fullerenes exhibits unique color and can be readily separated [34]. The C_{60} exhibits a purple color in toluene and C_{70} a red-wine color [23,34]. Pure C_{60} is obtained as the first fraction and pure C_{70} is obtained as the second fraction [31]. Using Soxhlet chromatography, which requires less solvent and is a combined process of extraction and chromatography, the fullerenes C_{76} to C_{96} that remain on the chromatographic column after elution of C_{60} and C_{70}, are washed out with hexane/toluene solvent mixtures and further separated by repeated HPLC [31]. The toluene extract of the soot material produced by resistive heating of graphite contains a variety of molecules larger than C_{60} and C_{70} in a total amount of 3% to 4% by weight. Enriched samples of higher fullerenes in milligram quantities were obtained by repeated chromatography on neutral alumina using a gradient elution with toluene/hexane and afforded five fractions of slightly different colors (brown to greenish) that gave stable crystalline solids upon evaporation of the solvents and led to the isolation of stable solid samples of C_{76}, C_{84}, C_{90}, C_{94}, and $C_{70}O$, an oxide of D_{5h}-C_{70} [26,28]. An efficient isolation method for the isolation of higher fullerenes (which are present in very small amounts in the raw extract) up to C_{96} using CS_2 as the eluent for HPLC resulted in isolation of C_{76}, C_{78}, C_{82}, C_{84}, C_{90}, and C_{96} by HPLC of CS_2 extracts of the carbon soot produced by the arc heating method [42]. Solid samples of purified fullerenes are obtained by crystallization [2,8].

2.5.3 Characterization of Fullerenes

Fullerenes can be characterized by various techniques such as mass spectroscopy, nuclear magnetic resonance (NMR) spectroscopy, infrared (IR) spectroscopy, Raman spectroscopy, ultraviolet-visible (UV-vis) spectroscopy, and x-ray diffraction (XRD) [10].

As mentioned earlier, mass spectroscopy of carbon clusters formed in the fullerene soot gives an idea about the different fullerene clusters present in the carbon soot [1,2,23]. Mass spectra of purified C_{60} or C_{70} fullerenes give rise to predominantly C_{60}^{+} or C_{70}^{+} molecular ion peaks by using field desorption, fast atom bombardment, or laser desorption techniques for evaporation and ionization. These techniques prevent generation of new carbon species from a pure sample or in other words there is not any significant fragmentation of the pure fullerene and therefore these techniques are useful to analyze purity of the fullerene solid sample [23].

Carbon-13 NMR spectroscopy is a useful technique to probe purity of C_{60} and C_{70} since solution of purified C_{60} gives a single resonance line at 143.2 ppm, whereas solution of lower symmetry purified C_{70} gives five distinct resonance lines at 130.8, 144.4, 147.8, 148.3, and 150.8 [2,9,10,23]. The five NMR lines are in the ratio 1:2:1:2:1 consistent with the elongated structure with D_{5h} symmetry [17]. The ^{13}C NMR spectra of chromatographically purified samples of soluble material extracted from arc-processed graphite are shown in Figure 2.4 [9].

Fullerenes exhibit unique and intense colors when dissolved in nonpolar solvents like hexane, toluene, benzene, xylene, and carbon disulfide [34]. C_{60} exhibits a purple color, C_{70} a red-wine color, and higher fullerenes (C_{76} to C_{94}) exhibit yellowish to green colors in solution due to electronic absorptions in the visible region of the electromagnetic spectrum [23]. The intense UV bands can be used to characterize pure fullerenes. Coupled with HPLC, it becomes a useful analytical tool for fullerene sample purity [23,34]. The

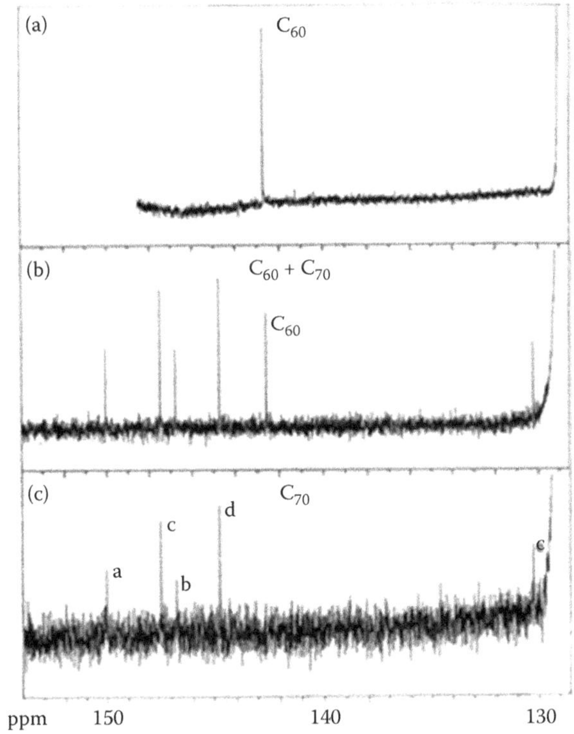

FIGURE 2.4
^{13}C NMR spectra of chromatographically separated C_{60} and C_{70}. (From R. Taylor, J.P. Hare, A.K. Abdul-Sada, and H.K. Kroto, *Journal of the Chemical Society, Chemical Communications* 20 (1990): 1423–1425. Reproduced by permission of The Royal Society of Chemistry.)

UV-visible spectra of chromatographically separated C_{60} and C_{70} have been studied [43,44]. Figure 2.5 [44] shows the UV-visible spectra of chromatographically separated C_{60} and C_{70} in hexane solution. The spectra of C_{60} shows bands at 213, 257, and 329 nm [2,44]. There is a weak band at 404 nm. The detailed spectra of the broad shoulder band from 440 to 670 nm has maxima at 500, 540, 570, 600, and 625 nm.

Following the discovery of C_{60} theoretical calculations focused on vibrational properties of C_{60} were performed [2,10]. The results showed that, because of its high symmetry, C_{60} should have only ten Raman active and four IR active modes [2,4,10]. Infrared spectrum of purified C_{60} shown in Figure 2.6 [8] shows four distinct IR absorption lines at 1429, 1183, 577, and 528 cm^{-1} [8,10,45].

The presence of only four strong IR bands in the spectrum agrees well with theoretical predictions which have shown that the free, truncated icosahedral C_{60} molecule with its unusually high symmetry will show only four IR active vibrational modes [4,8,10]. The presence of some other weak IR lines or any additional bands may be due to absorption by C_{70} or some other impurity in the sample. Both vibrational Raman and IR measurements on chromatographically separated C_{60} and C_{70} fullerene clusters were performed by Bethune et al. [45]. The vibrational frequencies for C_{60} (Raman lines and IR lines) agreed well with the calculated values. The C_{70} spectra were distinctive but complex.

XRD of solid crystalline C_{60} shows a distance of 10Å between the centers of adjacent C_{60} molecules and a molecular diameter of about 7.1Å consistent with a van der Waals interatomic distance of 2.9Å [2,4,8,34].

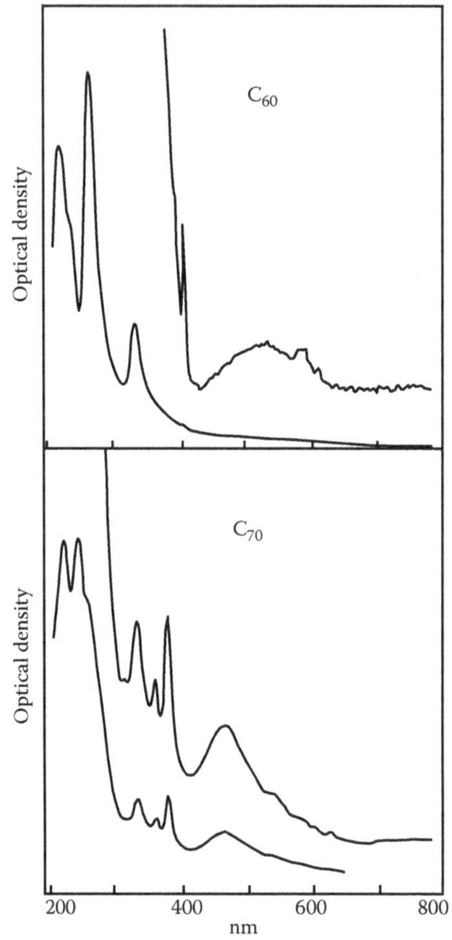

FIGURE 2.5
UV/visible spectra of chromatographically separated C_{60} (top) and C_{70} (bottom) in hexane solution. (From J.P. Hare, H.W. Kroto, and R. Taylor, *Chemical Physics Letters* 589 (2013): 57–60. With permission.)

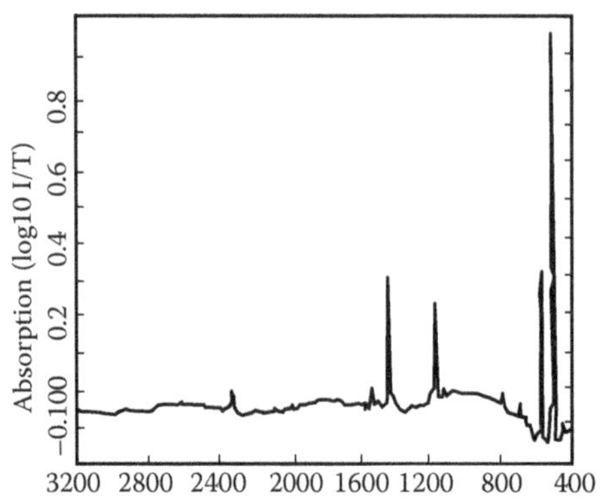

FIGURE 2.6
IR spectrum of pure C_{60}. (Reprinted by permission from Macmillan Publishers Ltd., *Nature*, W. Kratschmer, L.D. Lamb, K. Fostiropoulos, and D.R. Huffman, 347: 354–358, copyright 1990.)

The HPLC analyses, UV-visible spectra, mass spectral analyses, and IR analyses of toluene extracts of flame-derived soot are consistent with those reported for fullerenes obtained by graphite vaporization [30]. By analyzing the chromatogram of toluene extract using the technique of continuous acquisition of UV spectra as peaks elute from the HPLC, Howard et al. found that the chromatogram did not show the presence of peaks associated with the typical PAH commonly produced in flames [30].

2.6 Formation Mechanism of Fullerenes

The formation of fullerenes, especially C_{60}, is an interesting study and generated curiosity because of its abundance and stability shown during mass spectral analyses. The process of formation of fullerenes from graphite is recognized as a kinetically controlled process [25,27,34]. Graphite is the thermodynamically most stable form of carbon, however, vaporization of graphite under certain conditions yields fullerenes, a kinetically stable molecule [25,27,34]. In order to explain the formation of fullerenes, different mechanisms were put forward such as the fullerene road mechanism and the pentagon road mechanism [7]. High yields of fullerene C_{60} up to 40% of all the carbon vaporized is most satisfactorily explained by the pentagon road mechanism [25]. The presence of pentagons in the fullerenes provided a clue as to the formation mechanism of fullerenes [4]. According to the pentagon road mechanism, the fullerene formation starts with carbon atoms in the hot vapor forming linear carbon chains. The chains slowly condense and grow to form open graphitic sheets. Under the right reactive conditions of high temperature and high density of helium the free floating optimally open graphitic sheets with dangling bonds attain stability by curling up to the most favorable structure that inevitably involves formation of pentagons (pentagon rule) [4,10]. Closure by pentagon formation minimizes the number of dangling bonds in a given growing structure and eventually the closure reduces the number of dangling bonds to zero [34]. In addition, if the pentagons are isolated the structure is extra stable compared to fused or adjacent pentagons, which are strained [10]. This is also known as the isolated pentagon rule (IPR) [27,34]. By acquiring curvature or closure to a perfect sphere to avoid unstable and reactive dangling bonds, C_{60} becomes a stable and most favorable product during condensation of pure hot carbon. High temperature is important for the graphitic sheets to anneal effectively. In other words, the carbon vapor or hot carbon plasma created during arc discharge or laser vaporization cools and condenses slowly if the temperature is high enough for carbon clusters to get enough time to rearrange and adopt the most stable form. The inert gas helium in the vaporization zone is critical in the annealing process [4,10]. At high pressures of about 100 to 200 torr, helium provides the right annealing conditions for carbon clusters. At low buffer gas pressure carbon radicals diffuse far away from the hot zone and continue to grow in an area too cool to anneal to spherical molecules [10]. At very high pressure of buffer gas, a very high number of carbon radicals are produced in the hot reaction zone so that the carbon concentration is so high that the graphitic sheets form faster than annealing and particles growing beyond 60 carbon atoms [10,31]. An optimum high density of helium in the vaporization zone ensures a longer time for linear carbon chains to slowly grow to graphitic sheets, which inevitably curl up and condense to form stable fullerenes [4]. Helium controls the migration as well as cooling or condensation of high-energy carbon species in fullerene formation.

Ping and Da performed theoretical studies to explain fullerene formation, which were in good agreement with experimental results [46]. Isotope fractionation studies during fullerene synthesis by carbon arc evaporation also helped in understanding the mechanism of fullerene formation [47]. Isotope composition studies showed that while the raw fullerene soot, the solvent-extracted fullerene soot, C_{60} and C_{70} were enriched in C^{13} relative to graphite anode, the cathode deposit was depleted in C^{13}. The slower diffusing C^{13} containing carbon species have a higher residence time in the hottest zone of the carbon arc, allowing sufficient reaction time to form larger carbon intermediates and fullerene precursors, which eventually form C^{13}-enriched fullerenes.

A detailed mechanism for the formation of C_{60} and C_{70} fullerenes in flames has been given by Howard et al. [30]. The soot growth and evolution of the unique structural features of fullerenes is based on the transformation of polycyclic aromatic hydrocarbons produced in flames that undergo intramolecular rearrangements and other reactions such as addition of C_2 species or elimination of hydrogen, ring closure, and so forth to form fullerenes C_{60} and C_{70} [30].

2.7 Structure of Fullerenes

2.7.1 Structures of Diamond, Graphite, and C_{60}

Before the discovery of fullerenes, diamond and graphite were the only known allotropes of crystalline carbon [34]. Both diamond and graphite are network solids that consist of infinite network of C atoms [4]. Diamond crystallizes as a hard solid consisting of a continuous 3D covalent bonding network of sp^3 hybridized carbon atoms (C–C–C bond angle 109.5° of a regular tetrahedron) [34]. This results in a very strong 3D crystal lattice with cubic symmetry. Graphite crystallizes as a soft solid consisting of layers of 2D hexagonal lattice arrangement of carbon atoms. Each layer is made up of a continuous 2D covalent bonding network of sp^2 hybridized carbon atoms (C–C–C bond angle of 120° of a regular hexagon) with a delocalized π-bonding network within each layer [34]. The different layers in graphite are held together by weak van der Waals forces.

With the discovery of fullerenes there were three known forms of pure carbon: the network solids, diamond and graphite; and a new class of discrete molecules—the zero-dimensional (0D) fullerenes [4,10]. Euler proved in the 18th century that a closed cage or cluster made up of only hexagons and pentagons must always contain 12 pentagons, but can have any number of hexagons other than one [10,48]. The fullerenes are carbon clusters that can be described by this rule [7,23,48].

C_{60}, the most abundant representative of the fullerene family, has the closed cage structure of a truncated icosahedron—a polygon with 60 vertices and 32 faces of which 12 faces are pentagons, and 20 are hexagons [1,11,20]. An icosahedron is a 20-faced polyhedron with 12 vertices. Each face of an icosahedron is an equilateral triangle and at each of the 12 vertices of the icosahedron, five of the triangles meet. Truncating or cutting off these vertices replaces each of them by a pentagonal face and also converts each of the twenty former trianglular faces into a hexagon. The truncated icosahedron is one of the Archimedean solids [2]. The oldest known picture of a truncated icosahedron was painted by Pierodella Francesca (1420–1492), who was an Italian renaissance painter and mathematician. The hollow or skeleton model of truncated icosahedron painted by Leonardo da Vinci appeared in the book *De Divina Proportione* by Fra Luca Pacioli, published around 1498 [2,48]. The shape of icosahedron and that of truncated icosahedron can also be visualized in the art depicted in an Italian Cathedral (Figure 2.7). In C_{60} the pentagonal rings sit at the vertices of an icosahedron such that no two pentagonal rings are adjacent to each other [1,10]. Thus there are 20 hexagonal and 12 pentagonal rings forming a spherical shape with 60 carbon atoms at each of the 60 vertices, with the perfect spherical symmetry of a soccer-ball [1,11,19,48–50], as shown in Figure 2.8 [49].

Each carbon in the sphere is bonded to three other carbon atoms and is sp^2 hybridized and forms σ-bonding network [34]. But the strain in the structure resulting from the pentagons to effect closure of the cage introduces a mixture of $sp^2 + sp^3$ character in fullerenes. Due to this local curvature at each carbon site, the hybridization in C_{60} is not pure sp^2 as in a planar graphite sheet [10,34,51]. The pyramidalization of the sp^2-hybridized carbon atoms result in an excess of strain to C_{60} [11,52]. The fourth electron, that is, the π electron

FIGURE 2.7
The shape of C_{60} can also be visualized in the art depicted in an Italian cathedral. (Courtesy of Ole Krogh Andersen, http://www2.fkf.mpg.de/andersen/fullerene/vatican.html. With permission.)

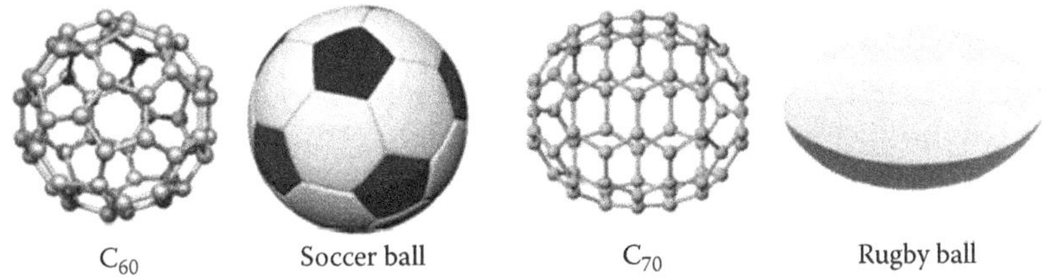

FIGURE 2.8
Structure of (a) C_{60} buckminsterfullerene resembles a soccer ball and (b) C_{70} resembles a rugby ball. (With kind permission from Springer Science+Business Media: *Structural Chemistry*, 21, 1155–1169, 2010, T.C. Dinadayalane and J. Leszczynski.)

of each carbon atom is localized in the hexagonal rings [11]. The C_{60} molecule thus has two bond lengths. The hexagon ring bonds, corresponding to 30 hexagon–hexagon edges or 6,6 junctions, shared by two adjacent hexagons, with increased double bond character can be considered "double bonds" and are shorter with a bond length of about 1.38Å [11,52]. The pentagon ring bonds corresponding to 60 hexagon–pentagon edges or 5,6 junctions, fusing a pentagon and a hexagon, are somewhat longer, about 1.45Å, since C_{60} tends to avoid double bonds in the pentagonal rings [11,34,52]. It is the release of excess strain in the curved structure of C_{60} that is responsible for the chemical reactivity of fullerene. A release of strain is in fact associated with the change of hybridization from sp^2 to sp^3 that accompanies most chemical reactions [11,52]. The chemical reactivity of C_{60} is that of a strained, electron-deficient polyalkene [52]. It reacts readily with nucleophiles [11]. The vast majority of reactants attack the 6,6 ring junctions of C_{60} [11]. The chemical reactivity of C_{60} is discussed in a later section.

C_{60} is highly strained due to the curvature in the structure but the geodesic and electronic bonding factors in the structure make it a kinetically stable structure [11,25,27,34,52]. The advantage of curvature and the loss of dangling bonds far exceeds the loss of stability due to inefficient π-bonding [10]. It is a perfectly symmetrical structure with no dangling bonds or adjacent pentagons, which can introduce further strain in the molecule [10]. C_{60} is the smallest fullerene with 12 isolated pentagons and is therefore the most stable fullerene [2,7]. All 60-carbon atoms are chemically equivalent or identical and indistinguishable [34,51]. This is based on its icosahedral symmetry and experimentally based on the single peak observed in the ^{13}C NMR at about 143 ppm [9].

2.7.2 Electronic Structure of C_{60} Fullerene

A knowledge of the electronic structure of C_{60} is useful in understanding its properties. To begin with, in a simple way, carbon has a ground state electronic configuration of $1s^22s^22p^2$ with two core 1s electrons and four valence electrons. To form a sp^2 bonded network of carbon atoms, one of the electrons in the 2s orbital is promoted to the empty 2p orbital resulting in three sp^2 hybridized orbitals containing three of the four valence electrons on each carbon atom. The fourth electron is in the unhybridized p orbital. In C_{60}, 60 carbon atoms combine to form a sphere of σ-bonded network of sp^2 hybridized orbitals while the 60 remaining p orbitals, which contain the fourth valence electron of each of the 60 carbon atoms and were not involved in sp^2 hybridization and the σ-bond formation, form the π-bonds [34]. In the Huckel molecular orbital approach, these 60 p orbitals form 60 molecular orbitals. The Huckel π-molecular orbital diagram of C_{60} is shown in Figure 2.9 [51]. There are 30 bonding molecular orbitals and 30 antibonding molecular orbitals. In the ground state all 60 electrons occupy the bonding orbitals and this represents a measure of the stability of the molecule [34]. As also mentioned in the previous section, due to the local curvature at each carbon site, the hybridization is not pure sp^2 in C_{60} but there is an intermediate rehybridization between pure sp^2 (graphite) and sp^3 (diamond), which leads to enhanced electron affinity of C_{60} [34,51].

The electronic structure of the C_{60} shows that the highest occupied molecular orbital (HOMO) is completely occupied and there are three low-lying degenerate lowest unoccupied molecular orbitals (LUMO) [34,52]. C_{60} can readily accept six additional electrons that can go into the three degenerate LUMOs in electrochemical reductions or, in chemical conversions, should be occupied by the electrons transferred by the lone pair of an attacking nucleophile [34,52].

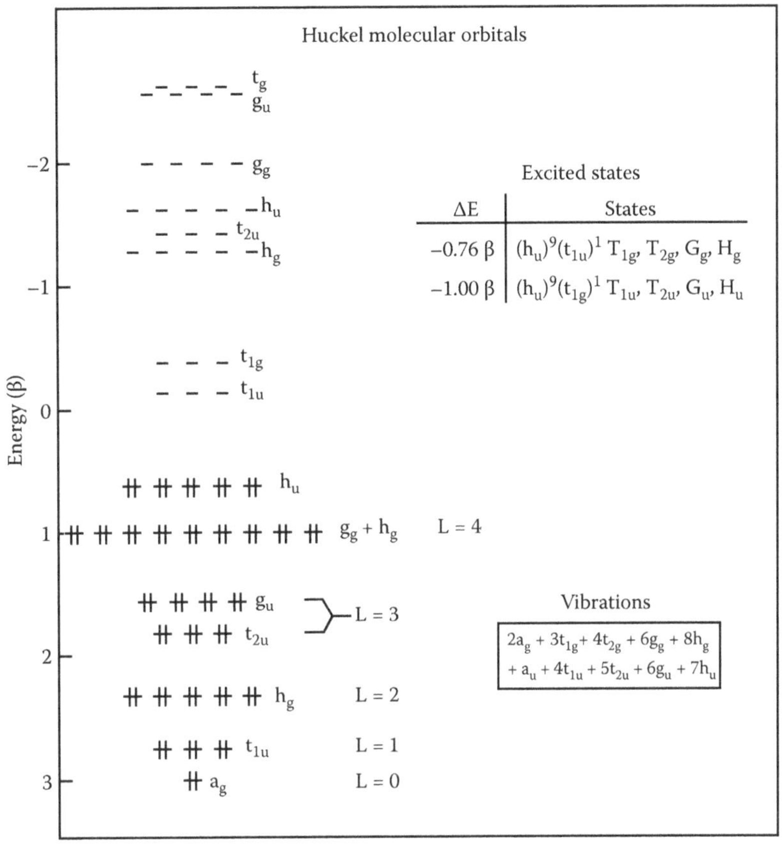

FIGURE 2.9
Energy-level diagram of Huckel π-molecular orbitals for C_{60}. (Reprinted from *Chemical Physics Letters* 125, no. 5, R.C. Haddon, L.E. Brus, and K. Raghavachari, 459–464, Copyright 1986, with permission from Elsevier.)

2.7.3 Structure of Other Fullerenes

The proposal that closed hollow cages form when carbon nucleates in the vapor phase and that C_{60} was a truncated icosahedron with enhanced stability due to chemical and geodesic factors was also confirmed by theoretical and experimental studies [2,53]. Theoretical studies based on chemical and geodesic factors and experimental studies based on mass spectrum predicted and showed a "magic" number series in the 20–80 carbon atoms range [2,10,50,53]. These "magic" fullerenes in the order of increasing stability were C_{20}, C_{24}, C_{28}, C_{32}, C_{36}, C_{50}, C_{60}, and C_{70} [2,10].

For carbon, hexagons and pentagons are likely structures to form a ball, since these are the only structures with small or negligible strain [10]. For a ball to close with only 5- and 6-sided faces, mathematically, 12 pentagonal faces are required with any number of hexagons other than one [7,10,23,48]. There is no ring or angle strain on an sp^2 hybridized carbon in a hexagon since the sp^2 hybridized carbon in a hexagon has planar bonds at 120° to each. There is an angle strain of 12° on each sp^2 hybridized carbon in a pentagon since an isolated sp^2 hybridized carbon in a pentagon has internal angles at 108°, which introduce a ring strain in the structure [10]. Two pentagons together produce 24° angle strain per carbon atom. Pentagons should therefore be isolated to avoid ring strain. For the IPR allowed fullerene C_n, applying Euler's theorem, there are 12 pentagons and (n/2) – 10 hexagon faces, that is, 20 hexagon faces for C_{60} [34]. The soccer ball C_{60} with I_h symmetry is the first structure that allows pentagons to be isolated from each other and is therefore the smallest fullerene that follows IPR [7,10,34]. Other isomers are possible for fullerene with 60 carbon atoms but the most stable and symmetric is the I_h isomer [7]. The ellipsoidal C_{70} with D_{5h} symmetry is the next smallest fullerene that follows the IPR [34]. For fullerenes with less than 60 carbon atoms there are no isolated pentagons leading to low stability of these fullerenes [7]. The C_{50} fullerene is the first to avoid triple pentagon corners; C_{28} is the first to avoid quartet pentagon corners. The smaller cages C_{24} and C_{20} can also exist, but they are even less stable [10].

The dominance of C_{60} and C_{70} was due to the fact that these were the smallest fullerenes in which no two pentagons share an edge [2,4,17]. Fullerenes with more than 70 carbon atoms have pentagons in strained positions and typically possess lower symmetry [4,7,10,34]. As discussed earlier the solvent extracted fullerene soot contains higher fullerenes C_{76}, C_{78}, C_{84}, C_{90}, and beyond up to giant fullerenes [24,26,28,29,42]. These fullerenes show the IPR with 12 pentagons in each fullerene structure and a corresponding increase in the number of hexagon faces [34]. Several IPR-allowed isomers are possible for higher fullerenes and the number of isomers increases exponentially as the number of carbon atoms increases [7,10]. Some isomers possess an isolated pentagon structure with an open cage structure [10,34]. The icosahedral C_{80} is an electronic open shell structure but it has not been isolated [34]. Open shell molecules are less stable than closed shell molecules, but they can be stabilized by metal encapsulation to form isolable endohedral fullerenes (e.g., $Sc_3N@C_{80}$) [34]. Endohedral fullerenes are discussed in a later section. The structure of "magic" number fullerenes and some of the higher fullerenes are shown in Figure 2.10 [54,55].

2.7.4 Structure of Fullerite

The discrete C_{60} molecules crystallize as a three-dimensional molecular solid, solid C_{60} (also called fullerite), held together primarily by weak dispersion forces [8,22,56–58].

The solid is highly compressible, whereas the molecular structure is essentially incompressible [58]. This solid-state crystalline form of C_{60} differs from other crystalline forms of carbon, namely, diamond and graphite. Diamond and graphite are atomic crystals with atoms as the repeating units; fullerenes are molecular crystals with fullerene molecules as the repeating units [7]. The fullerene molecules are held together by van der Waals forces and crystallize in a face-centered cubic structure (fcc lattice) with a lattice parameter of 14.17 Å at room temperature [23,56–58]. Figure 2.11 [57] shows the fullerite structure. The separation between nearest neighbor C_{60} clusters is 10 Å in this structure [2,23,34,56]. The distance between adjacent 7.1 Å diameter C_{60} molecules is thus 2.9 Å, consistent with a van der Waals interatomic distance [23,34].

For fcc hard-sphere packing in solid C_{60}, 26% of the available volume is empty and this interstitial space permits the insertion or intercalation of metal ions by exohedral metal doping between the C_{60} molecules [23,34,56]. There are two tetrahedral sites and one octahedral site per C_{60} molecule [23] in the solid C_{60}. The octahedral and tetrahedral holes of sizes 2.06 and 1.12 Å, respectively, in the crystal provide spaces that can be accommodated by a variety of species such as metal dopant species without any change in the basic structure of C_{60} [23]. Thus the intercalated solid has the stoichiometry A_3C_{60} when each of the available sites is filled with a single atom A [56] in the fcc lattice of solid C_{60}.

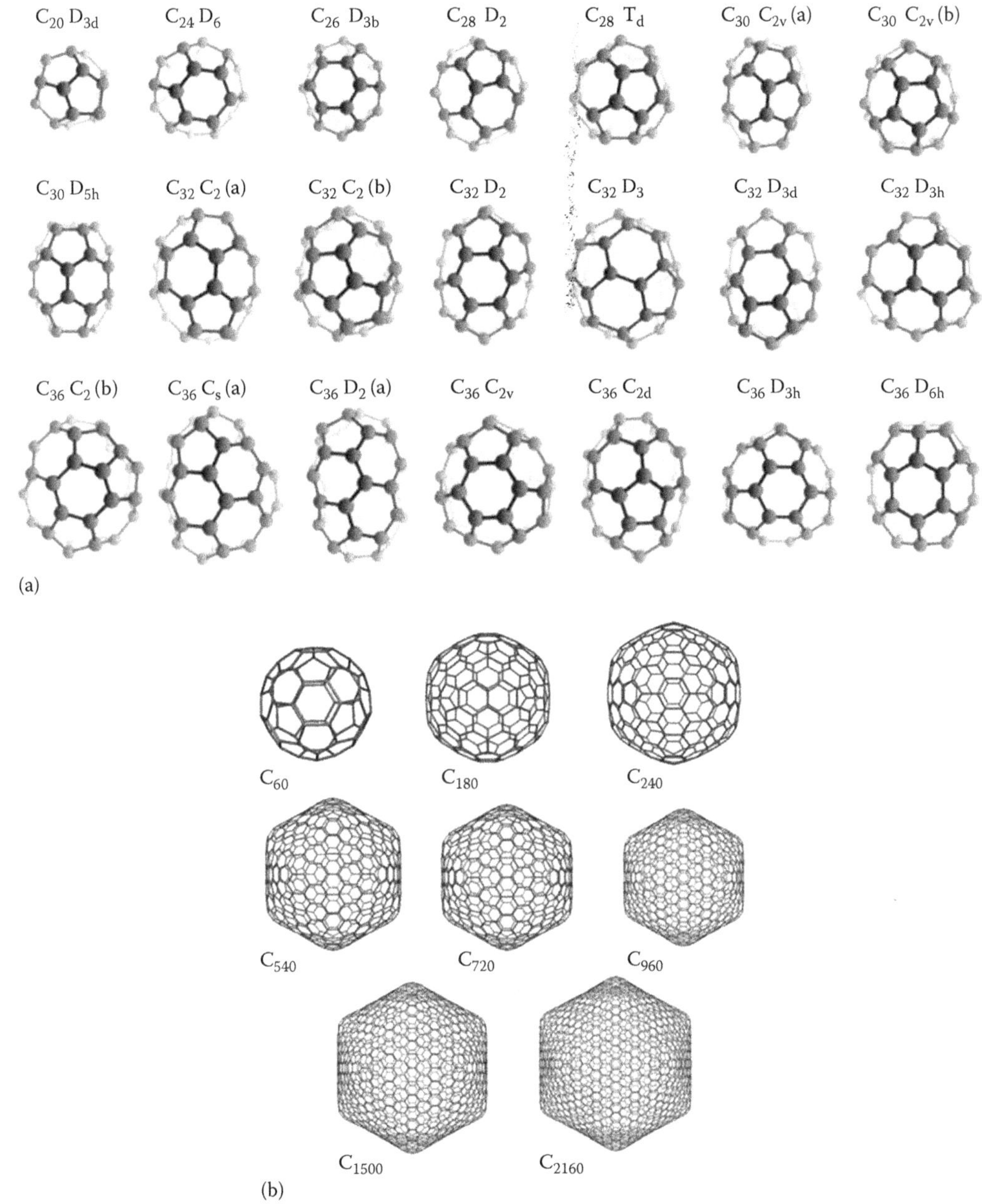

FIGURE 2.10
Structure of (a) small fullerenes and (b) giant fullerenes. ((a) Reprinted with permission from E. Małolepsza et al., *Journal of Physical Chemistry A* 111 (2007): 6649–6657. Copyright (2007) American Chemical Society; (b) Reprinted with permission from R.R. Zope et al., *Physical Review B* 77: 115452. Copyright (2008) by the American Physical Society.)

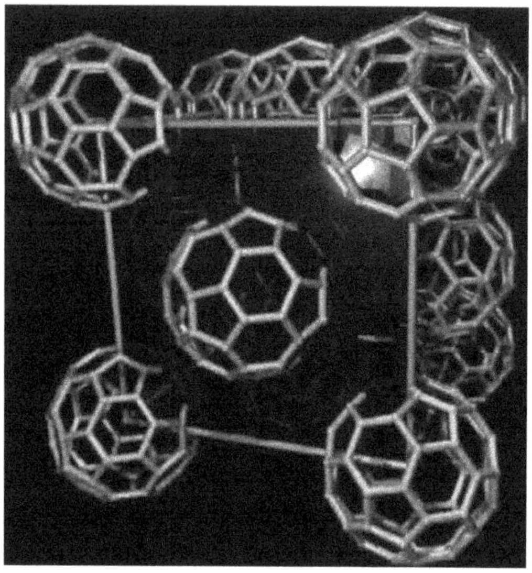

FIGURE 2.11
FCC crystal of C_{60} (fullerite). (Reprinted from H. Terrones and M. Terrones, *New Journal of Physics* 5 (2003): 126.1–126.37. © IOP Publishing & Deutsche Physikalische Gesellschaft. CC BY-NC-SA.)

2.8 Properties of Fullerenes

2.8.1 Physical Properties of Fullerenes

Fullerenes are extremely strong and resilient molecules with high bulk modulus and high impact strength [17,59,60]. The fcc crystals of C_{60}, held together only by van der Waals forces, are as soft as graphite [4]. Theoretical calculations suggest that the effective bulk modulus of a single C_{60} buckyball molecule is 668 GPa when compressed to 75% of its size in the closest packing of the crystals in the fcc lattice [59,60]. This property makes fullerenes become harder than steel and diamond, whose bulk moduli are 160 GPa and 442 GPa, respectively [4,59,60]. Fullerenes are extremely strong molecules, able to resist great pressures—they will spring back or bounce back to their original shape and normal volume when the pressure is relieved after being subject to over 3000 atmospheres [4,60]. Ion beam scattering experiments demonstrated the exceptionally high stability of fullerenes with respect to impact-induced fragmentation processes thus predicting high resilience for these molecules [17,61]. The fullerene C_{60} has remarkable impact strength being able to withstand high-speed collisions of up to 15,000 mph against stainless steel plate, merely bouncing back, unharmed, keeping their shapes, thus showing high stability [4,60,62].

2.8.2 Chemical Properties and Reactions of Fullerenes

The electronic structure of C_{60} (Figure 2.9) shows that the HOMO is completely occupied and there are three low-lying degenerate LUMOs. Thus C_{60} is more of an electron acceptor (electrophile) than an electron donor. C_{60} can readily accept six additional electrons that can go into the three degenerate LUMOs [34]. The C_{60} fullerene has just the right combination of size, chemical stability, and activity to serve as the building block for a large variety of solids with fascinating properties [63].

2.8.2.1 Metal-Doped Fullerenes

One of the most interesting properties of solid C_{60} or the bulk C_{60} are the electronic properties, where in various forms it can act as an insulator, conductor, semiconductor, or a superconductor [4]. Metal-doped fullerenes are the first three-dimensional organic electrical conductors [10]. Metal-doped fullerenes are solids that involve metal doping in the interstitial holes within the lattice of solid C_{60} [64,65].

The fcc lattice of C_{60} has octahedral and tetrahedral holes within the lattice, which can accommodate metal atoms to form metal-doped solids known as fullerides [23,56]. The process of metal doping in interstitial holes, also called metal intercalation, involves charge transfer from donor metal to acceptor C_{60} [34]. Vapor phase diffusion is one of the most important methods to prepare metal-doped C_{60} [23]. Metal doping or intercalation of C_{60} with alkali metals provides a way to alter the electronic properties of fullerenes since the smaller alkali metal ions should be able to fit into the interstices in the lattice without disrupting the network of contacts between the carbon spheroids [66]. Solid C_{60} is a semiconductor [4,7,58]. The addition of three electrons (half-filling) to the degenerate LUMO by exohedral metal doping with potassium leads to an fcc K_3C_{60} structure that is an electrical conductor at higher temperatures but is a superconductor at 18 K [4,34,67,68]. Figure 2.12 [68] shows the structure of metal-doped C_{60} (K_3C_{60}) with potassium ions filling the single octahedral hole and the two tetrahedral holes per C_{60} as per the stoichiometry requirements for fcc lattice. Both C_{60} and K_3C_{60} correspond to fcc lattices of nearly identical unit cell size: 14.11 Å for C_{60} and 14.24 Å for K_3C_{60} [68]. Potassium buckide, as it is also called, with potassium ions filling the cavities between the buckyballs, is a stable metallic

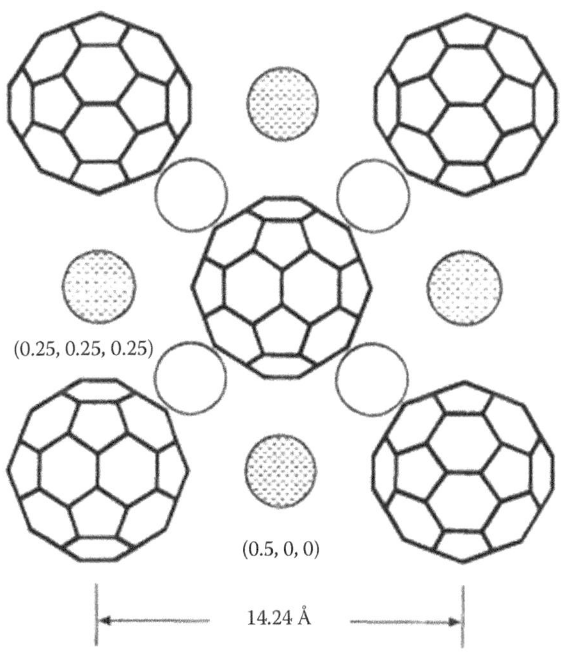

FIGURE 2.12
Structure of metal-doped C_{60} (K_3C_{60}). The open (0.25, 0.25, 0.25) and hatched (0.5, 0, 0) spheres represent the potassium at the tetrahedral and octahedral sites, respectively. (Reprinted by permission from Macmillan Publishers Ltd., *Nature*, P.W. Stephens, L. Mihaly, P.L. Lee, R.L. Whetten, S.-M. Huang, R. Kaner, F. Deiderich, and K. Holczer, 351: 632–634, copyright 1991.)

crystal and the first completely three-dimensional molecular metal [4,10]. The addition of six electrons by external doping yields the bcc (body-centered cubic) structure K_6C_{60} that has poor electrical conductivity [34]. In general, for the alkali metals A (Li, Na, K, Rb, Cs) only A_3C_{60} fcc exhibits superconductivity [23,34,58]. The A_6C_{60} bcc exhibit poor electrical conductivity. The superconducting temperature can be increased to 28 K when using Rb and can be further increased to 33 K by using Cs and Rb together [7]. A superconducting temperature of 40 K has been reported for Cs-doped C_{60} [69].

2.8.2.2 Endohedral Metallofullerenes

Endohedral fullerenes are fullerenes containing one or more atoms within its cage [31,70–76]. When metal atoms are trapped inside the fullerenes, the fullerenes are called endohedral metallofullerenes [34]. Figure 2.13 [76] shows a schematic illustration of endohedral metallofullerenes. The first endohedral structures isolated were the monometallofullerenes [70]. The discovery and proposal of the stable hollow sphere structure of C_{60} led Smalley et al. to wonder whether metal atoms could be trapped inside the fullerene cage [4,70]. The evidence that a single lanthanum atom can be trapped inside came within a week of the initial discovery of C_{60} when laser vaporization of a lanthanum-impregnated graphite disc produced concentrations of clusters at mass 859 corresponding to lanthanum plus C_{60} [70]. Further evidence for the existence of a metal atom inside the carbon cage came in a series of experiments called "shrink wrap" performed by Smalley's group [4,71]. This technique involves suspending carbon clusters in a magnetic field and blasting them with a laser just enough to knock pairs of carbon atoms out of the lattice, a method called "shrink-wrapping" [4]. The shrinking stops at C_{32} and the next blast bursts the brittle cluster into linear-chain fragments. These experiments were also performed for metal-encapsulated fullerenes causing the laser to shrink wrap the metal atom in a carbon envelope. It was found that the carbon cage bursts at a cluster size dictated by the ionic radius of the internal atom [4,71,72]. The potassium and lanthanum atom stops the shrinking process at C_{44}, while the slightly larger cesium atom stops at C_{48}. These studies not only proved the hollow-cage structure of fullerenes but also confirmed the existence of metal atoms inside the cage [71,72]. The general notation for fullerenes with metals inside their cage is $M@C_n$, where M is the metal inside the cage of a fullerene molecule with "n" carbon atoms [72]. For C_{60}, an endohedral fullerene will be represented as $M@C_{60}$ [72].

Laser vaporization of a lanthanum oxide–graphite composite rod heated to 1200°C in a stream of argon gas produced lanthanum containing fullerenes $La@C_{60}$, $La@C_{70}$, $La@C_{74}$ and $La@C_{82}$ in addition to the dominant products C_{60} and C_{70} [72]. $La@C_{82}$ was the only metallofullerene to be isolated by the dissolving and extracting in toluene in the presence

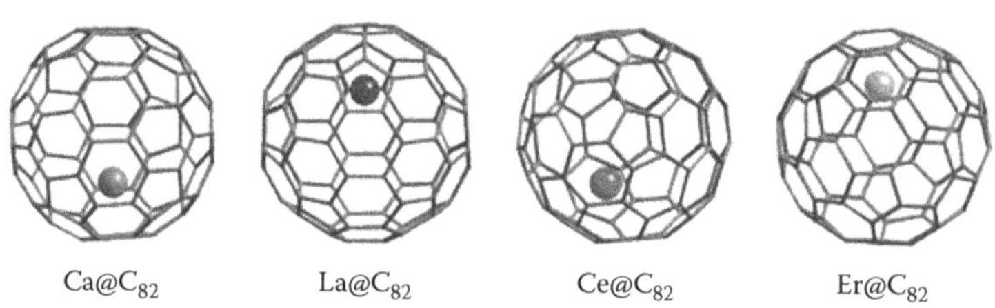

FIGURE 2.13
Schematic illustration of endohedral metallofullerenes. (Reprinted by permission from Macmillan Publishers Ltd., *Nature Chemistry*, K. Suenaga et al., *Nature Chemistry* 1: 415–418, copyright 2009.)

of air showing its unique stability [10,72]. With the development of the arc-discharge technique to generate macroscopic quantities of fullerenes [73], the metal encapsulated fullerenes were also prepared by arc vaporization method where metal–metal oxide is incorporated in the graphite rods prior to the arc vaporization process [60,74]. Generally, endohedral metallofullerenes yield 2% or less from the fullerene soot prepared by the arc discharge method [75]. Mass spectral analyses of the soot is a useful characterization technique to identify the presence of metal encapsulated fullerene clusters in the soot obtained from vaporization of metal–carbon rods. In studies done by Chai et al. [72] on lanthanum-containing fullerenes by arc discharge method, the La@C_{82} was found to be the most stable metallofullerene despite C_{82} not being the dominant species under normal conditions. The presence of a species within the cage promotes the production of larger fullerenes C_{82} in favor of C_{60} [72]. The scandium endohedral metallofullerenes have been isolated with scandium monomer, dimer, or trimer in a variety of fullerene sizes when scandium is vaporized with carbon in an inert gas atmosphere by the arc discharge technique [74]. Some of the representative scandium endohedral metallofullerenes that have been studied and isolated are Sc@C_{82}, Sc_2@C_{82}, Sc_3@C_{82}, Sc_2@C_{84}, and Sc_2@C_{86} with Sc_2@C_{82} being the most abundant [34,74].

The electronic shell of the fullerene is influenced by the encapsulated central atom [33]. Metal-encapsulated fullerenes involve charge transfer from metal to fullerene [34,75]. The formation of endohedral metallofullerenes upon arc discharge proceeds via the assembly of a carbon cage around endohedral atoms rather than random trapping of these atoms into the almost ready cages [75,77]. Consequently, the isomeric distribution of resulting molecules is influenced by donor–acceptor interactions between the forming carbon cage and the endohedral atoms. Thus, the structures of synthesizable endohedral compounds frequently differ from the available pristine fullerenes. The abundance of higher fullerene cages, such as C_{80}, C_{82}, and C_{84}, is much increased, and in many cases the products do not obey the isolated pentagon rule [75,77].

With the discovery of the dimetallofullerene, Sc_2@C_{66}, it was shown for the first time by Shinohara's group that a dimetallo structure is sufficient to break the isolated pentagon rule (IPR), which is the stability rule for conventional fullerenes [75].

A new class of cluster fullerenes called trimetallic nitride endohedral fullerenes A_3N@C_{80} (A = Sc, Y, Gd, Ho, Er) has been isolated and characterized [34,75]. This group of trimetallic nitride endohedral fullerenes, with Sc_3N@C_{80} being the first most abundant member, was obtained by a synthesis that involved introducing a small portion of nitrogen gas into the Krätschmer-Huffman generator during vaporization of graphite rods containing metal oxides [34,75]. By introducing NH_3 as the reactive gas, for the first time the nitride cluster fullerenes were produced as the dominant products in the soot, while the relative yield of the empty fullerene and conventional metallofullerenes was less than 5% [75]. It was thus shown for the first time that the endohedral fullerene is the main product in the soot mixture, making the isolation of the endohedral fullerenes much more facile as only one separation step even by a simple chromatographic technique was needed [75]. The carbon cage of such an endohedral fullerene has an isomeric form that has not been isolated as an empty cage so far. The C_{80} carbon cage has icosahedral symmetry and is an unstable open shell molecule in the absence of the trimetallic nitride cluster, but an important feature of these cages is that the internal size (~0.8 nm) of these cages is large enough to accommodate four atom molecular clusters like Lu_3N@C_{80} and Sc_3N@C_{80} [34,75]. The icosahedral C_{80} cage is stabilized by the electron transfer from the nitride cluster wherein the A_3N cluster donates six electrons to stabilize the carbon cage to form $(A_3N)^{+6}$@$(C_{80})^{-6}$ with enhanced stability [34].

The noble gases from He to Kr and nonmetal atoms like N and P also have been incorporated in fullerene cages [33,75]. These structures form preferably C_{60} endohedrals, which are prepared under high pressure or by ion bombardment of the preformed fullerene cages [33,75]. Noble gases have been encapsulated in the fullerene cages at high temperatures (600°C–1000°C) and pressures (40,000 psi). The endohedral fullerene He@C_{60} is the first noble gas–carbon compound [33]. However, unlike conventional metallofullerenes, these endohedral fullerenes are uncharged. Penetration of noble gas atoms into intact fullerene cages through a six-membered ring involves a very high activation energy and therefore must be accompanied by the breaking of at least one bond of the fullerene core [33]. According to a proposed window mechanism, the penetration involves thermal breaking of a 5–6 bond of fullerene and formation of a 9-membered ring large enough for atoms to pass in or out, which after being filled with a noble gas atom can reform by closing the opened bond [33].

The encapsulation of nitrogen (N) or phosphorus (P) atoms leads to endohedral fullerenes N@C_{60} and P@C_{60} [33,60]. There is an odd unpaired electron on both N and P, which are encapsulated in the carbon cage and hence shielded from the environment by the C_{60} fullerene leading to exceptional spin properties [33,60]. The incarceration of naked nitrogen atoms in a carbon cage leads to very interesting properties for these endohedral fullerenes and potential use in future quantum computing applications.

2.8.2.3 Functionalized Fullerenes

The enormous potential of fullerenes in practical materials science and biological applications is limited by hydrophobicity of the carbon cage and limited solubility of fullerenes in polar solvents combined with their aggregation tendency, which further reduces their solubility [11,78]. These processing difficulties can be overcome by organic derivatization of fullerenes through various reactions to obtain chemically modified fullerenes [78]. Functionalization can help to convert the hydrophobic surfaces of fullerenes to provide hydrophilic surfaces for diagnostic purposes and therapeutic purposes [34]. These reactions help to develop new fullerene products through introduction of useful extensions to the fullerene sphere [11,52]. These extensions alter their properties, for instance, solubility and electrochemical behavior, and therefore widen the range of applications in the field of materials science and technology [11,78]. The basic principles of organic chemistry of fullerenes have been summarized by Prato [11].

The C_{60} has conjugated double bonds but all the double bonds are localized in the hexagon rings [11]. It, therefore, is not a superaromatic compound but rather a polyene [11,52,78]. The reactivity of C_{60} is that of a strained electron-deficient alkene with high electron affinity [52,79]. C_{60} behaves as an electrophile and reacts readily with electron-rich species or nucleophiles and undergoes nucleophilic addition reactions [34]. Due to the closed cage structure of C_{60} the attacking agents can be added only to the exohedral surface of C_{60}, and most reactions occur at the highly reactive hexagon–hexagon edges between two pentagon rings [11,34]. The driving force for these reactions is the relief of steric strain [11,52,78]. Some of the important reactions of C_{60} are the Bingel reaction (nucleophilic cyclopropanation of C_{60} to a methanofullerene), [4 + 2] cycloaddition or the Diels Alder reaction, [3 + 2] cycloaddition or the Prato reaction and [2 + 2] cycloaddition [7,34,52,78,80]. Figure 2.14 [78] shows the important functionalization reactions of C_{60}.

Functionalized fullerenes containing 3-, 4-, 5-, and 6-membered rings built onto or across a 6,6 ring junction of C_{60} can be obtained by a variety of reactions [11]. The Bingel Hirsch reaction (Figure 2.14e) [78], or the Bingel reaction, is one of the most valuable

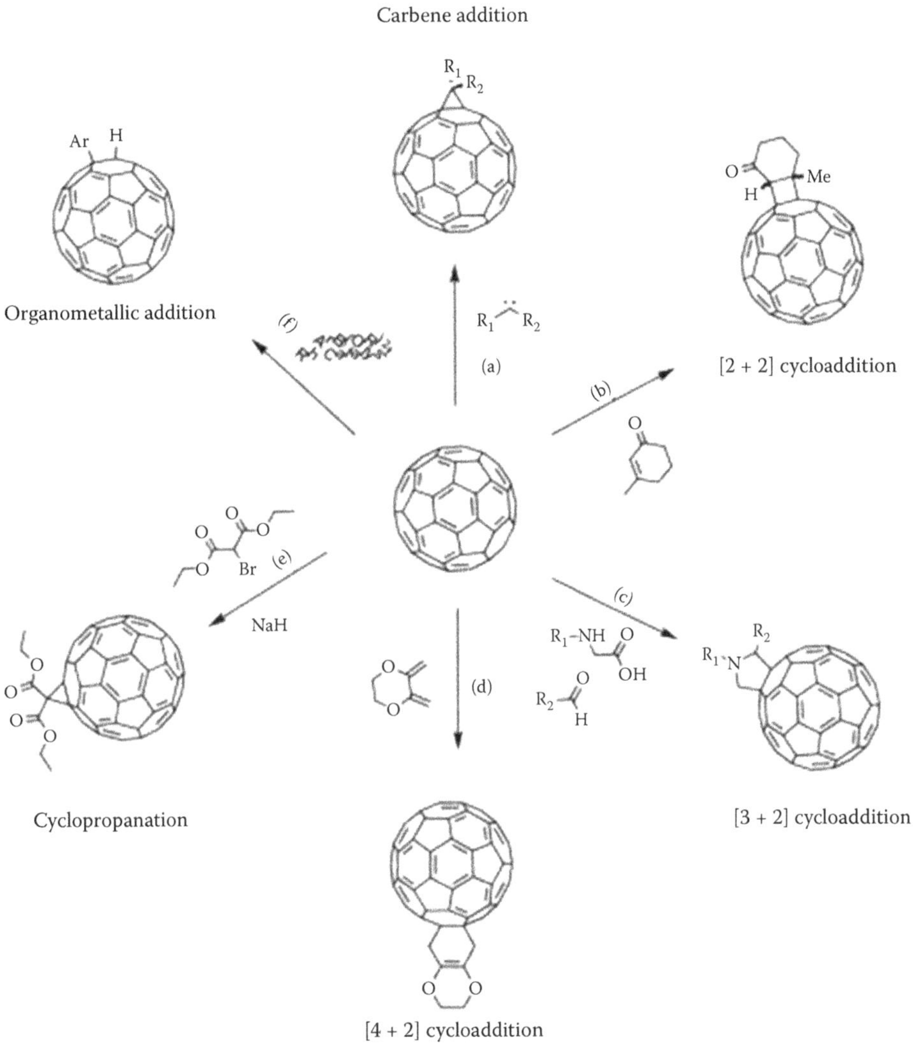

FIGURE 2.14
Covalent functionalization reactions of C_{60} (a) [1 + 2] cycloaddition; (b) [2 + 2] thermal cycloaddition; (c) [3 + 2] cycloaddition with azomethineylides; (d) [4 + 2] Diels–Alder cycloaddition; (e) Bingel–Hirsch reaction; and (f) Rh-catalyzed arylation. (From A.M. Lopez et al., *Journal of Materials Chemistry* 21 (2011): 1305–1318. Reproduced by permission of The Royal Society of Chemistry.)

functionalization reactions of fullerenes [78,80–83]. It involves a nucleophilic attack of a bromomalonate on a double bond of a fullerene in the course of which a cyclopropane ring (three-membered ring) is generated to give a methanofullerene with a bridged (6,6) bond. The Diels Alder reaction is another very important reaction of C_{60}. The 6,6 double bonds of C_{60} act as dienophiles and undergo [4 + 2] cycloaddition reactions with a variety of dienes (Figure 2.14d) [78]. As shown in Figure 2.14d, the classical Diels Alder reaction produces a six-membered ring [11]. The Prato reaction is the [3 + 2] cycloaddition reaction that was applied to fullerenes by Prato [84]. This fullerene functionalization reaction is based on the 1,3-dipolar cycloaddition of azomethineylides to C_{60} and is a powerful and widely used methodology (Figure 2.14c) [78] because of the versatility of the reactions and the stability of the resulting fulleropyrrolidines [78]. A mixture of amino acid and paraformaldehyde are refluxed in toluene to yield an ylide that reacts with the 6,6 double bond in C_{60} via 1,3-dipolar cycloaddition to yield a five-membered ring pyrrolidine derivative called N-methyl pyrrolidine. Azomethineylides can be obtained from a wide variety of easily accessible starting materials and react readily with C_{60}. The substituted pyrrolidine products can be further functionalized [84]. Cyclobutano fullerene derivatives are typically obtained by [2 + 2] cycloaddition reactions as shown in Figure 2.14b [11,78].

Fullerene C_{60} undergoes hydrogenation via Birch reduction using lithium in liquid ammonia in the presence of *tert*-butanol [2,73]. The reduction of C_{60} to $C_{60}H_{36}$ is fully reversible [73]. Oxidation of fullerene with Osmium tetroxide (OsO_4) gave the first confirmation of Buckminsterfullerene structure [2,85,86]. Osmium tetroxide is a powerful but selective oxidant and the osmylation of polycyclic aromatic hydrocarbons had been known for many years. Hawkins [86] explored various addition reactions including osmylation as a route to add functionality to the all-carbon C_{60} molecule. The idea was, that being a new organic compound with no elements other than carbon, C_{60} could serve as the parent and precursor to a whole class of novel organic compound [86]. Osmylation gave discrete stable products. The adduct, formed from a derivative of OsO_4, was $C_{60}(OsO_4)$(4-*tert*-butylpyridine)$_2$. The adduct broke the fullerene's symmetry allowing for crystallization and confirmation of the structure of C_{60} by x-ray crystallography [85,86]. This study yielded the first carbon–carbon bond lengths for the fullerene cage [2,85]. Fullerene epoxides of the type $C_{60}O_n$ such as $C_{60}O$, $C_{60}O_2$, or $C_{60}O_3$, obtained by epoxidation of fullerene and having epoxide groups directly attached to the fullerene cage, constitute an interesting class of fullerene derivatives [87]. A variety of fullerene derivatives are easily prepared from fullerene epoxides, in which the epoxy groups on the fullerene framework are efficiently activated to undergo chemical transformations in the presence of a Lewis acid. These chemical transformations of fullerene epoxides are important for development of functionalized fullerenes for material science applications [87].

Endohedral fullerenes have been functionalized to improve solubility in aqueous media especially for biological applications [34]. The Diels Alder reaction for the functionalization of $Sc_3N@C_{80}$ yields an adduct such as $Sc_3N@C_{80}C_{10}H_{10}O_2$ having a highly oriented crystal structure [34].

Another important and interesting area in the functionalization of fullerenes is the synthesis and applications of open-cage fullerene derivatives [78,88–92]. Vougioukalakis et al. [89] have reviewed the work on open-cage fullerenes. Hummelen et al. [88] first suggested controlled opening of the fullerene, encapsulation of the guest material and restoration of the original fullerene structure. This creative concept or the "molecular surgery" approach to introduce guests inside the fullerene sphere (Figure 2.15) [78] is an alternative strategy for synthesis of endohedral complexes. Open-cage fullerene derivatives, generated via consecutive cleavage of σ- and π- carbon–carbon bonds on the fullerene cage, allow small atoms or molecules to pass through their opening and be placed inside the cavity [89]. Restoration of the ruptured fullerene back to the pristine fullerene cage affords the corresponding endohedral complexes. Successful encapsulation experiments have been performed employing cage-opened fullerene derivatives using various techniques for the creation of an orifice, as well as the expansion of an existing one on the fullerene framework [89–92].

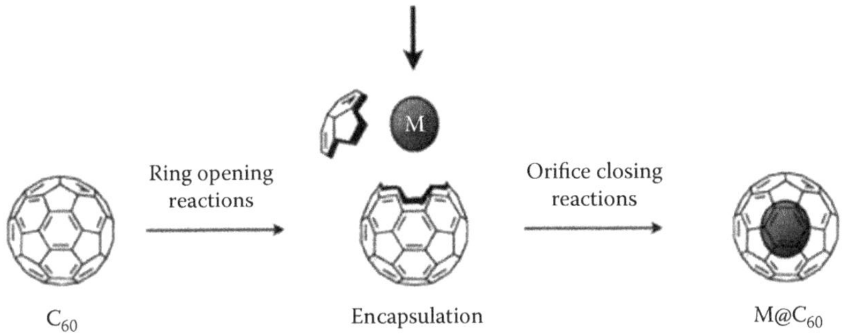

FIGURE 2.15
Ring opening and orifice closing reactions involving hole or cavity generation by provoking σ and π bond scission, insertion of the guest molecule, and reconstruction of the fullerene structure. (From A.M. Lopez et al., *Journal of Materials Chemistry* 21 (2011): 1305–1318. Reproduced by permission of The Royal Society of Chemistry.)

2.9 Applications of Fullerenes

The unique molecular architecture and interesting physicochemical properties of fullerenes find promising applications in various fields such as electronic, biomedical, health and personal care, and polymer composites [7,11,34,60,78,80,93–99]. This section reviews some of these applications. The use of fullerene chemistry to produce compounds useful in materials science and technology has been studied and reviewed by Prato [11,78].

2.9.1 Superconductors

Interest in research on organic superconductors has increased recently with the synthesis of Cs_3C_{60}, which exhibits superconductivity at superconducting transition temperature $T_c = 38$ K under pressure [93,94]. Unlike other metal-doped fullerenes such as K_3C_{60} or Rb_3C_{60} or C_{60} doped with Cs and Rb together discussed earlier, Cs-doped fullerene Cs_3C_{60} is an insulator at normal pressures but shows superconductivity at high pressure [93]. The superconducting transition temperature increases with the increasing size of the alkali metal [94]. As the size of alkali metals increases from Na to Cs (diameter of alkali metal Na = 0.429 nm, K = 0.533 nm, Rb = 0.559 nm, Cs = 0.614 nm) [94], the superconducting transition temperature changes as follows: Na_3C_{60} is not superconducting, $T_c = 18$ K for K_3C_{60}, 28–29 K for Rb_3C_{60}, 33 K by using Cs and Rb together, and 38 K for Cs_3C_{60} [94]. The interstitial atoms do not disrupt the fcc structure of solid C_{60}, thus allowing the overlap of π orbitals of neighboring C_{60} molecules resulting in conductivity [10], however, they do push the C_{60} molecules farther apart, which results in increasing T_c [94]. The Cs_3C_{60} has a bcc crystal structure in contrast to the normal fcc structure for other alkali metal-doped fullerenes. At normal pressure the space between C_{60} is so big due to the large size of the Cs atom that the electrons are pinned to the C_{60} balls and prevent them from relocating. When pressure is applied to Cs_3C_{60} the intermolecular distance between C_{60} is shortened and electrons start moving causing it to become metallic and simultaneously exhibit superconductivity with a high T_c thus causing the transition to a superconductive state.

2.9.2 Organic Solar Cells

Development of polymer-fullerene solar cells is an active area of research [78,95]. Thompson and Frechet [95] have comprehensively reviewed the work on polymer–fullerene composite solar cells. Solar energy has emerged as one of the frontrunners in the research field as an alternative to fossil fuel [95,96]. Use of sunlight for generation of electricity is made possible by a solar cell, or photovoltaic cell, which is an electrical device that converts the energy of light directly into electricity by the photovoltaic effect [95,96]. Semiconductor materials that exhibit photovoltaic effects are commonly used to make solar cells [96]. Polymer-based organic solar cells have the potential to provide an economical, flexible, and lightweight platform for solar energy conversion [60,95]. The function of such excitonic solar cells is based on photo-induced electron transfer from a donor to an acceptor [95]. Fullerenes have become the acceptors of choice because of their high electron affinity and ability to transport charge effectively [7,60]. The most effective solar cells have been made from bicontinuous polymer–fullerene composites, or so-called bulk heterojunctions (BHJs) in which a donor-type conjugated polymer is blended with an acceptor fullerene [7,78]. The BHJ solar cells are currently based on poly(3-hexylthiophene) (P3HT) and the fullerene derivative [6,6]-phenyl C_{61}-butyric acid methyl ester (PCBM), with power conversion efficiencies ranging from 5% to 7.4% [78,95]. Recently, two methanol-soluble fullerene surfactants have been developed as interfacial layers for cathodes in polymer solar cells and the performance of the BHJ solar cells based on using these surfactant-modified cathodes improved significantly to afford high power conversion efficiencies (as high as 6.63% for an Ag cathode) [97].

2.9.3 Biomedical and Personal Care Applications

Application of fullerenes in medicine and health is a growing area of research. The fullerenes and their derivatives have been extensively studied for toxicity, biocompatibility, and various applications such as molecular imaging, therapeutics, drug delivery, biosensing, and cosmetics [7,34,60,80,98,99]. Lalwani and Sitharaman [99] and Bakry et al. [98] have reviewed recent advances towards development of fullerenes and their derivatives for biomedical applications.

2.9.3.1 Biomedical Applications

Some of the medical applications of fullerenes include antiviral activity, antioxidant activity, radical scavenging, biocompatibility, drug delivery, therapeutics, and medical imaging [34,60,98,99]. Compounds with antiviral activities can be used for treatment of human immunodeficiency virus (HIV) infection and fighting acquired immunodeficiency syndrome (AIDS). Replication of HIV can be suppressed by several antiviral compounds, which are effective in preventing or delaying the onset of AIDS. Fullerenes (C_{60}) and their derivatives have potential antiviral activity, which has strong implications on the treatment of HIV infection [98]. Fullerene can sit inside the hydrophobic cavity of HIV proteases, inhibiting the access of substrates to the catalytic site of enzyme [98]. Fullerene-based systems for delivery of drugs and genes possessing selective tissue targeting, and with controlled release capabilities have been developed [99]. Water-soluble fullerenes can successfully pass through the cell membrane and serve as intracellular delivery agents. Endohedral metallofullerenes and other fullerene molecules with an atom of gadolinium inside and with chemical appendages that make them water soluble have been studied as

MRI contrast agents [34,60,99]. Compared to a typical MRI contrast agent where the metal gadolinium is lined to a nonfullerene molecule and is normally excreted quickly from the body, the fullerenes encapsulated with gadolinium might allow the contrast agent to remain in the body longer, allowing doctors to perform slower studies [60].

2.9.3.2 Personal Care Applications

Fullerenes find use in skin care applications as topical applications and therapeutic agents in cosmetics and sunscreens for free radical-scavenging, anti-aging, whitening, anti-inflammation, anti-wrinkle, and pore-tightening [98–104]. Use of fullerenes in cosmetics has been reviewed by Lens [100]. The fullerene C_{60} is a radical sponge due to its radical scavenging properties [101,104]. Strong antioxidant capacities and effective radical scavenging properties have made fullerenes suitable active compounds in the formulation of skin care products for the fast growing cosmetics industry [100]. Oxidative stress on skin resulting primarily from exposure to UV radiation can be prevented by utilizing antioxidant properties of fullerenes. Fullerenes inhibit oxidation by reacting with free radicals and reactive oxygen species [101,102]. Vitamin C60 BioResearch Corporation launched the world's first fullerene-contained cosmetic ingredient, the water-soluble fullerene Radical Sponge, in 2005 [103]. Radical Sponge is a blend of water-soluble polymer polyvinylpyrrolidone (PVP) and 1,3-butylene glycol (BG) that dissolves fullerene. Vitamin C60 BioResearch Corporation launched oil-soluble fullerene LipoFullerene in 2009 whereby fullerene blends with essential oils [103]. Many cosmetic products such as skin lotions, essential beauty oils, creams, cleansers, make-up bases and so forth contain Radical Sponge and LipoFullerene.

2.9.4 Biosensors

The unique electrochemical and physicochemical properties, together with the biocompatibility characteristics of fullerene, allow its wide use for designing the highly sensitive chemical/biosensors [80]. Advances in fullerene–C_{60} research for its use in biosensor applications has been reviewed by Pilehvar and De Weal [80]. A biosensor is an analytical device that consists of a biological recognition element in direct spatial contact with a transduction element, which ensures the rapid and accurate conversion of the biological events to measurable signals [80]. Fullerenes (C_{60}) are ideal nanomaterials for absorbing energy, taking up electrons and releasing them to the transducer.

Research on C_{60} over the years since its discovery has shown that C_{60} is a versatile carbon nanomaterial. Fullerenes and their derivatives are already showing promising and potential applications in varied fields like material science and technology, electronics, energy, biotechnology, health, and personal care, and will continue to play an important role in the future.

References

1. H.W. Kroto, J.R., Heath, S.C. O'Brien, R.F. Curl, and R.E. Smalley. C60: Buckminsterfullerene. *Nature* 318, no. 6042 (1985): 162–163.
2. H.W. Kroto, A.W. Allaf, and S.P. Balm. C60: Buckminsterfullerene. *Chemical Reviews* 91, no. 6 (1991): 1213–1235.

3. H.W. Kroto. Introduction. In *The Fullerenes*, edited by H.W. Kroto, J.E. Fischer, and D. Cox, 1–3. Pergamon Press, 1993.
4. R.F. Curl and R.E. Smalley. Fullerenes. *Vigyan Scientific American* (1991): 32–41.
5. B. Basu. Strange balls of carbon. *Science Reporter* (1991): 9–15.
6. W. Krätschmer, K. Fostiropoulos, and D.R. Huffman. The infrared and ultraviolet absorption spectra of laboratory-produced carbon dust: Evidence for the presence of the C60 molecule. *Chemical Physics Letters* 170, no. 2–3 (1990): 167–170.
7. H. Terrones. Fullerenes and beyond: Complexity, morphology, and functionality in closed carbon nanostructures. In *Springer Handbook of Nanomaterials*, edited by R. Vajtai, 83–104. Berlin: Springer-Verlag, 2013.
8. W. Kratschmer, L.D. Lamb, K. Fostiropoulos, and D.R. Huffman. Solid C60: A new form of carbon. *Nature* 347 (1990): 354–358.
9. R. Taylor, J.P. Hare, A.K. Abdul-Sada, and H.K. Kroto. Isolation, separation and characterisation of the fullerenes C60 and C70: The third form of carbon. *Journal of the Chemical Society, Chemical Communications* 20 (1990): 1423–1425.
10. M.A. Wilson, L.S. Pang, G.D. Willett, K.J. Fisher, and I.G. Dance. Fullerenes—Preparation, properties, and carbon chemistry. *Carbon* 30, no. 4 (1992): 675–693.
11. M. Prato. [60]Fullerene chemistry for materials science applications. *Journal of Materials Chemistry* 7, no. 7 (1997): 1097–1109.
12. S. Iijima. Helical microtubules of graphitic carbon. *Nature* 354, no. 6348 (1991): 56–58.
13. K.S. Novoselov, A.K. Geim, S.V. Morozov, D. Jiang, Y. Zhang, S.A. Dubonos, I.V. Grigorieva, and A.A. Firsov. Electric field effect in atomically thin carbon films. *Science* 306, no. 5696 (2004): 666–669.
14. D.E.H. Jones. Dreams in a charcoal fire: Predictions about giant fullerenes and graphite nanotubes. In *The Fullerenes: New Horizons for the Chemistry, Physics and Astrophysics of Carbon*, edited by H.W. Kroto and D.R.M. Walton, 9–18. Cambridge University Press, 1993.
15. R.E. Smalley. Discovering the fullerenes (Nobel Lecture). *Angewandte Chemie International Edition in English* 36, no. 15 (1997): 1594–1601.
16. J. Ginsberg. The discovery of fullerenes. In *A National Historic Chemical Landmark Brochure*. American Chemical Society, October 11, 2010.
17. P.W. Stephens. The C_{60} molecule. In *Physics & Chemistry of Fullerenes: A Reprint Collection*, vol. 1, edited by P.W. Stephens, 1–7. World Scientific, 1993.
18. E.A. Rohlfing, D.M. Cox, and A. Kaldor. Production and characterization of supersonic carbon cluster beams. *The Journal of Chemical Physics* 81, no. 7 (1984): 3322–3330.
19. S. Begly, R. Service, and D.J. Wilson. It's a buckyball. *Newsweek*, October 26, 1992, 56–58.
20. R.F. Curl. Pre-1990 evidence for the fullerene proposal. *Carbon* 30, no. 8 (1992): 1149–1155.
21. W. Krätschmer. The story of making fullerenes. *Nanoscale* 3, no. 6 (2011): 2485–2489.
22. W. Krätschmer and D.R. Huffman. Fullerites: New forms of crystalline carbon. *Carbon* 30, no. 8 (1992): 1143–1147.
23. C.M. Lieber and C.C. Chen. Preparation of fullerenes and fullerene-based materials. *Solid State Physics: Advances in Research and Applications* 48 (1994): 109–148.
24. D.H. Parker, K. Chatterjee, P. Wurz, K.R. Lykke, M.J. Pellin, L.M. Stock, and J.C. Hemminger. Fullerenes and giant fullerenes: Synthesis, separation, and mass spectrometric characterization. *Carbon* 30, no. 8 (1992): 1167–1182.
25. J.R. Heath. Fullerenes: C60's smallest cousin. *Nature* 393, no. 6687 (1998): 730–731.
26. C. Thilgen, F. Diederich, and R.L. Whetten. The higher fullerenes. In *Buckminsterfullerenes*, edited by W.E. Billups and M.A. Ciufolini, 59. Wiley-VCH, 1993.
27. A. Hirsch and M. Brettreich. Parent fullerenes. In *Fullerenes: Chemistry and Reactions*, edited by A. Hirsch and M. Brettreich, 1–48. Wiley-VCH, 2005.
28. F. Diederich, R. Ettl, Y. Rubin, R.L. Whetten, R. Beck, M. Alvarez, S. Anz, D. Sensharma, F. Wudl, K.C. Khemani, and A. Koch. The higher fullerenes: Isolation and characterization of C76, C84, C90, C94, and C70O, an oxide of D5h-C70. *Science* 252, no. 5005 (1991): 548–551.

29. F. Diederich and R.L. Whetten. Beyond C60: The higher fullerenes. *Accounts of Chemical Research* 25, no. 3 (1992): 119–126.
30. J.B. Howard, A.L. Lafleur, Y. Makarovsky, S. Mitra, C.J. Pope, and T.K. Yadav. Fullerenes synthesis in combustion. *Carbon* 30, no. 8 (1992): 1183–1201.
31. A. Krueger. Fullerenes—Cages made from carbon. In *Carbon Materials and Nanotechnology*, edited by A. Krueger, 33–122. Wiley-VCH, 2010.
32. G. Goodman, M.E Gershwin, and D. Bercovich. Fullerene and the origin of life. *The Israel Medical Association Journal* 14, no. 10 (2012): 602–606.
33. A. Hirsch. The parent fullerenes. In *The Chemistry of the Fullerenes*, edited by A. Hirsch, 1–36. Georg Thieme Verlag, 1994.
34. H.C. Dorn and J.C. Duchamp. Fullerenes. In *Introduction to Nanoscale Science and Technology*, edited by M.D. Ventra, S. Evoy, and J.R. Heflin, Jr., 119–134. Springer Science and Business Media, 2004.
35. R. Taylor, G.J. Langley, H.W. Kroto, and D.R. Walton. Formation of C60 by pyrolysis of naphthalene. *Nature* 366, no. 6457 (1993): 728–731.
36. X. Armand, N. Herlin, I. Voicu, and M. Cauchetier. Fullerene synthesis by laser pyrolysis of hydrocarbons. *Journal of Physics and Chemistry of Solids* 58, no. 11 (1997): 1853–1859.
37. J.S. Qiu, Y. Zhou, Z.G. Yang, D.K. Wang, S.C. Guo, S.C. Tsang, and P.J.F. Harris. Preparation of fullerenes using carbon rods manufactured from Chinese hard coals. *Fuel* 79, no. 11 (2000): 1303–1308.
38. L.S. Pang, A.M. Vassallo, and M.A. Wilson. Fullerenes from coal: A self-consistent preparation and purification process. *Energy & Fuels* 6, no. 2 (1992): 176–179.
39. A. Chuvilin, U. Kaiser, E. Bichoutskaia, N.A. Besley, and A.N. Khlobystov. Direct transformation of graphene to fullerene. *Nature Chemistry* 2, no. 6 (2010): 450–453.
40. L.P.F. Chibante, A. Thess, J.M. Alford, M.D. Diener, and R.E. Smalley. Solar generation of the fullerenes. *The Journal of Physical Chemistry* 97, no. 34 (1993): 8696–8700.
41. D. Laplaze, P. Bernier, C. Journet, J.L. Sauvajol, D. Bormann, G. Flamant, and M. Lebrun. The use of solar energy for the production of fullerenes and porous silicon. *Journal de Physique III* 7 no. 3 (1997): 463–472.
42. K. Kikuchi, N. Nakahara, T. Wakabayashi, M. Honda, H. Matsumiya, T. Moriwaki, S. Suzuki, et al. Isolation and identification of fullerene family: C76, C78, C82, C84, C90 and C96. *Chemical Physics Letters* 188, no. 3 (1992): 177–180.
43. H. Ajie, M.M. Alvarez, S.J. Anz, R.D. Beck, F. Diederich, K. Fostiropoulos, D.R. Huffman, W. Kraetschmer, and Y. Rubin. Characterization of the soluble all-carbon molecules C60 and C70. *Journal of Physical Chemistry* 94, no. 24 (1990): 8630–8633.
44. J.P. Hare, H.W. Kroto, and R. Taylor. Reprint of: Preparation and UV/visible spectra of fullerenes C60 and C70. *Chemical Physics Letters* 589 (2013): 57–60.
45. D.S. Bethune, G. Meijer, W.C. Tang, H.J. Rosen, W.G. Golden, H. Seki, C.A. Brown, and M.S. de Vries. Vibrational Raman and infrared spectra of chromatographically separated C60 and C70 fullerene clusters. *Chemical Physics Letters* 179, no. 1 (1991): 181–186.
46. Z. Zhen-Ping and G. Yong-Da. Structure of carbon caps and formation of fullerenes. *Carbon* 34, no. 2 (1996): 173–178.
47. K.M. Thomas, J.M. Lewis, S.H. Bottrell, S.P. Dean, and J. Foulkes. Isotope fractionation in the synthesis of fullerenes. *Carbon* 32, no. 5 (1994): 991–1000.
48. M.S. Dresselhaus, G. Dresselhaus, and P.C. Eklund. Historical introduction. In *Science of Fullerenes and Carbon Nanotubes: Their Properties and Applications*, 1–14. Academic Press, 1996.
49. T.C. Dinadayalane and J. Leszczynski. Remarkable diversity of carbon–carbon bonds: Structures and properties of fullerenes, carbon nanotubes, and graphene. *Structural Chemistry* 21 (2010): 1155–1169.
50. H.W. Kroto. C60B buckminsterfullerene, other fullerenes and the icospiral shell. *Computers & Mathematics with Applications* 17, no. 1 (1989): 417–423.
51. R.C. Haddon, L.E. Brus, and K. Raghavachari. Electronic structure and bonding in icosahedral C60. *Chemical Physics Letters* 125, no. 5 (1986): 459–464.

52. F. Diederich and C. Thilgen. Covalent fullerene chemistry. *Science* 271, no. 5247 (1996): 317.
53. H.W. Kroto. The stability of the fullerenes Cn, with n = 24, 28, 32, 36, 50, 60 and 70. *Nature* 329, no. 6139 (1987): 529–531.
54. E. Małolepsza, H.A. Witek, and S. Irle. Comparison of Geometric, electronic, and vibrational properties for isomers of small fullerenes C_{20}–C_{36}. *Journal of Physical Chemistry A* 111 (2007): 6649–6657.
55. R.R. Zope, T. Baruah, M.R. Pederson, and B.I. Dunlap. Static dielectric response of icosahedral fullerenes from C_{60} to C_{2160} characterized by an all-electron density functional theory. *Physical Review B* 77 (2008): 115452.
56. A.F. Hebard. Superconductivity in doped fullerenes. *Physics Today* 45, no. 11 (1992): 26–32.
57. H. Terrones and M. Terrones. Curved nanostructured materials. *New Journal of Physics* 5 (2003): 126.1–126.37.
58. K. Holczer and R.L. Whetten. Superconducting and normal state properties of the A3C60 compounds. *Carbon* 30, no. 8 (1992): 1261–1276.
59. R.S. Ruoff and A.L. Ruoff. The bulk modulus of C60 molecules and crystals: A molecular mechanics approach. *Applied Physics Letters* 59, no. 13 (1991): 1553–1555.
60. E. Ulloa. Fullerenes and their applications in science and technology, 2013. http://web.eng.fiu.edu/~vlassov/EEE-5425/Ulloa-Fullerenes.pdf.
61. R.D. Beck, P. St. John, M.M. Alvarez, F. Diederich, and R.L. Whetten. Resilience of all-carbon molecules C60, C70, and C84: A surface-scattering time-of-flight investigation. *The Journal of Physical Chemistry* 95, no. 21 (1991): 8402–8409.
62. K. Huseman, T. Jackson, W. Johnson, and D. Howsmon. High speed collisions. In "Fullerenes and 'Buckyballs': The State Molecule of Texas!," research.che.tamu.edu/groups/Seminario/materials/G04_Buckyballs.pdf. (With screenshots taken from http://www.nanoed.org/courses/carbon_nanotube/ballv02.jpg.)
63. L. Forro and L. Mihály. Electronic properties of doped fullerenes. *Reports on Progress in Physics* 64, no. 5 (2001): 649–699.
64. R.C. Haddon, A.F. Hebard, M.J. Rosseinsky, D.W. Murphy, S.H. Glarum, T.T.M. Palstra, A.P. Ramirez et al. Conductivity and superconductivity in alkali metal doped C60. In *Fullerenes Synthesis, Properties, and Chemistry of Large Carbon Clusters,* ACS Symposium Series, vol. 481, edited by G.S. Hammond and V.J. Kuck, 71–89. ACS, 1992.
65. R.C. Haddon. Electronic structure, conductivity and superconductivity of alkali metal doped (C60). *Accounts of Chemical Research* 25, no. 3 (1992): 127–133.
66. R.C. Haddon, A.F. Hebard, M.J. Rosseinsky, D.W. Murphy, S.J. Duclos, K.B. Lyons, B. Miller et al. Conducting films of C60 and C70 by alkali-metal doping. *Nature* 350, no. 6316 (1991): 320–322.
67. A.F. Hebard, M.J. Rosseinky, R.C. Haddon, D.W. Murphy, S.H. Glarum, T.T.M. Palstra, A.P. Ramirez, and A.R. Karton. Superconductivity at 18K in potassium-doped C60. *Nature* 350 (1991): 600–601.
68. P.W. Stephens, L. Mihaly, P.L. Lee, R.L. Whetten, S.-M. Huang, R. Kaner, F. Deiderich, and K. Holczer. Structure of single-phase superconducting K_3C_{60}. *Nature* 351 (1991): 632–634.
69. T.T.M. Palstra, O. Zhou, Y. Iwasa, P.E. Sulewski, R.M. Fleming, and B.R. Zegarski. Superconductivity at 40K in cesium doped C 60. *Solid State Communications* 93, no. 4 (1995): 327–330.
70. J.R. Heath, S.C. O'Vrien, Q. Zhang, Y. Liu, R.F. Curl, F.K. Tittel, and R.E. Smalley. Lanthanum complexes of spheroidal carbon shells. *Journal of the American Chemical Society*, 107, no. 25 (1985): 7779–7780.
71. F.D. Weiss, J.L. Elkind, S.C. O'Brien, R.F. Curl, and R.E. Smalley. Photophysics of metal complexes of spheroidal carbon shells. *Journal of the American Chemical Society* 110, no. 13 (1988): 4464–4465.
72. Y. Chai, T. Guo, C. Jin, R.E. Haufler, L.P.F. Chibante, J. Fure, L. Wang, J.M. Alford, and R.E. Smalley. Fullerenes with metals inside. *Journal of Physical Chemistry* 95, no. 20 (1991): 7564–7568.
73. R.E. Haufler, J. Conceicao, L.P.F. Chibante, Y. Chai, N.E. Byrne, S. Flanagan, M.M. Haley, S.C. O'Brien, and C. Pan. Efficient production of C60 (buckminsterfullerene), C60H36, and the solvated buckide ion. *Journal of Physical Chemistry* 94, no. 24 (1990): 8634–8636.

74. C.R. Wang, M. Inakuma, and H. Shinohara. Metallofullerenes Sc2@C82(I,II) and Sc2@C86(I, II): Isolation and spectroscopic studies. *Chemical Physics Letters* 300, no. 3 (1999): 379–384.
75. L. Dunsch and S. Yang. The recent state of endohedral fullerene research. *The Electrochemical Society Interface* 15, no. 2 (2006): 34–39.
76. K. Suenaga, Y. Sato, Z. Liu, H. Kataura, T. Okazaki, K. Kimoto, H. Sawada et al. Visualizing and identifying single atoms using electron energy-loss spectroscopy with low accelerating voltage. *Nature Chemistry* 1 (2009): 415–418.
77. A.I. Gusev, I.E. Kareev, and N.I. Ioffe. Fullerene, endohedral. *Glossary of Nanotechnology and Related Terms,* n.d. eng.thesaurus.rusnano.com/wiki/article1933.
78. A.M. Lopez, A. Mateo-Alonso, and M. Prato. Materials chemistry of fullerene C_{60} derivatives. *Journal of Materials Chemistry* 21 (2011): 1305–1318.
79. F.N. Diederich. Covalent fullerene chemistry. *Pure and Applied Chemistry* 69, no. 3 (1997): 395–400.
80. S. Pilehvar and K. De Wael. Recent Advances in electrochemical biosensors based on fullerene-C60 nano-structured platforms. *Biosensors* 5, no. 4 (2015): 712–735.
81. C. Bingel. Cyclopropanierung von fullerenen. *Chemische Berichte* 126, no. 8 (1993): 1957–1959.
82. X. Camps and A. Hirsch. Efficient cyclopropanation of C60 starting from malonates. *Journal of the Chemical Society, Perkin Transactions* 1, no. 11 (1997): 1595–1596.
83. A. Hirsch and M. Brettreich. Nucleophilic additions. In *Fullerenes: Chemistry and Reactions,* edited by A. Hirsch and M. Brettreich, 73. Wiley-VCH, 2005.
84. M. Maggini, G. Scorrano, and M. Prato. Addition of azomethineylides to C60: Synthesis, characterization, and functionalization of fullerene pyrrolidines. *Journal of the American Chemical Society* 115, no. 21 (1993): 9798–9799.
85. J.M. Hawkins, A. Meyer, T.A. Lewis, S. Loren, and F.J. Hollander. Crystal structure of osmylated C60: Confirmation of the soccer ball framework. *Science* 252, no. 5003 (1991): 312–313.
86. J.M. Hawkins. Osmylation of C60: Proof and characterization of the soccer-ball framework. *Accounts of Chemical Research* 25, no. 3 (1992): 150–156.
87. Y. Tajima, K. Takeshi, Y. Shigemitsu, and Y. Numata. Chemistry of fullerene epoxides: Synthesis, structure, and nucleophilic substitution-addition reactivity. *Molecules* 17, no. 6 (2012): 6395–6414.
88. J.C. Hummelen, M. Prato, and F. Wudl. There is a hole in my bucky. *Journal of the American Chemical Society* 117, no. 26 (1995): 7003–7004.
89. G.C. Vougioukalakis, M.M. Roubelakis, and M. Orfanopoulos. Open-cage fullerenes: Towards the construction of nanosized molecular containers. *Chemical Society Reviews,* 39, no. 2 (2010): 817–844.
90. N. Xin, X. Yang, Z. Zhou, J. Zhang, S. Zhang, and L. Gan. Synthesis of C60(O)3: An open-cage fullerene with a ketolactone moiety on the orifice. *The Journal of Organic Chemistry* 78, no. 3 (2013): 1157–1162.
91. L. Xu, S. Liang, J. Sun, and L. Gan. Open-cage fullerene with a stopper acts as a molecular vial for a single water molecule. *Organic Chemistry Frontiers* 2, no. 11 (2015): 1500–1504.
92. A. Krachmalnicoff, R. Bounds, S. Mamone, M.H. Levitt, M. Carravetta, and R.J. Whitby. Synthesis and characterisation of an open-cage fullerene encapsulating hydrogen fluoride. *Chemical Communications* 51, no. 24 (2015): 4993–4996.
93. Y. Takabayashi, A.Y. Ganin, P. Jeglič, D. Arčon, T. Takano, Y. Iwasa, Y. Ohishi et al. The disorder-free non-BCS superconductor Cs3C60 emerges from an antiferromagnetic insulator parent state. *Science* 323, no. 5921 (2009): 1585–1590.
94. C. Binns. Carbon nanostructures: Bucky balls and nanotubes. In *Introduction to Nanoscience and Nanotechnology,* edited by C. Binns, 53. Wiley, 2010.
95. B.C. Thompson, and J.M. Frechet. Polymer–fullerene composite solar cells. *Angewandte Chemie International Edition,* 47, no. 1 (2008): 58–77.
96. T.M. Razykov, C.S. Ferekides, D. Morel, E. Stefanakos, H.S. Ullal, and H.M. Upadhyaya. Solar photovoltaic electricity: Current status and future prospects. *Solar Energy* 85, no. 8 (2011): 1580–1608.

97. C.Z. Li, C.C. Chueh, H.L. Yip, K.M. O'Malley, W.C. Chen, and A.K.Y. Jen. Effective interfacial layer to enhance efficiency of polymer solar cells via solution-processed fullerene-surfactants. *Journal of Materials Chemistry* 22, no. 17 (2012): 8574–8578.
98. R. Bakry, R.M. Vallant, M. Najam-ul-Haq, M. Rainer, Z. Szabo, C.W. Huck, and G.K. Bonn. Medicinal applications of fullerenes. *International Journal of Nanomedicine* 2, no. 4 (2007): 639–649.
99. G. Lalwani and B. Sitharaman. Multifunctional fullerene- and metallofullerene-based nano-biomaterials. *Nano Life* 3, no. 3 (2013): 1342003.
100. M. Lens. Recent progresses in application of fullerenes in cosmetics. *Recent Patents on Biotechnology* 5, no. 2 (2011): 67–73.
101. Fullerene: Topical application for acne treatment. *Personal Care Magazine*, Feb. 28, 2012. http://www.personalcaremagazine.com/story/9443/fullerene-topical-application-for-acne-treatment.
102. S. Inui, H. Aoshima, A. Nishiyama, and S. Itami. Improvement of acne vulgaris by topical fullerene application: Unique impact on skin care. *Nanomedicine: Nanotechnology, Biology and Medicine* 7, no. 2 (2011): 238–241.
103. Vitamin C60 BioResearch Corporation. https://www.ulprospector.com/en/asia/PersonalCare/Suppliers/13314/Vitamin-C60-BioResearch-Corporation.
104. C.N. McEwen, R.G. McKay, and B.S. Larsen. C60 as a radical sponge. *Journal of the American Chemical Society* 114, no. 11 (1992): 4412–4414.

3

Single-Walled Carbon Nanotubes

3.1 History

Roger Bacon at Union Carbide observed images of straight and hollow nanowhiskers with carbon layers that were separated by the same spacing as the planar layers of graphite when he was studying carbon fibers in the late 1950s. In the 1970s, Morinobu Endo, who was earning his PhD at the University of Orleans, France, observed carbon fibers with a hollow core and catalytic particles at the end under a high-resolution transmission electron microscope. These fibers were produced by the pyrolysis of benzene and ferrocene at 1000°C. Russian scientists, Bochvar and Gal'pern, did Huckel calculations showing that graphene sheets can curl up to make hollow molecules with a relatively large band gap. The worldwide interest for carbon-based nanomaterials came after Sumio Iijima synthesized nanocarbons made of hollow tubes by the electric arc technique in 1991 [1].

Two years later, a breakthrough occurred when single-walled carbon nanotubes (SWCNTs) were discovered simultaneously and independently by Sumio Iijima and Donald Bethune. More attention was paid to these materials when the Nobel Prize in Chemistry was awarded in 1996 for the discovery of fullerenes or a caged molecule of 60 carbon atoms (C_{60}).

3.2 Structure of Single-Walled Carbon Nanotube (SWCNT)

SWCNTs can be considered to be formed by rolling of a single layer of graphite (called graphene) into a seamless cylinder, schematically represented in Figure 3.1. Similarly, a multiwalled carbon nanotube (MWCNT) is a coaxial assembly of cylinders of SWCNTs, one within another. Generally, nanotubes have a length-to-diameter ratio of more than 1000. Therefore, these can be considered as nearly one-dimensional structures and accordingly their properties vary with structure [2].

The diameter and chirality is determined by the *m, n* indices, which are the key parameters of nanotubes. Depending on the chirality SWCNTs can either be metal like conductors or semiconductors. As grown, nanotubes are closed at both ends by the hemispherical cap, which is formed by the replacement of hexagons with pentagons in the graphite sheet leading to the curvature [3].

A SWCNT consists of two separate sections with different physical and chemical properties. The first section is the sidewall of the tube and the second is the end cap of the tube. Again the

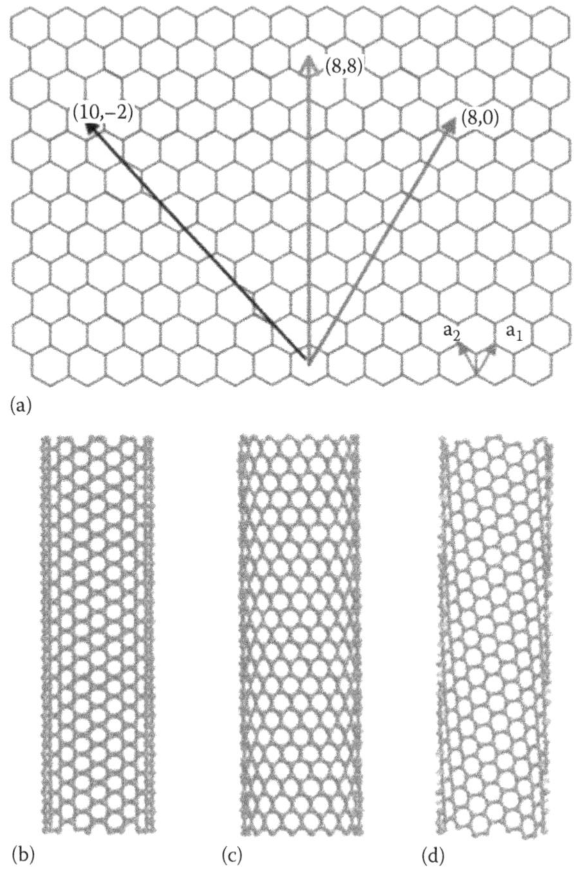

FIGURE 3.1
(a) Schematic honeycomb structure of a graphene sheet. SWCNTs can be formed by folding the sheet along lattice vectors. The two basis vectors a_1 and a_2 are shown. Folding of the (8,8), (8,0), and (10,–2) vectors lead to (b) armchair, (c) zigzag, and (d) chiral tubes, respectively. (Reprinted with permission from H. Dai. Carbon nanotubes: Synthesis, integration, and properties. *Accounts of Chemical Research* 35, no. 12: 1035–1044. Copyright (2002) American Chemical Society.)

sidewall may be with different chiralities according to the rolling of the graphite sheet in different directions (Figure 3.1). The end cap structure is similar to fullerene, such as C_{60}.

3.3 Synthesis of Carbon Nanotubes

A variety of synthesis methods exist for producing CNTs. The three main production methods used for the synthesis of CNTs are arc discharge, laser ablation, and chemical vapor deposition (CVD).

3.3.1 Arc Discharge

The arc discharge method, initially used for producing fullerenes (C_{60}), is the most common and perhaps easiest way to produce CNTs. In this technique two carbon rods are separated by approximately 1 to 2 mm, in an enclosure that is usually filled with inert gas (helium, argon) at low pressure (between 50 and 700 mbar) as shown in Figures 3.2 and 3.3. A direct current of 50 to 100 A driven by approximately 20 V is applied, which creates a high temperature (~4000 K) discharge between the two electrodes. The discharge vaporizes one of the carbon rods (anode) and forms a small rod shaped deposit on the other rod (cathode) [4]. In order to increase the yield of SWCNTs, the anode is filled with a metal catalyst, such as Fe, Co, Ni, Y, or Mo. In the absence of any catalyst, the main product will be MWCNTs. Typical dimensions of MWCNTs are an inner diameter of 1 to 3 nm and an

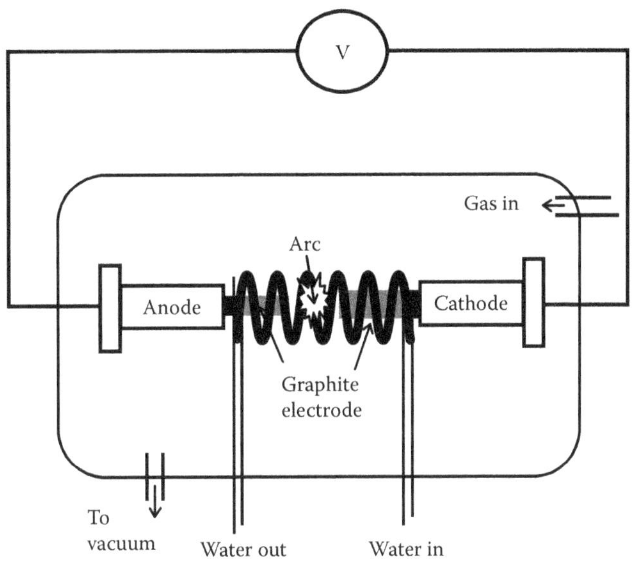

FIGURE 3.2
Schematic diagram of direct current arc discharge setup. (From V. Choudhary, B.P. Singh, and R.B. Mathur, Carbon nanotubes and their composites, in *Syntheses and Applications of Carbon Nanotubes and Their Composites*, edited by S. Suzuki, 953–978, InTech, 2013.)

Production of carbon nanotubes by dc-Arc discharge

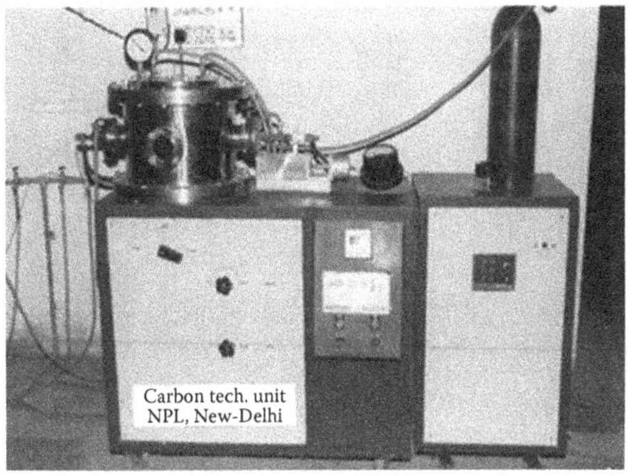

FIGURE 3.3
Photograph of direct current arc discharge setup for the synthesis of CNTs at CSIR-NPL, India.

outer diameter of approximately 10 to 20 nm. Because no catalyst is involved in this process, there is no need for any complex purification step. This means MWCNTs having a low amount of defects can be synthesized this way. Generally, SWCNTs are synthesized using the direct current arc discharge process by filling the catalyst + graphite powder into a hole drilled in a graphite electrode acting as anode. Arcing takes place between the anode and a pure graphite-based cathode under optimized chamber conditions [4].

In one of Mathur et al.'s earlier studies [5], both SWCNTs and MWCNTs were selectively synthesized simultaneously in a single experiment. In the experiment, instead of filling the catalyst powder into a hole drilled in the graphite electrode, a catalyst–graphite composite electrode was synthesized in the laboratory. These electrodes were used as anodes in the arcing process and a high-density graphite block was used as the cathode. The yield of SWCNTs was more than double in this case due to uniform distribution of catalyst in the whole graphite anode.

3.3.2 Laser Ablation

In the laser ablation technique, a pulsed laser vaporizes a graphite target containing a metal catalyst. A schematic diagram of laser ablation set up for the production of CNTs is shown in Figure 3.4. The graphite target is placed in a furnace at ~1200°C in an inert atmosphere. The CNTs deposited on the cooler surface of the reactor, as the vaporized carbon condenses. The yield of nanotube synthesis by this process is around 70% [4].

3.3.3 Chemical Vapor Deposition

Chemical vapor deposition (CVD) is the most accepted method for the bulk production of CNTs. In CVD, decomposition of hydrocarbons takes place in the presence of a particular transition metal catalyst such as Fe, Ni, or Co. It is a low-cost and scalable technique for mass production of CNTs. In this process, hydrocarbon vapors pass through a tubular reactor that consists of a high temperature furnace in the presence of catalyst material at sufficiently high temperature (600°C–1200°C) to decompose the hydrocarbon. A schematic diagram of CVD is shown in Figure 3.5. CNTs grow on the catalyst in the reactor, which

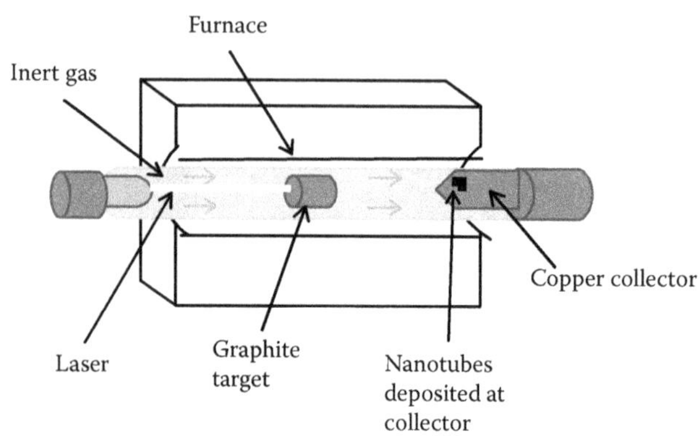

FIGURE 3.4
Schematic diagram of laser ablation set-up for CNTs synthesis. (From V. Choudhary, B.P. Singh, and R.B. Mathur, Carbon nanotubes and their composites, in *Syntheses and Applications of Carbon Nanotubes and Their Composites*, edited by S. Suzuki, 953–978, InTech, 2013.)

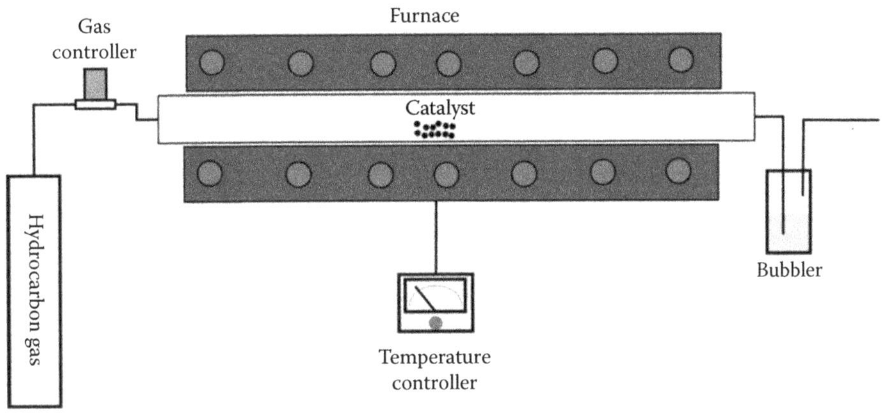

FIGURE 3.5
Schematic diagram of the CVD setup.

are collected upon cooling the system to room temperature. Generally a low-temperature CVD (600°C–900°C) growth condition yields MWCNTs, whereas a high-temperature (900°C–1200°C) reaction favors SWCNT growth. This indicates that SWCNTs need higher energy of formation due to small diameters means small curvature bears high-strain energy. Therefore, growth of MWCNTs is easier than SWCNTs from most of the hydrocarbons, whereas SWCNTs grow from selected hydrocarbons, namely, methane and carbon monoxide, which have a stability in the temperature range of 900°C to 1200°C [6].

Compared to other methods of synthesizing CNTs, several parameters, including temperature, feed gases, flow rate, catalyst components, and heating rate, are available to control the growth rate in CVD. By changing the growth conditions, the properties of the produced CNTs (length, orientation, and diameter) to some extent can be controlled. It has been observed that the gas phase processes produce CNTs with fewer impurities and are most amenable to large-scale production. Therefore, the gas phase techniques such as CVD offer the greatest potential for scaling up CNT production. Two important processes that have proved capable of industrial-scale production of CNTs are the high-pressure carbon monoxide (HiPCO) process and fluidized bed CVD.

3.3.4 High-Pressure Carbon Monoxide Process

The high-pressure carbon monoxide (HiPCO) process was first developed at Rice University for the synthesis of SWCNT and demonstrated for lab-scale production. The HiPCO process is a type of CVD where an organometallic compound, such as $Fe(CO)_5$, is fed into a reaction zone at high pressures (up to 50 atm) and high temperatures (900°C–1100°C), as shown in Figure 3.6. In this process organometallic compounds, such as ferrocene, decompose into nanoscale metal clusters that catalyze the CO disproportionate reaction to form by-products of CO_2 and SWCNTs.

3.3.5 Fluidized Bed

An important process based on CVD technique for large-scale production of CNTs is the fluidized bed CVD (FBCVD) process. In this process, as shown in Figure 3.7, feedstock gases and catalyst material are fed into the reaction zone where CNTs form and

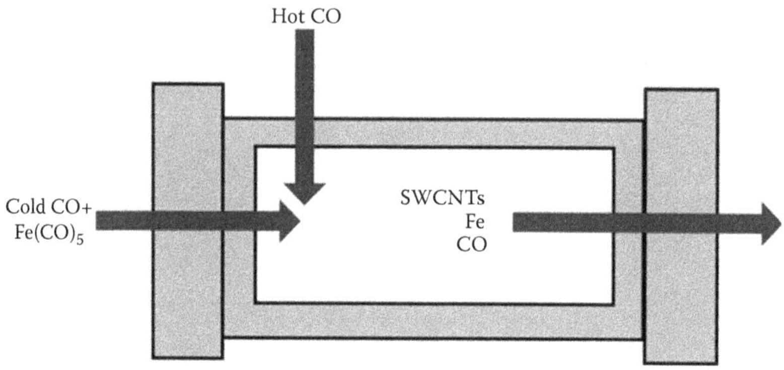

FIGURE 3.6
Schematic diagram of the HiPCO setup. (From Chen et al. *Int. J. Environ. Res. Public Health* 2013, 10(4), 1231–1249.)

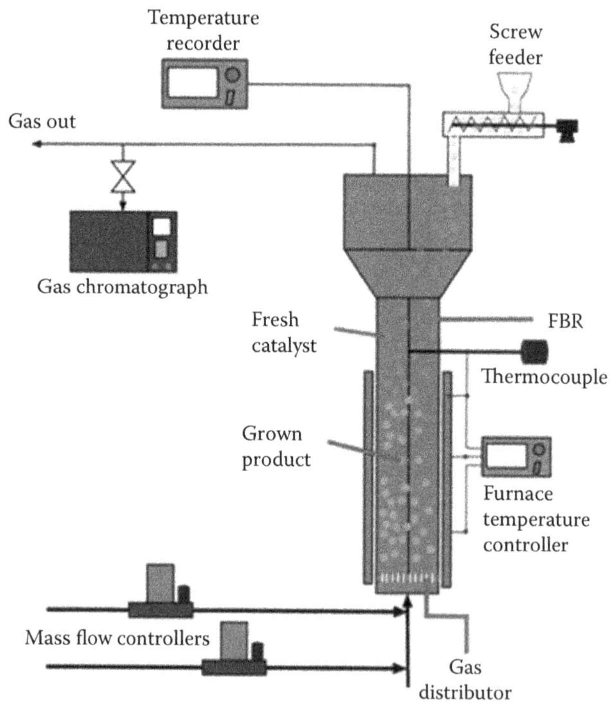

FIGURE 3.7
Schematic diagram of a fluidized bed chemical vapor deposition system. (Y. Liu, W.Z. Qian, Q. Zhang, G.Q. Ning, G.H. Luo, Y. Wang, D.Z. Wang, and F. Wei: Synthesis of high-quality, double-walled carbon nanotubes in a fluidized bed reactor. *Chemical Engineering & Technology*. 2009. 32. No. 1. 73–79. Copyright Wiley-VCH Verlag GmbH & Co. KGaA. Reproduced with permission.)

then collected in a continuous fashion. Basically, the FBCVD process involves two distinct reaction zones: In the first zone, the catalyst precursor reacts and forms the catalytic materials and then this flows downstream to the second reaction zone where the CNTs form through reaction with this catalytic material. The advantage of this process is that the catalyst source is continuous, which enables the development of highly scalable CNTs with controllable properties. However, the drawback to FBCVD is often the excess metal present in the CNTs that can be difficult to remove, and the difficulty achieving the level of accuracy in growth processes that can be comparable to fixed substrate-based CNT growth techniques [7,8].

3.4 Growth Mechanism

The growth mechanism of CNT has been debatable since its discovery. Based on the reaction conditions and postdeposition product analyses, various methods of synthesis proposed by researchers are often contradicting in nature. Therefore, no single mechanism for CNT growth is well established to date. Nevertheless, the most general, widely accepted mechanisms are discussed based on the method of synthesis.

3.4.1 Growth Mechanism Based on Chemical Vapor Deposition

For the synthesis of CNTs, normally, nanometer-size metal particles are required to decompose hydrocarbon at lower temperature. The most common metals used are Fe, Co, and Ni, which are used because of two main reasons: (1) high solubility of carbon in these metals at high temperatures and (2) high carbon diffusion rate in these metals. In addition, the high melting point and low equilibrium vapor pressure of these metals provide a wide temperature window of CVD for a broad range of carbon precursors. Recently it has been found that Fe, Co, and Ni have stronger adhesion with the growing CNTs (than other transition metals) and therefore they are more efficient in forming high-curvature or low-diameter CNTs such as SWCNTs [6,9]. Organometallocenes such as ferrocene, cobaltocene, and nickelocene are widely used as a CNT catalyst, because they release metal nanoparticles in situ, which catalyzes the hydrocarbon decomposition in an efficient way. Generally, the nanotube diameter depends on the size of the catalyst particles. Hence, metal nanoparticles of controlled size are used to grow CNTs of controlled diameter.

In one of the most widely accepted general mechanisms, first, a hydrocarbon vapor comes in contact with the hot metal nanoparticles and it decomposes into carbon and hydrogen species in which hydrogen flies away and carbon gets dissolved into the metal. When the carbon-solubility limit is reached in the metal at that temperature, as-dissolved carbon precipitates out in the form of a cylindrical network, which is called CNT [6].

Now there are two cases. In the first case (Figure 3.8a (1)), when the catalyst–substrate interaction is weak (metal has an acute contact angle with substrate), decomposition of hydrocarbon on the top surface of the metal takes place and carbon diffuses through the metal. The carbon in the form of CNT precipitates out of the bottom of the metal particle, which pushes up the whole metal particle off the substrate (Figure 3.8a (2)). Till the top of the metal is free for fresh hydrocarbon decomposition, the concentration gradient exists in the metal and allows the diffusion of carbon, and CNT continues to grow longer and longer (Figure 3.8a (3)). When the metal is fully covered with excess carbon, its catalytic activity ceases and the growth of CNT is stopped (Figure 3.8a (4)). This is generally known as the tip-growth model.

In the second case (Figure 3.8b), when the catalyst–substrate interaction is strong (metal has an obtuse contact angle with substrate), hydrocarbon decomposition and carbon diffusion take place initially similar to the tip-growth mechanism, but the precipitation of CNT fails to push the metal particle up; therefore the precipitation is compelled to emerge out from the apex of metal having minimum interaction with the substrate. First, carbon crystallizes

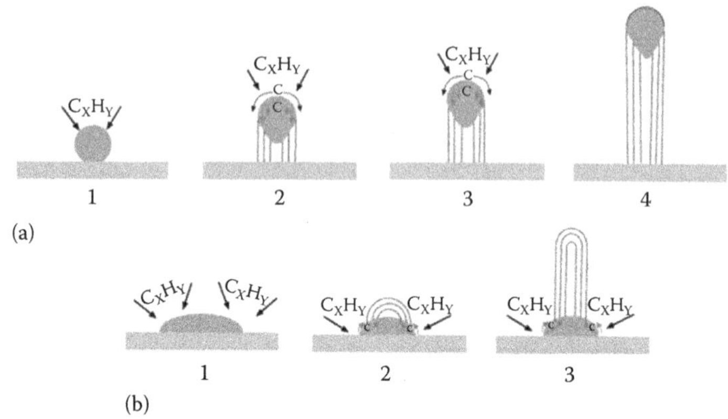

FIGURE 3.8
Widely accepted growth mechanisms for CNTs: (a) tip-growth model and (b) base-growth model.

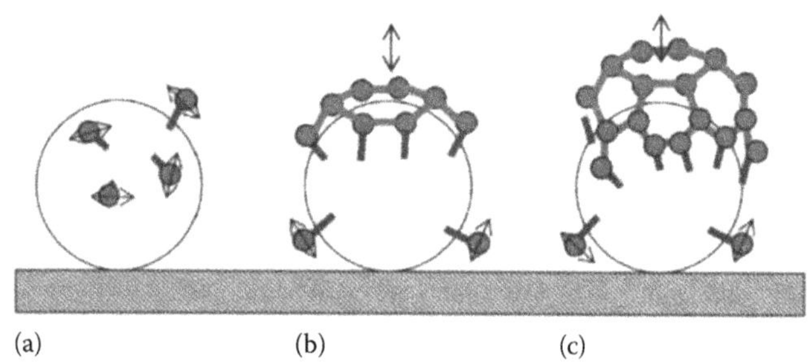

FIGURE 3.9
Schematic representation of the basic steps of SWCNT growth on a Fe catalyst, as observed in *ab initio* simulations. (a) Diffusion of single C atoms (spheres) on the surface of the catalyst. (b) Formation of a sp^2 graphene sheet floating on the catalyst surface with edge atoms covalently bonded to the metal. (c) Root incorporation of diffusing single C atoms or dimer. (Reprinted with permission from J.-Y. Raty, F. Gygi, and G. Galli, *Physical Review Letters*, 95, 096103. Copyright 2005 by the American Physical Society.)

out in the form of hemispherical dome, which is the most favorable closed-carbon network on a spherical nanoparticle. Then this extends up in the form of seamless graphitic cylinder as CNT. Successive hydrocarbon decomposition takes place on the lower side surface of the metal and as-dissolved carbon in metal diffuses upward. Hence CNT grows up with the catalyst particle rooted on its base and this is known as the base-growth model [6].

Another growth mechanism was also suggested by Raty et al. [10] based on a molecular dynamics simulation study for the early stages growth of SWCNT on metal nanoparticles. They explained that carbon atoms diffuse only on the outer surface of the metal cluster. First, a graphene cap is formed which floats over the metal and the border atoms of the cap remain anchored to the metal. Subsequently, more carbon atoms join the border atoms pushing the cap up and thus constituting a cylindrical wall as CNT (Figure 3.9) [10].

3.4.2 Mechanism of SWCNT Growth Using Arc Evaporation and Laser Ablation

There are good reasons for assuming that the mechanisms of SWCNT formation in the arc-evaporation and laser-vaporization processes are generally similar. In both processes, similar

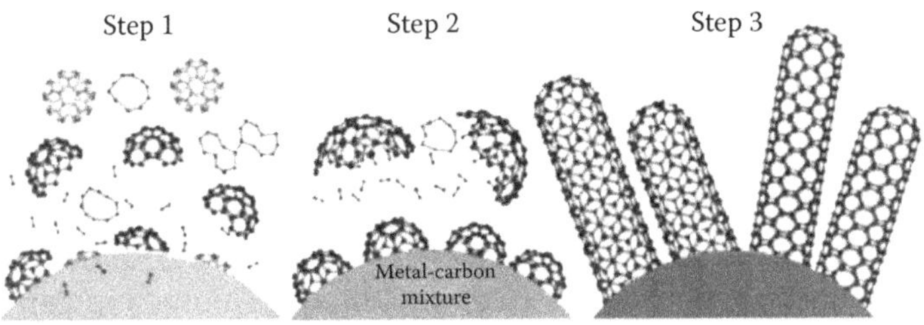

FIGURE 3.10
Illustration of a model for SWCNT growth proposed by Kataura and colleagues [11]. In step 1, carbon clusters with fullerene-like structures form, and some of this carbon is dissolved in metal particles. When the metal particles are saturated with carbon, their surfaces become covered with carbon fragments (step 2). Finally, in step 3 SWCNTs grow out from the particles. (Reprinted from *Carbon* 38, no. 11, H. Kataura, Y. Kumazawa, Y. Maniwa, Y. Ohtsuka, R. Sen, S. Suzuki et al. Diameter control of single-walled carbon nanotubes, 1691–1697. Copyright 2000, with permission from Elsevier.)

starting materials, namely, a graphite–metal mixture are used and both involve the vaporization of this mixture followed by condensation in an inert atmosphere. The nanotube-containing soot produced by both processes is almost the same in appearance and contains bundles of SWCNTs with disordered or amorphous carbon and metal particles. It is assumed that mechanisms proposed for one process are applicable to both as discussed next [12].

Although various models have been proposed for the growth of SWCNTs by the arc or laser methods, it is mainly accepted that the mechanism probably involves root growth more than tip growth. This means that the tubes grow away from the metal particles, with carbon. This is supported by the fact that metal particles are not found at the tips of SWCNTs produced by arc discharge or laser ablation [12].

Kataura et al. proposed a model for SWCNT growth that is similar to the Geohegan mechanism. In the model, it emphasized that fullerene-like carbon fragments play an important role in nucleating growth, as illustrated in Figure 3.10. In the first stage, which occurs at very high temperatures (i.e., ~2000°C–3000°C), small carbon clusters nucleate and these are mainly fullerene-like structures that are rich in pentagons. The metal atoms are in the gas phase at this stage. As the system cools, metal atoms condense and form particles or droplets, which become supersaturated with carbon. This takes place around the eutectic temperature. These particles then become covered with fullerene-like carbon fragments. The open edges of these fragments tend to attach to the particles to remove dangling bonds. These fragments then act as precursors for SWCNT growth and carbon being supplied by precipitation from the particles or from the available disordered carbon with particles [11].

As discussed earlier several mechanisms have been proposed for the growth of SWCNTs, but there is still no single growth model that has been unambiguously accepted by researchers. Moreover, how to control the broad diameter distribution among CNTs and their chirality is still a big challenge.

3.5 Properties of SWCNTs

SWCNTs are endowed with exceptionally high material properties, very close to their theoretical limits, such as electrical and thermal conductivity, strength, stiffness, and low density.

3.5.1 Mechanical Properties

The impressive mechanical properties of SWCNTs originate from their geometrical structure and C–C bond. The carbon–carbon bond in the basal plane of graphite is one of the strongest bonds in nature. Small diameter SWCNTs are very stiff and exceptionally strong, which corresponds to having a high Young's modulus and tensile strength. This would imply that SWCNTs are very stiff and difficult to bend. However, SWCNTs seem to be very resilient when bent and can buckle like straws without breaking and can be straightened back without any damage. Most of the materials fracture at bending due to presence of defects or dislocations, but SWCNTs have very few defects on their sidewalls, which is why they have very high mechanical properties and resilience (i.e., strain to failure).

Various efforts have been made to investigate the mechanical properties of CNTs, experimentally and theoretically. Nanotubes are the stiffest fibers, with a measured Young's modulus of 1.4 TPa. Additionally, they have an elongation to failure of 20% to 30%, and tensile strength around 100 GPa. These values are much higher comparatively with high-strength steel (Young's modulus ~200 GPa and its tensile strength is 1 to 2 GPa). The exceptional stiffness and strength with low density implies that CNT could be used as an ideal reinforcement in composite materials.

3.5.2 Specific Gravity

Specific strength of any material plays a key role in designing of structural materials. The hollow structure of CNTs makes them very light (specific gravity of 0.8), and this is very useful for a variety of lightweight high strength applications. The specific gravity of steel, on the other hand, is 7.8. Therefore SWCNTs have specific strength value at least two orders of magnitude greater than steel. Similarly, the specific strength of traditional carbon fibers (specific gravity ~1.78) is only 40 times that of steel.

3.5.3 Thermal Conductivity

The specific heat and thermal conductivity of CNT systems are determined primarily by phonons. Nanotubes are very conductive for phonons in their axial direction, whereas in the transverse direction their conductivity is very low. Theory predicts an unusually high value of thermal conductivity, that is, 6000 W/mK at room temperature; however thermal conductivities as high as 3000 W/mK at room temperature have been measured [13] which is equivalent to the thermal conductivity of natural diamond and basal plane of graphite.

3.5.4 Electrical Conductivity

SWCNTs can either be a metal or semiconductor, depending on their diameters and helical arrangement. The condition whether a CNT is metallic or semiconducting can be obtained based on the band structure of a 2D graphite sheet and periodic boundary conditions along the circumference direction. The differences in conductivity can easily be derived from properties of the graphene sheet [14]. It was shown that a (n, m) nanotube is metallic if $n = m$ or $(n - m) = 3i$, where i is an integer and n and m are parameters defining the nanotubes structure or chiral vector. Due to their one-dimensional nature, charge carriers can travel through nanotubes without scattering, resulting in ballistic transport. The absence of scattering means that joule heating is minimized and thus nanotubes can carry very large current densities (up to 100 MA/cm^2) [15]. In addition, carrier mobility as high as 10^5 cm^2/Vs has been observed in semiconducting nanotubes [16].

3.6 Purification of SWCNTs

At the time of synthesis of SWCNTs, a lot of impurities remained embedded in the material. These impurities are metallic catalyst particles, amorphous carbon, and graphitic nanoshells. Removal of these metallic impurities without degradation of the structure of CNTs is a big challenge. Several techniques have been used to purify the SWCNTs.

3.6.1 Acid Treatment

The treatment with different acids such as HCl and HNO_3 is the simplest route to separate the transition metal from the SWCNT soot.

3.6.2 Air Oxidation

Oxidation of SWCNTs soot at ~350°C removes the amorphous carbon because it is unstable at this temperature, but SWCNTs are stable at this temperature.

3.6.3 High Temperature Annealing

For some applications of SWCNTs, such as their use as biomaterials, complete removal of metal particles is of particular importance. However, it is very difficult to achieve this by acid treatment because most of the metal particles are encapsulated by carbon layers that form a barrier to acid attack. The physical properties of carbon and metals are different at high temperature (>1400°C) under inert atmosphere. It is well known that graphite is stable even at 3000°C, whereas metal evaporates at temperatures higher than their boiling point (1300°C–1400°C). Therefore, high temperature annealing can effectively remove metal particles. A detailed purification route is discussed by Mathur et al. [5].

3.7 Characterization of SWCNTs

After the synthesis of SWCNTs, their thorough characterization is required and depending upon the properties their application can be decided. Different types of SWCNTs can be used for different applications. Some of the important characterization techniques used for SWCNTs are discussed next.

3.7.1 Raman Spectroscopy

Raman spectroscopy has become an indispensible tool for the characterization of CNTs and especially for SWCNTs. Study of vibrational modes in CNTs has been useful for understanding their structural makeup. A CNT vibrational spectrum is interesting due to its intricate dependence on diameter and chirality, and Raman spectroscopy has been a useful tool to study these vibrational modes. Resonant Raman spectroscopy has been found more suitable for the characterization of SWCNTs compared to normal Raman spectroscopy. The resonant feature comes from the use of excitation laser energy that coincides with actual optical transitions of SWCNT, resulting in a resonant enhancement effect that increases the Raman scattering probability from about 1 in 10 million to approximately 1 in 1000.

Raman spectrum of SWCNTs shows three important bands: (1) radial breathing mode (RBM) in the low frequency region (100–600 cm^{-1}), (2) D-band (due to disorder, peak around 1350 cm^{-1}) and (3) G-band (corresponding to the tangential vibrations of carbon atoms, peak around 1580 cm^{-1}). The RBM frequency is inversely proportional to the tube diameter. RBM cannot be detected in CNTs having tube diameter >2 nm. Apart from the aforementioned, some other bands have also been reported at around 2600 cm^{-1} (second-order D-band) and some low intensity peaks in the 1700–1800 cm^{-1} range. The presence of RBM in the Raman spectra of SWCNTs is a unique feature and is widely used for the estimation of tube diameter [17].

It has been mentioned by Araujo et al. [18] that for SWCNTs having diameter <1.2 nm, the following relation can be used for the estimation of diameter of SWCNTs: $\omega_{rbm} = 227/d_t$, where ω_{rbm} is the Raman shift (cm^{-1}), and d_t is the tube diameter of CNT (in nanometers).

G-band vibration is an in-plane C–C stretching mode that is isotropic in graphene but anisotropic in carbon nanotubes because of its curvature, which causes the phonons to split into two G^+ and G^- peaks. Overall the line shape and intensity of these split peaks give the information regarding the metallicity of SWCNTs; either semiconducting or metallic. The line shapes of the G band for both types of SWCNTs are shown in Figure 3.11 [19].

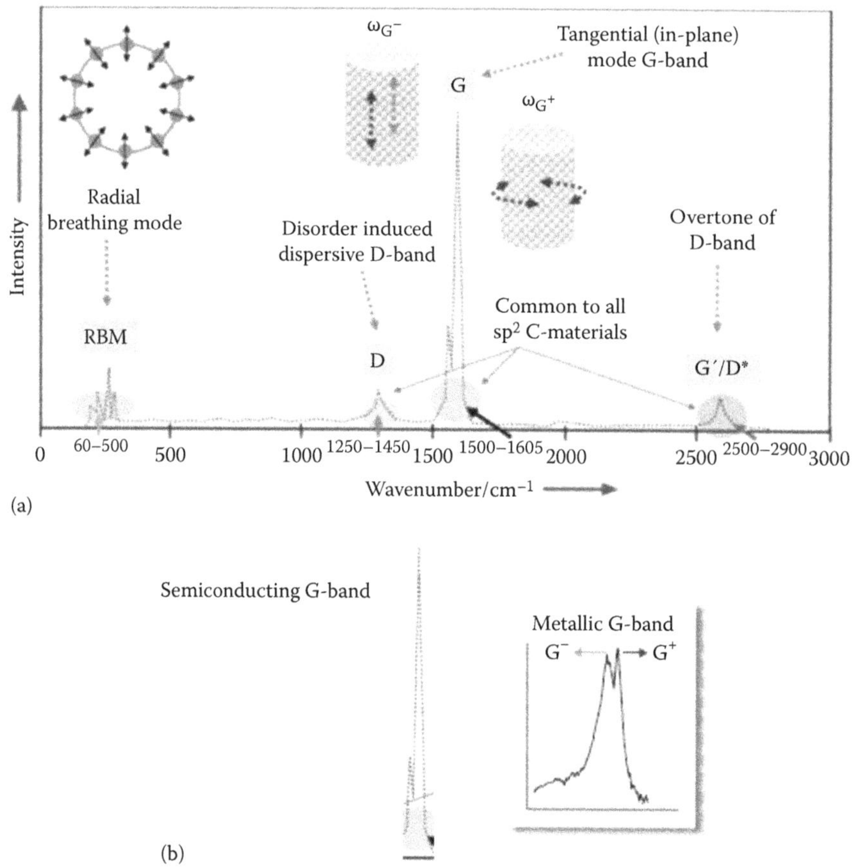

FIGURE 3.11
(a) Raman spectrum of SWCNT. (b) G-band splitting of Raman spectra for semiconducting and metallic SWCNTs. (Reprinted with permission from S. Basu-Dutt, M.L. Minus, R. Jain, D. Nepal, and S. Kumar. Chemistry of carbon nanotubes for everyone. *Journal of Chemical Education* 89, no. 2: 221–229. Copyright 2012 American Chemical Society.)

The D-band is the out of plane bending mode that is generally Raman inactive in pristine or perfect sp^2 carbon due to symmetry. When this symmetry is broken in structures of graphite, graphene, CNT, this D-band becomes Raman active and its intensity increases with the increase of disorder.

3.7.2 X-Ray Diffraction

Basically, SWCNTs are synthesized in the form of ropes as shown in Figure 3.12 [20]. These ropes have some specific structural configuration. X-ray diffraction (XRD) of pure individual SWCNTs does not provide any significant information but the XRD of SWCNT ropes give significant information. The XRD of SWCNT ropes is shown in Figure 3.13.

X-ray diffraction of small flake from ground collar material after background subtraction is shown in Figure 3.13, and data show an intense peak near diffusion vector $Q = 0.43\ Å^{-1}$ and four weaker peaks up to $1.8\ Å^{-1}$ (Figure 3.13, trace a). These low-Q scattering peaks indicate the existence of a two-dimensional lattice of SWNTs organized in bundles. The width of the peak at $Q = 0.43\ Å^{-1}$ yields a coherence length of the order of 100 Å, in very good agreement with the average transverse size of the bundles. The region above $Q = 3\ Å^{-1}$ (Figure 3.13, inset) is dominated by two peaks at 3.1 and $3.6\ Å^{-1}$, which correspond to the (111) and (200) reflections of pure face-centred cubic nickel. These x-ray diffraction data are very consistent with the observations made by HRTEM (Figure 3.12b) [20]. More details on this aspect can be seen in work done by Journet et al. [20].

3.7.3 UV/Vis/NIR Spectroscopy of SWCNTs

UV/Vis/NIR absorption spectroscopy has been demonstrated to be a powerful and inexpensive tool for providing large information regarding physical and electronic properties of SWCNT. In addition to the information regarding the type of SWCNT or its composition, the UV/Vis/NIR technique is also a very useful tool for ascertaining the degree of dispersion and purity of SWCNTs.

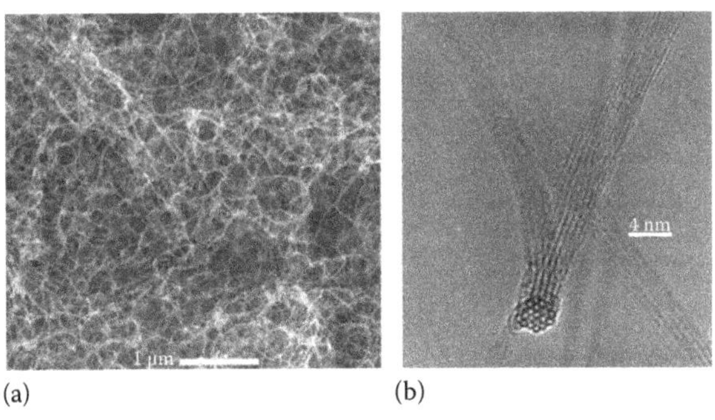

FIGURE 3.12
(a) Scanning electron microscopy image of the light and porous material that formed as a collar around the cathode deposit in our apparatus, showing a high density of entangled carbon filaments. Scale bar, 1 mm. (b) High-resolution transmission electron microscopy view of one bundle from the collarette bent in such a way that it is seen edge-on. The bundle has a uniform width and consists of, 20 SWNTs of almost uniform diameter, packed according to a triangular lattice. The focusing conditions are such that the lattice is imaged as a triangular arrangement of strong white dots (which are located at the centers of the SWNTs) on a dark background. Scale bar, 4 nm. (Reprinted by permission from Macmillan Publishers Ltd., *Nature*, C. Journet, W. K. Maser, P. Bernier, A. Loiseau, M. Lamyde la Chapelle, S. Lefrant et al., 388, no. 6644 (1997): 756–758, copyright 1997.)

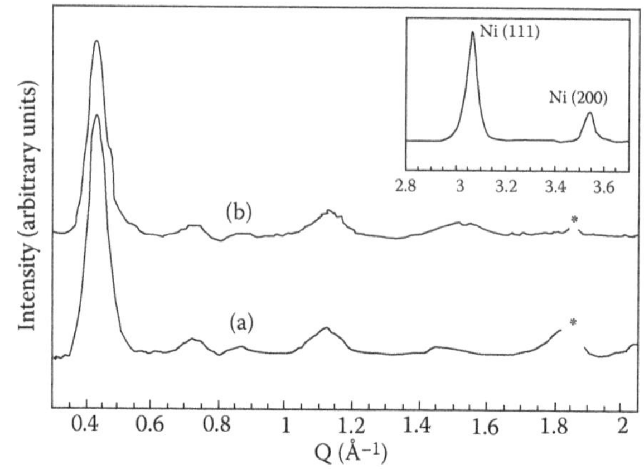

FIGURE 3.13
X-ray diffraction patterns at low angle of collar sample (trace a) and of the sample obtained by the laser ablation technique by Thess et al. (trace b). The graphite peak, due to remaining graphitic particles, has been removed for clarity; its position is shown by an asterisk. Inset: diffraction pattern associated with the Ni nanoparticles. (Reprinted by permission from Macmillan Publishers Ltd., *Nature*, C. Journet, W. K. Maser, P. Bernier, A. Loiseau, M. Lamyde la Chapelle, S. Lefrant et al., 388, no. 6644: 756–758, copyright 1997.)

The electronic structure of a SWCNT is unique for (n, m) indices of nanotubes with significantly high density of states, so-called van Hove singularities, caused by rolling the 2D graphene sheet into a 1D CNT. However, these electronic states are different for semiconducting and metallic CNT as shown in Figure 3.14. As shown in the figure in case of semiconducting SWCNTs the electronic levels in the valence and conduction band lead to S_{11} and S_{22} transitions. In case of metallic SWCNTs these energy levels are separated farther leading to M_{11} transitions [21].

The absorbance and emission spectrum of highly dispersed and debundled SWCNTs in an aqueous solution of lysozyme are shown in Figure 3.15. The absorbance (gray) spectrum

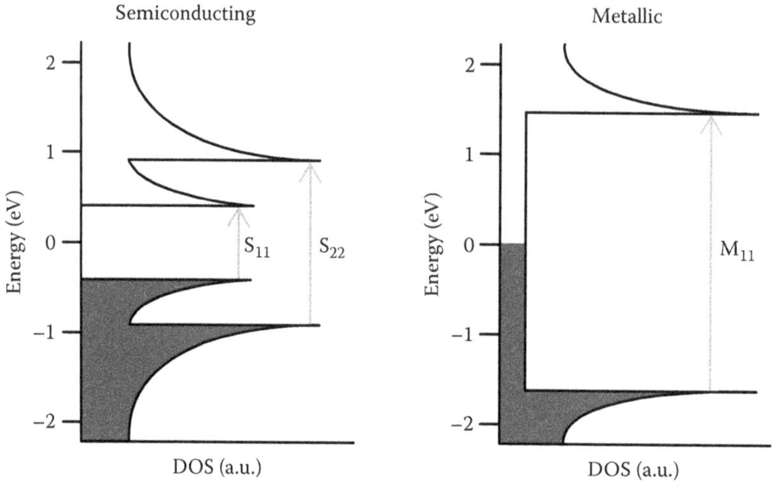

FIGURE 3.14
The electronic density of states of semiconducting and metallic SWCNTs. (From R. Saito, M. Fujita, G. Dresselhaus, and M.S. Dresselhaus, *Physical Review B* 46, no. 3 (1992): 1804–1811.)

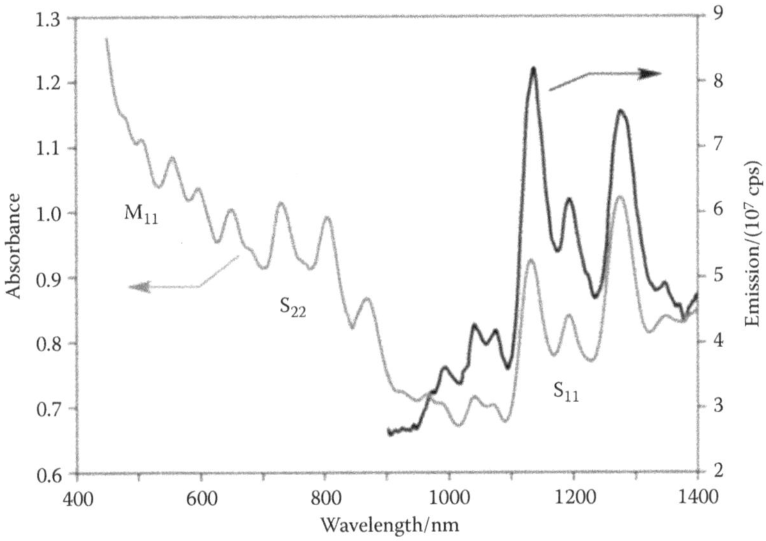

FIGURE 3.15
Absorbance (gray) and emission (black) spectra of SWCNT dispersed in lysozyme. (D. Nepal and K.E. Geckeler: pH-sensitive dispersion and debundling of single-walled carbon nanotubes: Lysozyme as a tool. *Small*. 2002. 2. No. 3. 406–412. Copyright Wiley-VCH Verlag GmbH & Co. KGaA. Reproduced with permission.)

shows peaks in the near-IR and far-IR absorption regions (600–1500 nm) corresponding to the first (S_{11}) and second (S_{22}) interband transitions for semiconducting SWCNT, whereas the first transition (M_{11}) for metallic SWCNT falls in the UV region. In addition, debundled semiconducting SWCNTs also display a near-IR fluorescence corresponding to the S_{11} transition between 900 and 1400 nm, as shown in the emission (dark) spectrum. The interactions between adjacent SWCNTs in bundles quench the luminescence [19,22,23].

3.8 Applications of Single-Walled Carbon Nanotubes

3.8.1 Microelectronics

Single-walled carbon nanotubes are a model 1D electronic system. The electronic properties of SWCNTs are closely linked to their structure, which is defined by the chiral indices (n, m). Whereas the relationship between n and m defines whether they are metallic and semiconducting, the absolute value of (n, m) defines the magnitude of the band gap in semiconducting SWCNTs. Based on their electronic band-gap (Eg), SWCNTs can be broadly classified into metallic (Eg = 0), quasi-metallic (Eg ≈ kT), and semiconducting (Eg >> kT) [24]. It is easy to see that semiconducting SWCNTs could find application as field-effect transistors (FETs) and metallic SWCNTs could be used as interconnects in nanoelectronic circuits, but not vice versa. Indeed, any and all components in nanoelectronics could be constructed out of specific types of SWCNTs, allowing for an all-CNT electronic world, making SWCNTs one of the most versatile electronic materials.

The performance of SWCNTs in electronic devices ensures extremely high carrier mobility and ballistic transport, ambipolar carrier transport, ultrafast switching, zero hysteresis, and extremely high current-carrying capacity. Lately these have all been demonstrated on individual SWCNT devices.

3.8.1.1 First SWCNT-Based Computer

Miniaturization of electronic apparatus/devices/components has been the major motivation behind the semiconductor industry, and the main improvements in computational power and energy efficiency have been seen in the last two decades. Although advancement in this technology will be continued, there is also a requirement to explore alternative technologies. Digital circuits of transistors based on CNTs have the potential to surpass silicon-based technology by improving the energy efficiency (more than an order of magnitude). Therefore, CNTs are an exciting accompaniment to existing semiconductor technologies [25]. Researchers at Stanford University announced the world's first CNT-based computer with a one-bit 178-transistor proof-of-concept processor in September 2013. This CNT computer runs an operating system with multitasking capability. It is a substantial advancement in this field because CNTs are prominent among a variety of emerging technologies that are being considered for next-generation highly energy-efficient electronic devices.

Smaller, faster, and cheaper electronic devices bring with them the problem of heat dissipation due to large circuitry in small volume, which might effect the efficiency of the device. Due to the very high electrical and thermal conductivities of CNTs, they are extremely efficient for conducting and controlling electricity and heat. They are so small that it takes very small energy to switch them off. The combination of efficient conductivity and low energy to switch off make CNTs excellent candidates to serve as smart electronic transistors. Energy dissipation in silicon-based systems has been a major concern and is a major benchmark in pushing CNTs toward practical use.

Before these CNTs we used in practical devices, it is necessary to overcome the drawback of inherent imperfaction. First, CNTs do not necessarily grow in neat parallel lines as so-called aligned CNTs. In due course, researchers have developed new processes and technologies for the growth of up to 99.5% of aligned CNTs. Though a billion tubes can come on a tiny chip, a tiny degree of misaligned tubes could cause faults. A second type of imperfection has also obstructed CNT technology, which depends on how the CNTs grow; a fraction of these CNTs can end up behaving like metallic wires that always conduct electricity, instead of being as semiconductors that can be switched on/off as per the requirement. Since mass production is the main goal, researchers had to find efficient ways to sort out the problem of misaligned and/or metallic CNTs. For the elimination of the wire-like or metallic nanotubes, the Stanford team used a novel technique that switched off all the good (semiconductor) CNTs and then they pumped the electricity in the semiconductor circuit. The electricity was concentrated in the metallic nanotubes, which make the metallic CNTs too hot and vaporized into carbon dioxide (Figure 3.16). By using this technique all of the metallic CNTs were eliminated from the circuit.

The solution for the problem of misaligned nanotubes is very difficult for that bypassing the misaligned nanotubes required even greater delicacy. Therefore the Stanford researchers developed a powerful algorithm that maps a circuit layout with a guarantee to work no matter whether or where CNTs might be off center. The Stanford team assembled a basic computer with 178 transistors by using this imperfection-immune design.

In this work, the SWCNTs were grown on a quartz substrate to yield highly aligned CNTs, and were transferred onto the target SiO_2 wafer. Before CNT transfer, the wafer

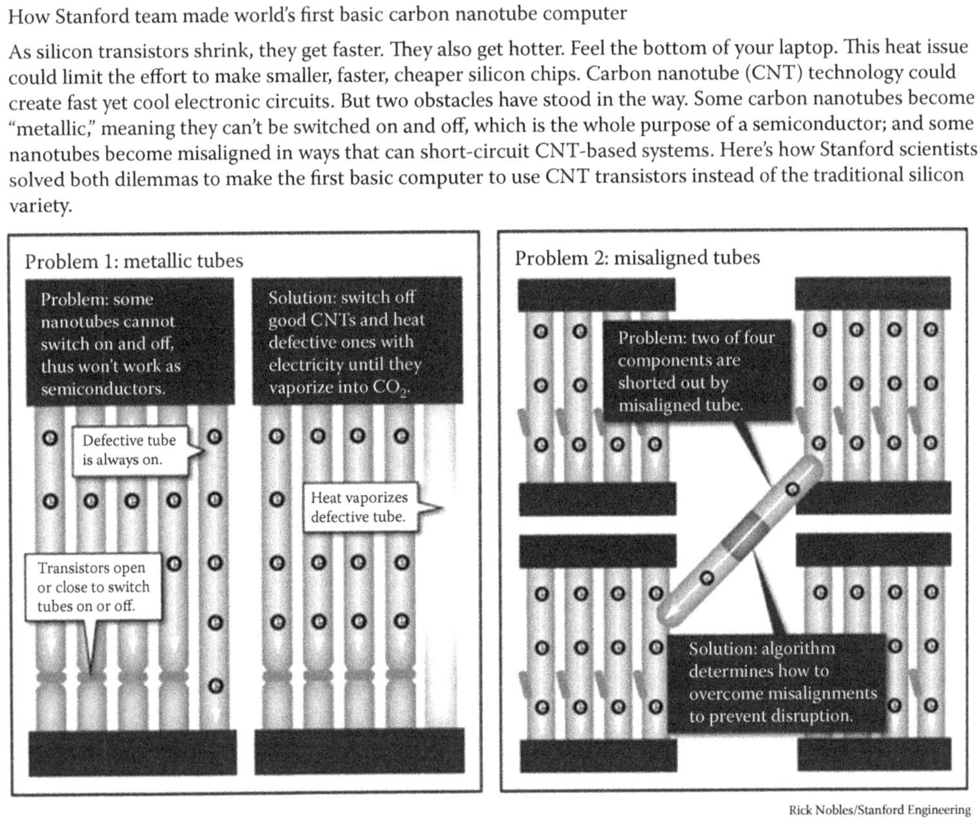

FIGURE 3.16
A scheme for the problems and their solution in designing a CNT-based computer. (From Stanford Electrical Engineering/Computer Science, Max Shulaker. http://engineering.stanford.edu/news/stanford-engineers-build-first-computer-using-carbon-nanotube-technology.)

undergoes processing to define bottom-layer wires and the local back gates of the transistors [25,26].

3.8.2 Solar Cells

One of the promising applications of SWCNTs is their use in solar panels, due to their strong characteristic of UV/Vis/NIR absorption. Research has shown that they can provide a sizable increase in efficiency. Solar cells developed at the New Jersey Institute of Technology use a CNT complex, formed by a mixture of CNTs and fullerenes to form snake-like structures. Fullerenes trap electrons, but they cannot make electrons flow. Sunlight excites the polymers, and the buckyballs will grab the electrons. CNTs, behaving like copper wires, will then be able to make the electrons or current flow [27].

Additional research has been conducted on creating SWCNT hybrid solar panels to increase the efficiency further. These hybrids are created by combining SWCNTs with photoexcitable electron donors to increase the number of electrons generated. It has been observed that the interaction between the photoexcited porphyrin and SWCNT generates electron–hole pairs at the SWCNT surfaces. This phenomenon has been observed experimentally and contributes practically to an increase in efficiency.

3.8.3 Hydrogen Storage

A lot of research is going on for using CNTs to store hydrogen to be used as a fuel source. By taking benefit of the capillary effects of the small CNTs, it is possible to condense gases in high density inside SWCNT. This allows hydrogen gas to be stored in high densities without being condensed into a liquid by high pressure. There is a possibility to use this storage method, which could be used on vehicles in place of gas fuel tanks for hydrogen-powered vehicles. Presently, regarding hydrogen-powered vehicles there is an issue of on-board storage of the fuel. Available technologies for storage methods involve cooling and condensing of the hydrogen gas to a liquid state for storage, which causes a loss of energy compared to the energy associated with the gaseous state. The storage efficiency can be increased by using SWCNTs, which would allow one to keep the H_2 in its gaseous state. This method allows for a volume-to-energy ratio smaller than that of current gas-powered vehicles.

3.8.4 Biological and Biotechnological Applications

SWCNTs with diameters of 1~2 nm, and length ranging from a few nanometers to several micrometers, is a 1D nanomaterial that may behave differently from spherical nanoparticles in biological environments. This offers new opportunities in the area of biomedical research.

The flexible CNT may bend to assist multiple binding sites of a functionalized nanotube to one cell. This improved affinity of nanotubes conjugated with targeting ligands. Due to very high surface area (theoretically 1300 m^2/g), SWCNTs permit efficient loading of multiple molecules along the length of the nanotube [28].

CNT-based sensors have been developed for the detection of biological species including proteins and DNA. Due to unique optical properties of SWCNTs, these can be utilized as optical tags or contrast agents for a variety of biological imaging techniques. Additionally, researchers have revealed that properly functionalized CNTs are able to enter cells without toxicity, transporting various biological molecular cargoes into cells.

Interesting optical properties of semiconducting SWCNTs can be exploited for sensitive and selective biomolecule detection. Band gap fluorescence of SWCNT has been explored as a method for near-IR imaging of both *in vitro* and *in vivo* biological systems. This showed the promise for *in vitro* biomolecule detection assays.

Small drug molecules can be covalently conjugated to SWCNTs for *in vitro* delivery. Fluorescent dyes and drug cargoes can be simultaneously linked to 1,3-dipolar cycloaddition functionalized CNTs via amide bonds for the anti-cancer drug delivery into cells.

3.8.5 Space Elevator

A space elevator is proposed as a space transportation system. For decades, space elevators have been touted by futurists and science-fiction writers as an alternative to expensive rocket launches. But so far, nobody has managed to bring the concept into reality. The main component of it is a cable anchored to the surface and extending into space. It will be designed in such a way that permits vehicle transport along the cable from the surface of a planet, such as Earth, into space or orbit, without the use of any rockets. A space elevator for Earth would consist of a cable with one end attached near the surface of the equator

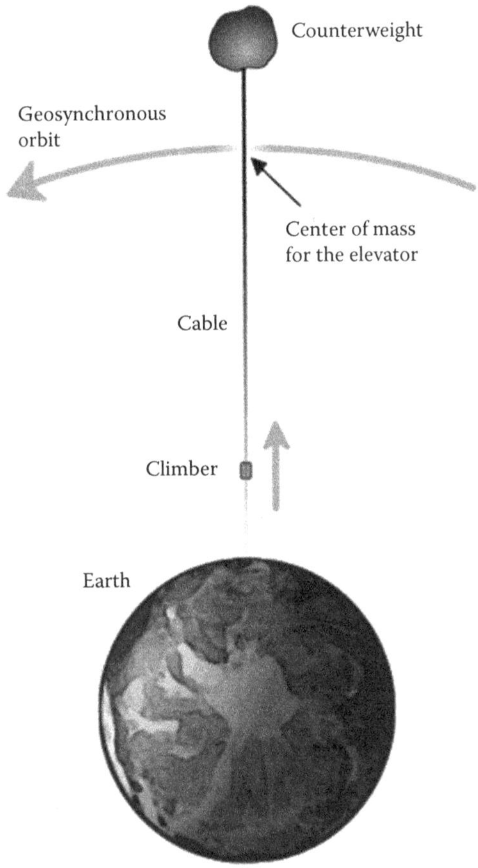

FIGURE 3.17
Space elevator.

and the other end in space beyond geostationary orbit (Figure 3.17). The cable for the space elevator must be so strong such that it can carry not only its own weight but also carry the additional weight of climbers.

One of the biggest obstacles is material that should be strong and light enough to build the incredibly long cable. Several metals like titanium, steel, or aluminum alloys have breaking lengths of only 20–30 km. Present fiber materials such as Kevlar, glass fiber, and carbon fiber have breaking lengths of 100–400 km. CNTs are expected to have breaking lengths of 5000–6000 km at sea level [29]. Being the strongest material with low density SWCNT ropes are being envisaged by NASA as cable for space elevators in the future.

References

1. S. Iijima. Helical microtubules of graphitic carbon. *Nature* 354, no. 6348 (1991): 56–58.
2. H. Dai. Carbon nanotubes: Synthesis, integration, and properties. *Accounts of Chemical Research* 35, no. 12 (2002): 1035–1044.

3. V. Mittal. *Polymer Nanotube Nanocomposites Synthesis, Properties, and Applications.* Scrivener Publishing LLC, 2010.
4. V. Choudhary, B. Singh, and R. Mathur. Carbon nanotubes and their composites. In *Syntheses and Applications of Carbon Nanotubes and Their Composites,* edited by S. Suzuki, 953–978. InTech, 2013.
5. R. Mathur, S. Seth, C. Lal, R. Rao, B. Singh, T. Dhami et al. Co-synthesis, purification and characterization of single-and multi-walled carbon nanotubes using the electric arc method. *Carbon* 45, no. 1 (2007): 132–140.
6. M. Kumar. *Carbon Nanotube Synthesis and Growth Mechanism.* InTech, 2011.
7. S. Nanot, N.A. Thompson, J.-H. Kim, X. Wang, W.D. Rice, E.H. Haroz, Y. Ganesan, C.L. Pint, and J. Kono. Single walled carbon nanotubes. In *Springer Handbook of Nanomaterials,* edited by R. Vajtayi, 105–146. Springer-Verlag, 2013.
8. Y. Liu, W.Z. Qian, Q. Zhang, G.Q. Ning, G.H. Luo, Y. Wang, D.Z. Wang, and F. Wei. Synthesis of high-quality, double-walled carbon nanotubes in a fluidized bed reactor. *Chemical Engineering & Technology* 32, no. 1 (2009): 73–79.
9. F. Ding, P. Larsson, J.A. Larsson, R. Ahuja, H. Duan, A. Rosen et al. The importance of strong carbon-metal adhesion for catalytic nucleation of single-walled carbon nanotubes. *Nano Letters* 8, no. 2 (2008): 463–468.
10. J.-Y. Raty, F. Gygi, and G. Galli. Growth of carbon nanotubes on metal nanoparticles: A microscopic mechanism from ab initio molecular dynamics simulations. *Physical Review Letters* 95, no. 9 (2005): 096103.
11. H. Kataura, Y. Kumazawa, Y. Maniwa, Y. Ohtsuka, R. Sen, S. Suzuki et al. Diameter control of single-walled carbon nanotubes. *Carbon* 38, no. 11 (2000): 1691–1697.
12. P.J.F. Harris. Solid state growth mechanisms for carbon nanotubes. *Carbon* 45, no. 2 (2007): 229–239.
13. S. Berber, Y.K. Kwon, and D. Tomanek. Unusually high thermal conductivity of carbon nanotubes. *Physical Review Letters* 84, no. 20 (2000): 4613–4616.
14. P. Avouris. Carbon nanotube electronics. *Chemical Physics* 281, no. 2–3 (2002): 429–445.
15. B.Q. Wei, R. Vajtai, and P.M. Ajayan. Reliability and current carrying capacity of carbon nanotubes. *Applied Physics Letters* 79, no. 8 (2001): 1172–1174.
16. T. Durkop, B.M. Kim, and M.S. Fuhrer. Properties and applications of high-mobility semiconducting nanotubes. *Journal of Physics-Condensed Matter* 16, no. 18 (2004): R553–R580.
17. R. Gupta, B.P. Singh, V.N. Singh, T.K. Gupta, and R.B. Mathur. Origin of radial breathing mode in multiwall carbon nanotubes synthesized by catalytic chemical vapor deposition. *Carbon* 66 (2014): 724–726.
18. P.T. Araujo, I.O. Maciel, P.B.C. Pesce, M.A. Pimenta, S.K. Doorn, H. Qian et al. Nature of the constant factor in the relation between radial breathing mode frequency and tube diameter for single-wall carbon nanotubes. *Physical Review B* 77, no. 24 (2008): 241403.
19. S. Basu-Dutt, M.L. Minus, R. Jain, D. Nepal, and S. Kumar. Chemistry of carbon nanotubes for everyone. *Journal of Chemical Education* 89, no. 2 (2012): 221–229.
20. C. Journet, W. K. Maser, P. Bernier, A. Loiseau, M. Lamyde la Chapelle, S. Lefrant et al. Large-scale production of single-walled carbon nanotubes by the electric-arc technique. *Nature* 388, no. 6644 (1997): 756–758.
21. R. Saito, M. Fujita, G. Dresselhaus, and M.S. Dresselhaus. Electronic structure of graphene tubules based on C_{60}. *Physical Review B* 46, no. 3 (1992): 1804–1811.
22. M.J. O'Connell, S.M. Bachilo, C.B. Huffman, V.C. Moore, M.S. Strano, E.H. Haroz et al. Band gap fluorescence from individual single-walled carbon nanotubes. *Science* 297, no. 5581 (2002): 593–596.
23. D. Nepal and K.E. Geckeler. pH-sensitive dispersion and debundling of single-walled carbon nanotubes: Lysozyme as a tool. *Small* 2, no. 3 (2002): 406–412.
24. A. Vijayaraghavan. Applications of chirality-sorted individual single-wall carbon nanotube devices. *Journal of Materials Chemistry* 22, no. 15 (2012): 7083–7087.

25. M.M. Shulaker, G. Hills, N. Patil, H. Wei, H.-Y. Chen, H.S.P. Wong et al. Carbon nanotube computer. *Nature* 501, no. 7468 (2013): 526–530.
26. Stanford Electrical Engineering/Computer Science, Max Shulaker, http://engineering.stanford.edu/news/stanford-engineers-build-first-computer-using-carbon-nanotube-technology.
27. New flexible plastic solar panels are inexpensive and easy to make, http://www.sciencedaily.com/releases/2007/07/070719011151.htm.
28. Z. Liu, X. Sun, N. Nakayama-Ratchford, and H. Dai. Supramolecular chemistry on water-soluble carbon nanotubes for drug loading and delivery. *ACS Nano* 1, no. 1 (2007): 50–56.
29. Breaking length, http://wordpress.mrreid.org/2013/08/10/breaking-length/.

4

Multiwalled Carbon Nanotubes

4.1 Introduction

Carbon nanotubes (CNTs) were originally discovered as by-products of fullerenes and now are considered to be the building blocks of future nanoscale electronic and mechanical devices. The credit for the discovery of multiwalled carbon nanotubes (MWCNTs) is given to Sumio Iijima. His work was published in *Nature* in 1991 [1]. Much of the historical development of CNTs including single-walled and multiwalled carbon nanotubes is covered in Chapter 2.

4.2 Difference between Multiwalled Carbon Nanotubes (MWCNTs), Single-Walled Carbon Nanotubes (SWCNTs), and Carbon Nanofibers (CNFs)

Multiwalled carbon nanotubes (MWCNTs) can easily be distinguished from single-walled carbon nanotubes (SWCNTs) on the basis of their diameter and Raman characteristics. MWCNTs are assemblies of several SWCNTs arranged in a co-axially Russian doll structure, and each SWCNT corresponds to one wall of MWCNT. The wall-to-wall distance in MWCNTs is 0.340 nm, which is slightly higher than the sheet-to-sheet difference in graphite (0.335 nm) [2]. Although CNTs and carbon nanofibers (CNFs) are both hollow and of nanometer dimensions and have similar production methods, there are significant differences that affect their performance and ability to be processed. The first difference is based on their morphology. CNFs, also known as stacked cup CNTs (SCCNTs), have a unique morphology in that graphene planes are canted from the fiber axis resulting in exposed edge planes on the interior and exterior surface of the fiber [3]. On the other hand, MWCNTs are typically an assembly of concentric cylinders made of carbon sheet. Another difference is based on their sizes. CNTs typically have diameters ranging from 1 to 100 nm, whereas CNFs or SCCNTs have diameters from 50 to 200 nm depending on the type of CNF and synthesis technique. Figure 4.1 shows a comparative structure and dimensions of different forms of these carbons.

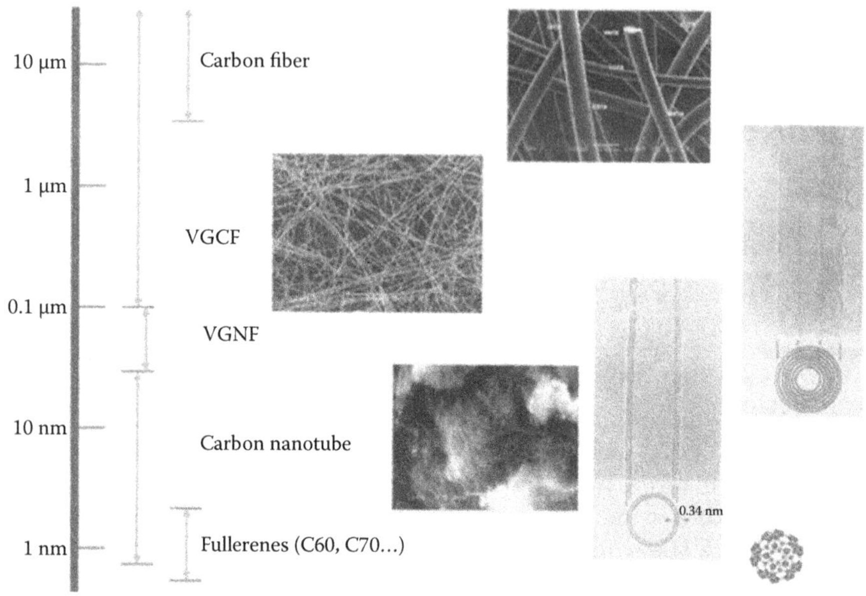

FIGURE 4.1
Comparative structure and dimensions of different forms of carbons. (From Unitek Int. Co. Ltd.)

4.3 Synthesis of MWCNTs

MWCNTs and SWCNTs are generally produced by similar techniques. Several techniques have been developed in the last 25 years, and now synthesis of MWCNTs has been well established and several industries are producing them on a large scale. The most commercially viable technique for the production of MWCNTs is chemical vapor deposition (CVD) and is being followed by most of the commercial producers with slight modifications. Besides CVD, arc discharge and laser ablation are two other techniques for the synthesis of MWCNTs.

4.3.1 Arc Discharge

Arc discharge is the electrical breakdown of a gas to generate the plasma. The chamber consists of two electrodes: one of which is called anode, which is filled with powdered carbon precursor, mainly graphite particles, along with the catalysts; and the other electrode is called cathode, which is usually a pure graphite rod. The chamber is generally filled with a gas or sometimes with liquid environment. An electric field is applied between the two electrodes, which are brought close enough to generate an arc, and a gap of 1 to 2 mm is maintained to attain a steady discharge. The arc current generates plasma of very high temperature ~4000°C, which sublime the carbon precursor inside the arc and form the CNTs. A schematic of a typical arc discharge apparatus is shown in Figure 4.2.

Generally arc discharge is used for the synthesis of SWCNTs, MWCNTs, and fullerenes. The operating conditions, carbon precursor, and catalyst in the arc chamber determine the type and purity of carbon nanomaterials. It has been found that high current density and low gas pressure results into the formation of fullerenes while at low current density

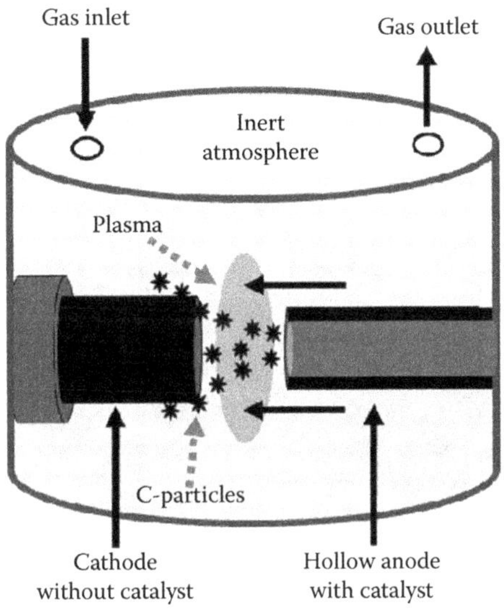

FIGURE 4.2
Schematic for the synthesis of CNTs by arc discharge technique.

and high pressure nanotubes are formed. There are primarily two deposits in the arc discharge technique: One is the chamber deposits, where the growth takes place in the presence of catalysts and the product is in the form of webs of SWCNT ropes. The second is the cylindrical mass called the cathode deposit, which contains mainly MWCNTs in its core. Interestingly this does not contain any metallic impurities.

4.3.2 Laser Ablation

Laser ablation is very similar to the arc discharge technique due to a very similar reaction condition and on the fact that both reactions probably occur with the same mechanism. It was initially used for the synthesis of fullerenes. In a typical laser ablation experiment, the laser pulse is focused onto a graphite target containing a metal catalyst. The target is kept in inert gas flow and the system is heated at around 1000°C by an electric furnace. The laser evaporates the target into plasma in which CNTs are formed. Laser ablation is not suitable for the bulk production of CNTs, but it produces high purity of CNTs (>90%) and narrower size distribution than arc discharge and CVD technique.

4.3.3 Chemical Vapor Deposition

CVD is a very important and industrially viable technique for the large-scale production of MWCNTs. In CVD, the hydrocarbon precursor is decomposed on the metal catalyst particle available at substrate in the presence of inert gas. The hydrocarbon can either be liquid state as toluene, benzene, xylene, and ethanol, or in the gas form as acetylene and ethylene. The catalyst can be any transition metal such as Ni, Co, and Fe. The photograph of the two zone furnace setup used for the CVD process for the synthesis of MWCNTs is shown in Figure 4.3. In this method, MWCNTs are synthesized using toluene as the carbon source and ferrocene as the catalyst source. Ferrocene is a good precursor for producing

FIGURE 4.3
Photograph of the CVD setup used for the synthesis MWCNTs at CSIR-NPL, New Delhi.

iron particles used for nanotube growth and is soluble in the hydrocarbon (toluene). A known quantity of ferrocene is dissolved in the solvent toluene and the mixture is sonicated for 60 min to achieve uniform solution of ferrocene in toluene. The solution of precursor raw material is injected into a reaction zone of a tubular quartz reactor of 50 mm in diameter, kept inside the furnace, with a flow rate of 10 ml/hr through the fine capillary. The length of the capillary is adjusted so that the solution drops inside the reactor in the first zone where a temperature of 200°C is maintained. The mixture of two components—a catalyst and hydrocarbon—is transported into the main reaction zone of the furnace, maintained at around 800°C, with the help of argon as the carrier gas. The flow rate of carrier gas is so adjusted that the maximum amount of precursor hydrocarbon is consumed inside the desired temperature zone.

The temperature of the reaction zone is maintained at 750°C to 800°C during the experiment. Inside the reactor, the Fe particles decomposed out of ferrocene in the first zone get deposited on the quartz substrate and then become fragmented in Fe nanoparticles. The decomposition of the toluene on the surface of Fe nanoparticles results into the formation of carbon, and growth of CNTs occurs via diffusion of carbon through Fe nanoparticles. After injecting all the solution into the reactor, the furnace is allowed to cool down with continuous flow of inert gas until it cools to room temperature. The deposited CNTs from the quartz reactor are scraped out with the help of a brush and a good amount of CNT is harvested in the form of black fluffy powder.

The CVD process cannot only be used for the bulk synthesis of MWCNTs but also in perfectly aligned form in different architecture. This has also been previously demonstrated by several researchers and used for a variety of applications. The schematic for the synthesis of aligned CNTs by CVD technique is shown in Figure 4.4. A substrate coated with the aluminum buffer layer is placed inside the reaction zone of the CVD furnace and a mixture of hydrocarbon/catalyst precursor is injected. At the end of the reaction the substrate is covered with aligned MWCNTs [4].

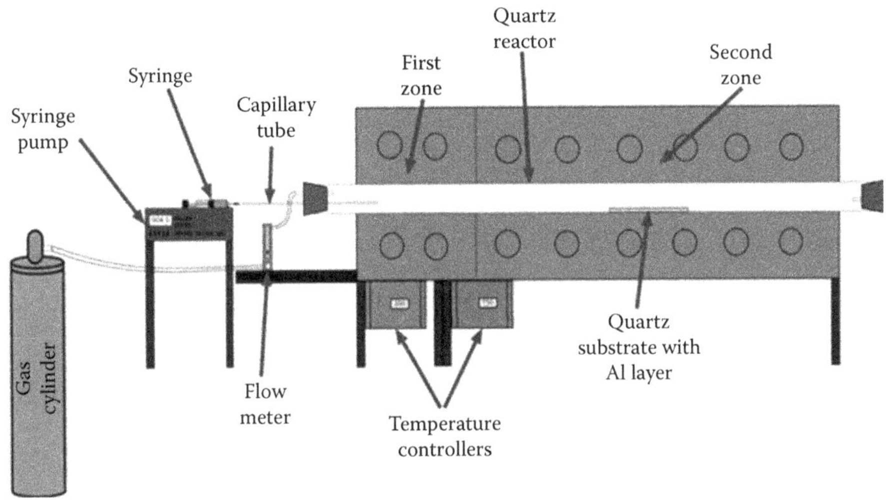

FIGURE 4.4
Schematic of CVD setup used for the synthesis of vertically aligned MWCNTs. (From A.K. Arya, B.P. Singh, J. Jyoti, S. Pati, and S. Dhakate, *Advanced Material Letter* 6, no. 12 (2015): 1094–1097.)

4.3.4 Fluidized Bed

The fluidized bed process is the most promising route for the large-scale production of the CNTs. The schematic of the fluidized bed for the growth of CNTs is shown in Figure 4.5. The CNT growth mechanism is based on three independent phenomenons: catalytic dissociation of hydrocarbon molecules, atomic carbon saturation in catalyst nanoparticles, and carbon precipitation from the catalyst to form CNTs [5]. The processing conditions in fact decides various features of resulting CNTs such as dimensions, chirality, number of walls, and diameter. The catalyst also plays a very important role in the growth of CNTs by fluidized bed and influences the quality and the quantity of the CNTs produced. Mostly, transition metals in elemental form (Fe, Mo, Co, Ni) or in compound form are used as

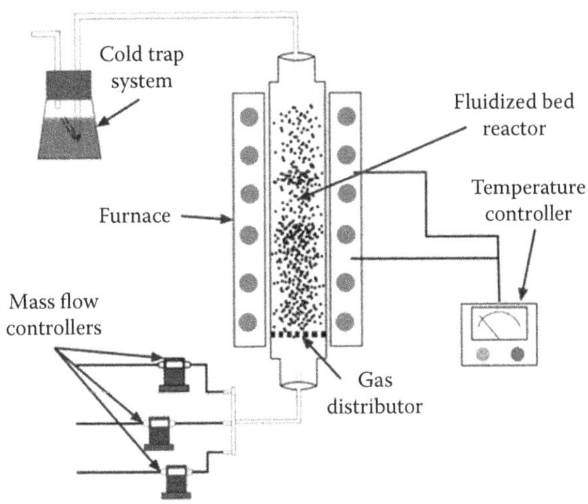

FIGURE 4.5
Schematic diagram of the fluidized bed reactor for the production of CNTs.

catalysts in CNT synthesis, but iron is found to be more active metal for the synthesis of CNTs. In addition to the catalyst, catalyst support also plays an important role. The catalyst is dispersed on a supported material with a nanosized pore (<50 nm), for example, magnesia, alumina, zeolite, silica, and calcium oxide. The small-sized supporting catalysts are very cohesive in nature. The production efficiency of CNTs can be increased by fluidizing the catalyst particles in the presence of carbon source gas heated to the reaction temperature in the reactor [5]. A steady flow of the catalyst is maintained inside the reactor along with the flux of mixed gases (methane/acetylene as the carbon source plus argon as the carrier gas) for a specific optimized time.

4.4 Purification and Functionalization

As-produced MWCNTs have impurities in the form of catalyst and amorphous carbon, and purification of these MWCNTs is very important. Generally 10–12 wt/% of a catalyst is mostly found as impurity along with the MWCNT bundle. A detailed procedure for purification of CNTs was discussed in Chapter 3. Additionally, these CNTs do not have efficient surface functionality to make chemical bonds with the polymer matrix for the development of composite materials. To realize the outstanding mechanical properties in a fruitful way, it is essential to use surface-modified MWCNTs as reinforcing materials in a polymer matrix.

A number of studies on CNT–polymer composites have focused on improving the dispersion and load transfer efficiency by improving the compatibility between CNTs and the polymer matrix through covalent chemical functionalization of the CNT surface [6,7]. The surface modification involves attaching different functional groups on the surface and the end caps of CNTs. A major challenge in the emerging area of nanotubes chemistry was the development of an oxidation process for CNTs. It can either be carried out by refluxing it in nitric acid or in the combination of nitric and sulfuric acid. These drastic conditions led to the opening of the tube caps as well as the defects in the sidewalls due to oxidative etching and associated release of carbon dioxide [8]. The final products provide the nanotubes of short lengths whose ends and sidewalls are decorated with a high density of various oxygen-containing groups (mainly carboxyl groups). Under less vigorous conditions, such as refluxing in nitric acid, the shortening of the tubes can be minimized. The chemical functionalization is then limited mostly to the opening of the tube caps and the formation of functional groups at defect sites along the sidewalls [8].

These carboxylic functional groups can be further replaced by more reactive groups like –COCl or $–CORNH_2$, as shown in Figure 4.6. The addition of these functional groups on CNTs cause intermolecular repulsion between functional groups on the surface, which overcomes weak van der Waals attraction between CNTs. It is also vital to stabilize the dispersion to prevent reagglomeration of CNTs. A wide range of surface functionality can thus be created on CNTs.

It has been observed by the researchers that use of amine functionalized CNTs resulted in a significant improvement in the mechanical properties of the CNT epoxy composites due to the formation of covalent bond by the opening of epoxide ring with amine groups available on the surface of CNTs, as shown in Figure 4.7 [9].

A host of literature is available on the subject of the chemical modification of CNTs.

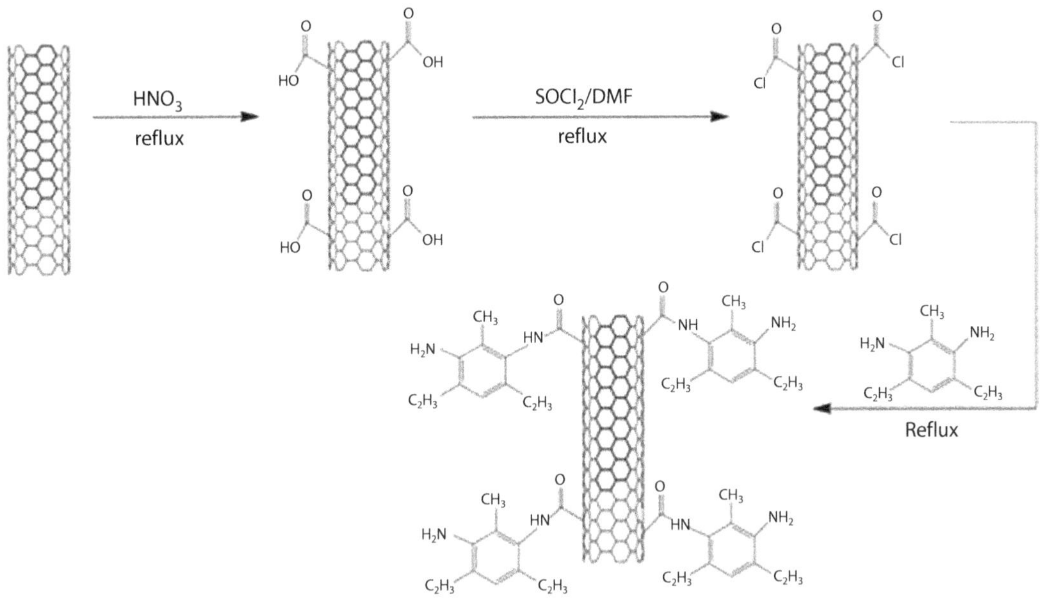

FIGURE 4.6
Chemical modifications of nanotubes through chemical oxidation followed by subsequent chlorination and amidation.

4.5 Characterization

4.5.1 Scanning Electron Microscopy

Scanning electron microscopy (SEM) is useful for imaging surface features of bulk samples of nanotube. Generally as-produced CNTs by CVD on the quartz reactor are in the form of bundles. SEM study (Figure 4.8a–d) of as-produced CVD-CNTs reveals that these bundles of CNTs are relatively straight. The high magnification SEM images of the CNT bundles (Figure 4.8e and f) shows a very clean, catalyst-free surface of individual nanotubes.

4.5.2 High-Resolution Transmission Electron Microscopy

Transmission electron microscopy (TEM) is generally used to study the diameter distribution; and number of layers and type of CNTs, either SWCNTs or MWCNTs. Several TEM images of CVD-CNTs are shown in Figure 4.9. Figure 4.9a and b show the TEM micrographs which reveal the presence of catalyst within MWCNTs. Some of the synthesized MWCNTs were found to be filled with catalyst particles in the form of wires (Figure 4.9b and c) and the tip of CNTs, surrounded by several graphitic layers (Figure 4.9d).

4.5.3 Raman Spectroscopy

Raman spectroscopy has become an indispensable tool for the characterization of MWCNTs. A typical spectrum of MWCNTs shows three important bands: (1) a low

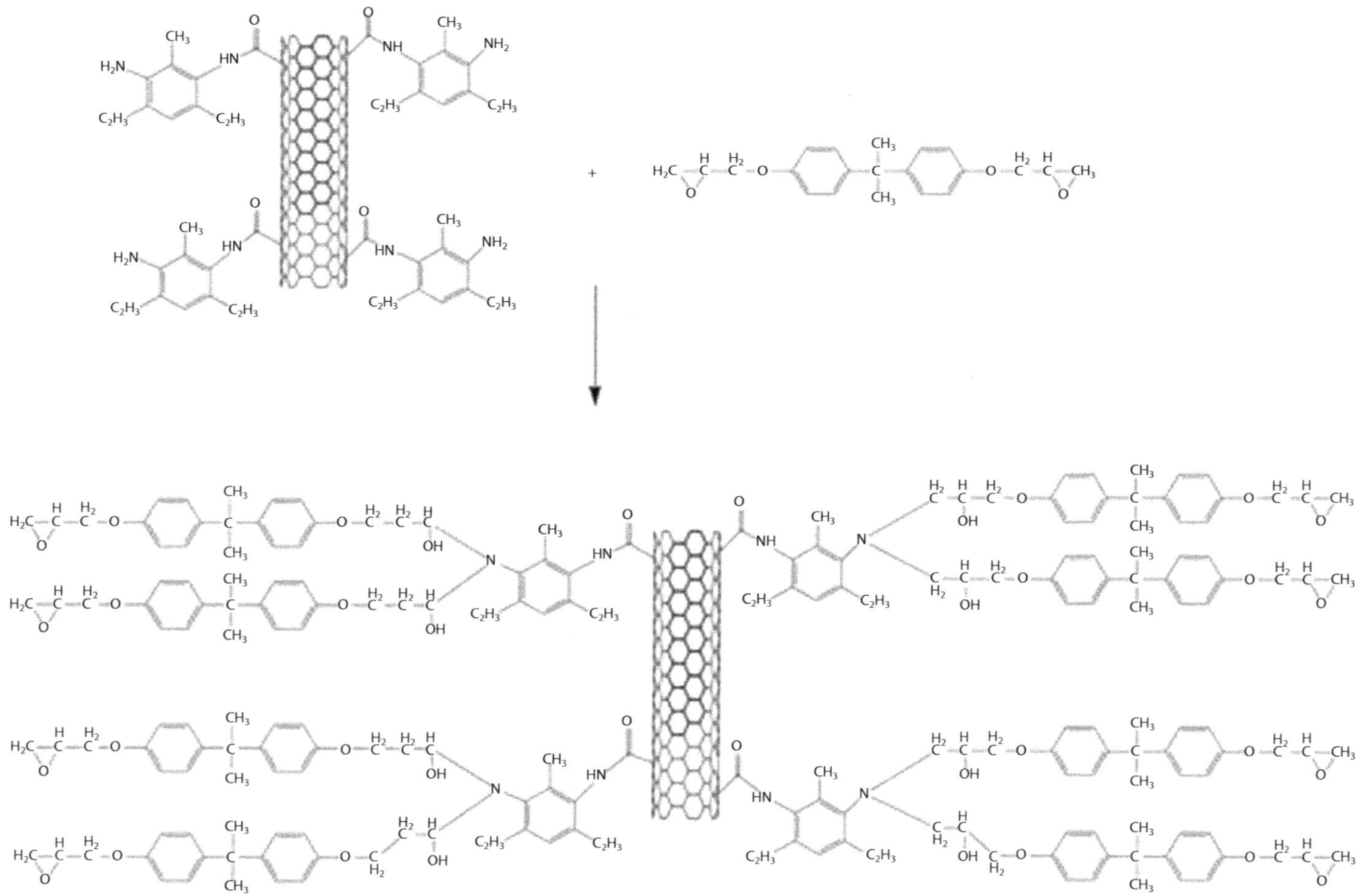

FIGURE 4.7
Pictorial view for the formation of amine-modified MWCNTs and their interaction with the epoxy matrix. (From B.P. Singh, V. Choudhary, S. Teotia, T.K. Gupta, V.N. Singh, S.R. Dhakate, and R.B. Mathur, *Advanced Materials Letters* 6, no. 2 (2015): 104–113.)

FIGURE 4.8
SEM images of CVD-MWCNTs at different magnification: (a) 70×, (b) 350×, (c) 500×, (d) 3000×, (e) 10,000×, and (f) 50,000×.

frequency region (100–600 cm^{-1}), also called radial breathing mode (RBM), (2) the band around 1340 cm^{-1} called the D band (due to disorder), and (3) the band around 1580 cm^{-1} called the G band (corresponding to the tangential vibrations of carbon atoms). The RBM frequency is inversely proportional to the tube diameter and RBM cannot be detected for CNT having tube diameter >2 nm. Other than these three main features, some other bands have also been reported at around 2600 cm^{-1} (second-order D band) and some low intensity bands in the range of 1700–1800 cm^{-1}.

Moreover, the ratio of peak intensity of the G band and D band (I_G/I_D) indicates the quality of CNTs, that is, a higher ratio corresponds to good quality of CNTs. The surface modification of CNTs during functionalization can also be investigated using Raman spectroscopy.

The Raman spectrum of MWCNTs in Figure 4.10 shows the G band position at ~1573 cm^{-1} and the D band at 1347 cm^{-1}. The I_G/I_D intensity ratio of >2.5 confirms reasonably good quality of MWCNTs.

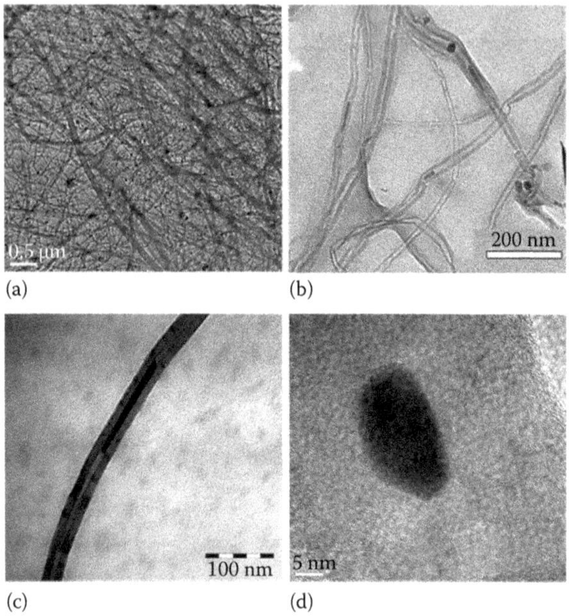

FIGURE 4.9
TEM images of CVD-MWCNTs. (a) Low magnification image showing highly entangled long length tubes; (b) image showing uniform diameter tubes with nanosized dispersed catalyst particles; (c) high magnification image of single MWCNT showing encapsulated iron catalyst inside the cavity; and (d) high resolution image of single MWCNT showing graphitic fringes surrounding the catalyst particle.

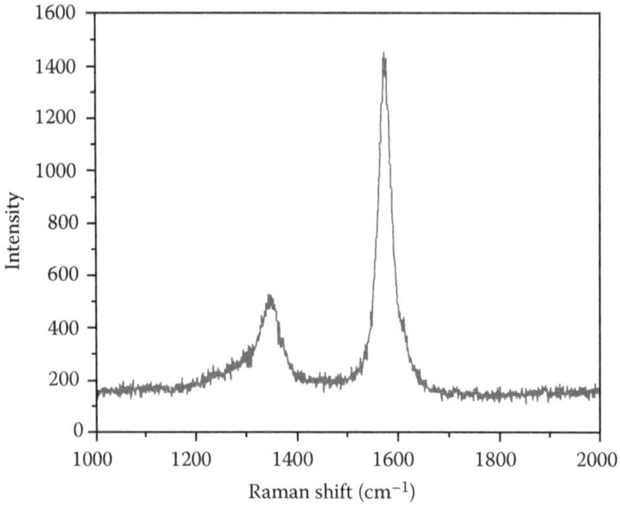

FIGURE 4.10
Raman spectrum of CVD-MWCNTs.

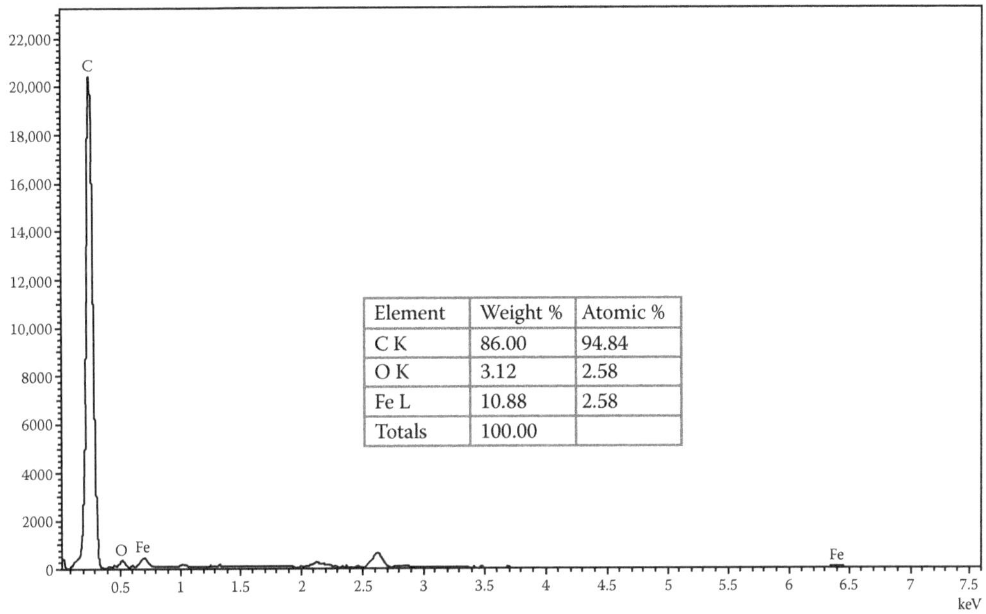

Element	Weight %	Atomic %
C K	86.00	94.84
O K	3.12	2.58
Fe L	10.88	2.58
Totals	100.00	

FIGURE 4.11
EDX of CVD-MWCNTs.

4.5.4 Energy-Dispersive X-Ray Spectroscopy

Energy-dispersive x-ray spectroscopy (EDX or EDS) is an analytical technique used for the elemental analysis or chemical composition of the material using x-ray energy spectra. Besides the other applications, it can determine the type of impurities present in the sample, that is, metallic impurities in as-produced MWCNTs. The elemental composition of a typical MWCNT sample is given in the inset table of Figure 4.11. The EDX of as-produced MWCNTs contains ~10.88 wt. % Fe, which comes from the ferrocene used as a catalyst for the growth of MWCNTs.

4.5.5 Thermal Gravimetric Analysis

Thermal gravimetric analysis (TGA) is a thermal analysis technique that measures the weight change in a material as a function of temperature and time in a controlled environment. It is very useful to investigate the thermal stability of a material or to investigate its behavior in different atmosphere (inert or oxidizing). It can also be used for ascertaining the catalyst contents as well as the other carbon impurities present in the as-produced soot [10]. The catalytic impurities present in MWCNTs are evaluated by recording the thermogravimetric traces in oxygen atmosphere. Figure 4.12 shows the TG/DTG traces for CVD-MWCNTs. There is no mass loss up to 440°C, after which a sharp mass loss is observed with the DTG peak at 550°C. The mass loss in this region is due to the oxidation of CNTs present in the sample. The mass loss is stabilized beyond 680°C. The total mass loss up to 1000°C is ~87%, which is due to the oxidation or burn off of CNTs. The remaining 13% mass represents the remaining iron catalyst or its oxide, present in the as-produced sample as an impurity.

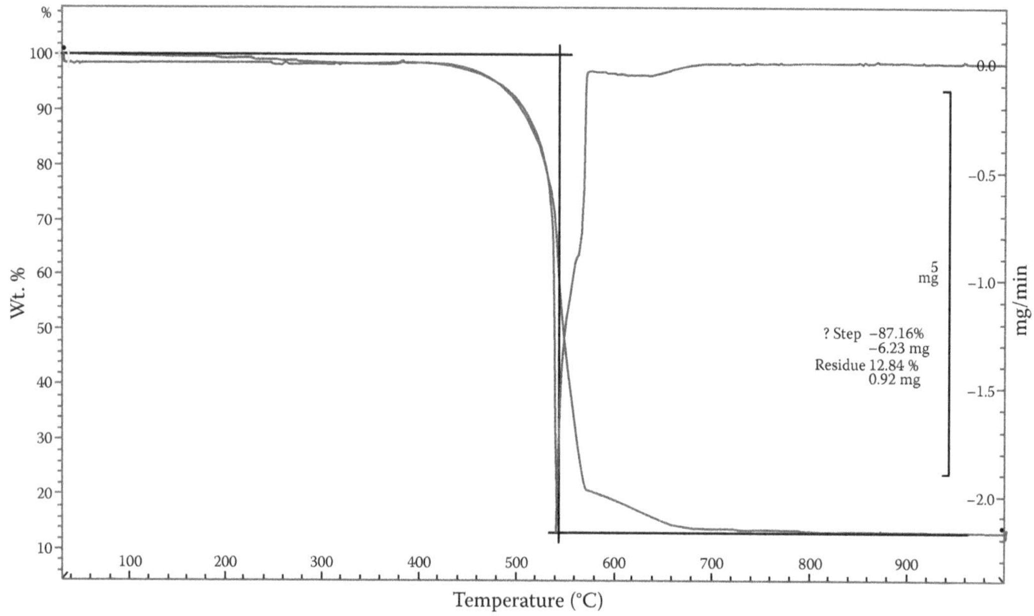

FIGURE 4.12
A typical TGA run of CVD-MWCNT batch carried out in air atmosphere.

4.6 Applications

The outstanding properties of CNTs have opened a new age of advanced multifunctional materials for a variety of applications such as reinforcement in plastic to make them high-strength conducting plastics for electromagnetic interference (EMI) shielding and electrostatic discharge (ESD) applications, CNT-based yarn for structural composites, anodes for lithium ion batteries, fuel cell components, armor materials, self-healing composite materials, and conducting cables for motors.

4.6.1 Polymer Nanocomposites for Structural Applications

Polymer composites consisting of polymers reinforced with various additives such as carbon fibers, graphite fibers, glass fibers or Kevlar fibers [11], and carbon black are increasingly being used in defense, aerospace, automobile, sports, and electronic sectors as lightweight, high-strength and high electrical and thermal conducting materials [12–16]. In recent years, nanocomposites, in which the reinforcing additive has nanoscale dimensions have attracted both scientific and technological interest to meet the growing demands for materials with improved properties for challenging applications [17,18]. CNTs have particularly led to research in the development of advanced polymer nanocomposites due to their unique material properties as discussed earlier, which is comprised of extraordinarily high mechanical, electrical and thermal properties [1,19,20]. CNTs, both SWCNTs with diameters 1–2 nm and MWCNTs with diameters 10–100 nm, have a very high aspect ratio and therefore very large surface area [7,21]. This suggests that small amounts of CNTs as compared to conventional reinforcements can provide a large volume of interface that

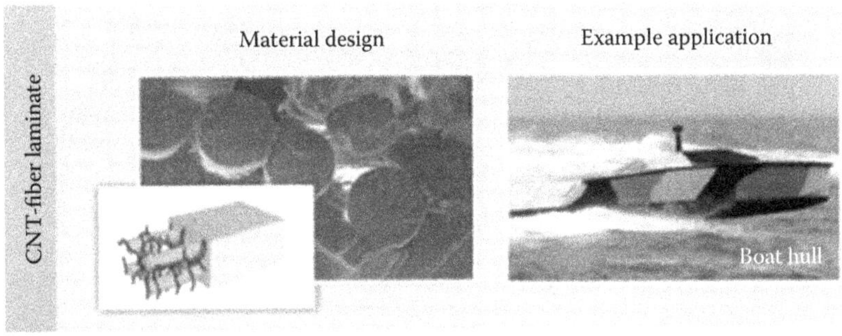

FIGURE 4.13
Emerging CNTs composites. (a) A micrograph showing the cross section of carbon fiber laminates with CNTs dispersed in epoxy resin and (b) a lightweight CNT–fiber composite boat hull for maritime security boats. (Images courtesy of Zyvex Technologies.)

will significantly enhance the properties of polymers in terms of imparting strength and conductivity to a composite system [22].

In recent years, most of studies have been carried out to enhance the mechanical properties of CNT–polymer composites [7,23–27]. Different polymers, both thermoplastics (e.g., polymethylmethacrylate [PMMA] [28–35], polystyrene [36–38], polycarbonate, and polypropylene [7,39–41]) and thermosets (e.g., phenolics [42], polyimide [23,43] and epoxy [44–51]) have been investigated as matrices to make CNT-reinforced polymer composites.

Although neat CNT–polymer composites have so far not realized the desired results, an alternate approach is initiated in terms of using a CNT-dispersed resin matrix for fiber reinforced composites.

The high mechanical strength of these nanocomposites could be utilized to make high-end sporting goods, such as tennis rackets and baseball bats, thus delivering superior performance. Recent examples include strong, lightweight wind turbine blades and hulls for maritime security boats and composite wind turbine blades [52] that were made by using carbon fiber composites with CNT-enhanced resin (Figure 4.13) [52]. This can also be useful for the aerospace industry because the main driving force for materials substitution in the aerospace industry is weight reduction at a reasonable cost while maintaining reliability and safety standards. Reducing the weight of the structure allows lifting a greater payload and/or reducing fuel consumption.

4.6.2 CNT Yarn for Structural Composites

As discussed earlier, it is very difficult to properly align CNT in a polymer matrix through conventional composite processing techniques; however, several techniques have been used to draw fibers out of a CNT-dispersed polymer. Not much success has been achieved so far. Recently, dry spinning of CNT yarn is suggested as an alternative form of an aligned bundle of MWCNT. In the long run, it is believed that CNT yarn and laminated sheets made by direct CVD or spinning from CNT forest or other drawing methods will only compete with the carbon fiber-based technologies for high-end applications, especially in weight-sensitive applications. Several researchers have made multiwalled CNT-based yarn by various techniques such as dry spinning, wet spinning, and melt spinning with polymers. A stiffness value of 357 GPa and a strength of 8.8 GPa has been achieved for a few millimeters long gauge length. For centimeters long gauge length, strength values of

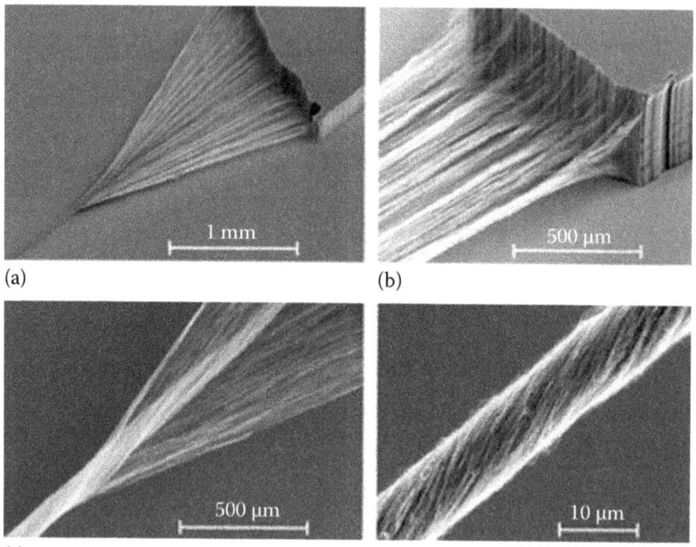

(a) (b) (c) (d)

FIGURE 4.14
SEM images showing a web being drawn from a CNT forest and then twisted into a yarn. (a) Overall view of the process clearly showing the spinning triangle that forms; (b) CNT detachment from the forest edge to form a web; (c) convergence point of the spinning triangle; and (d) the final yarn. (Reprinted from *Carbon*, 48, K. Sears, C. Skourtis, K. Atkinson, N. Finn, and W. Humphires. Focused ion beam milling of carbon nanotube yarns to study the relationship between structure and strength, 4450–4456. Copyright 2010, with permission from Elsevier.)

2 GPa have been achieved, which equals the specific strength of commercially available Kevlar (DuPont) fibers [52]. Recent advances in preparation allow CNTs to grow up to several millimeters in length, which could be aligned to continuous macroscopic CNT fibers (Figure 4.14) [53,54]. This provides an opportunity for preparation of superstrong continuous nanotube reinforced composites.

4.6.3 Electrostatic Discharge and Electromagnetic Interference Shielding Applications

The growing demand for electrostatic discharge (ESD) protection and electromagnetic interference (EMI) shielding in the plastic industry for the electronics sector has increased the research on developing new materials, particularly on electrically conductive polymer composites using carbon-based filler [40].

EMI is becoming a serious problem nowadays due to the rapid growth in electronics and instrumentation. It can disturb normal functioning of equipment or may lead to its complete breakdown. Therefore, some shielding mechanism must be developed to prevent the appliances from the harmful effect of this electromagnetic (EM) noise.

Compared to conventional metal-based EMI shielding materials, carbon-based conductive polymer composites are attractive due to their light weight, resistance to corrosion, flexibility, and processing advantages [55,56]. Carbon black is commonly used as a conducting filler in polymer composites [56,57]. However, in using carbon black as filler, a major disadvantage is the requirement of a high amount of carbon black (up to 30%–40%) to achieve desired conductivity, which results in deterioration of the mechanical properties of polymer [58]. Recently, CNT as alternative to carbon black is used to overcome the above drawback. A major advantage of using CNTs is that the conductive composite can be formed at low CNT loading, due to low percolation thresholds. The small diameter, high aspect ratio, high conductivity, and mechanical strength of CNTs make them an excellent option for creating conductive composites for high-performance EMI shielding materials [59,60].

Several studies have been reported on the ESD and EMI shielding properties of CNT–polymer composites. Different thermoplastic and thermoset polymer matrices have been investigated in order to realize the superior properties of CNTs for the development of lightweight, strong, conductive material.

4.6.4 CNTs as Anode Materials for Lithium-Ion Batteries

CNTs have been approved to be a good anode material for lithium batteries [61–67] due to their unique structure (one-dimensional cylindrical tubule of graphite sheet), high electrical conductivity (10^6 S m^{-1} at 300 K for SWCNTs and >10^5 S m^{-1} for MWCNTs) [68], low density, high rigidity (Young's modulus of the order of 1 TPa) [69,70], and high tensile strength (up to 60 GPa) [71]. The prediction of electronic properties of MWCNTs, which contain multiple layers of graphene sheets that may have different chirality, is more complicated [72]. However, due to the multiple rolled layers, MWCNTs are able to insert lithium ions in a way similar to graphite, making them a promising candidate as an anode material for lithium-ion batteries (LIBs) [73]. Along with interplanar spacing of graphitic sheets that allows for the intercalation/deintercalation of lithium ions, the hollow cores of MWCNTs are also available for lithium-ion intercalation. Therefore, the unique structures of MWCNTs should result in a higher capacity than that of graphite. Much work has been done to verify the excellent electrochemical performance of MWCNT-based anode materials in LIBs applications. In a study by Welna et al. [74], the electrochemical properties of vertically aligned MWCNTs and that of nonaligned MWCNTs for comparison were investigated. The configuration of the aligned MWCNTs provides more sites at which lithium can be adsorbed, and the intimate contact of the MWCNTs in the aligned structure helps to maintain electrical continuity to the current collector. Both of these help to improve the electrochemical performance of MWCNT-based anode. A high lithium storage capacity of 980 mAh/g in the first cycle was obtained and stabilized at about 750 mAh/g after more than 10 cycles, whereas for the nonaligned MWCNTs, the values reduced to 158 and 58 mAh/g, respectively. The vertically aligned MWCNTs also displayed a substantially greater rate capability than the nonaligned ones, which emphasizes the important role that structure order plays in electrode performance. Applying different synthesis methods and treatment methods, different forms of CNTs have been developed to be used as anodes in LIBs.

In addition there is currently an enormous research effort aimed at developing ultrathin, flexible, and soft batteries to cater for the bendable modern gadgets. The goal is improved rate capability without a penalty in charge capacity and sufficient electrochemical cycling characteristics. Flexible batteries are not only needed, for example, for rolled-up displays, active radio-frequency identification tags, integrated circuit smart cards, and implantable medical devices, but there is also the intention to place large flexible batteries in hollow spaces of the auto body of future hybrid and electric vehicles. Needless to stress that high power and high energy density are expected. Of course, the battery performance is closely related to the structural and electrochemical properties of the applied electrodes. Hence, the development of flexible electrodes, free standing with high energy and power density, good rate capability that can function safely for many years becomes important.

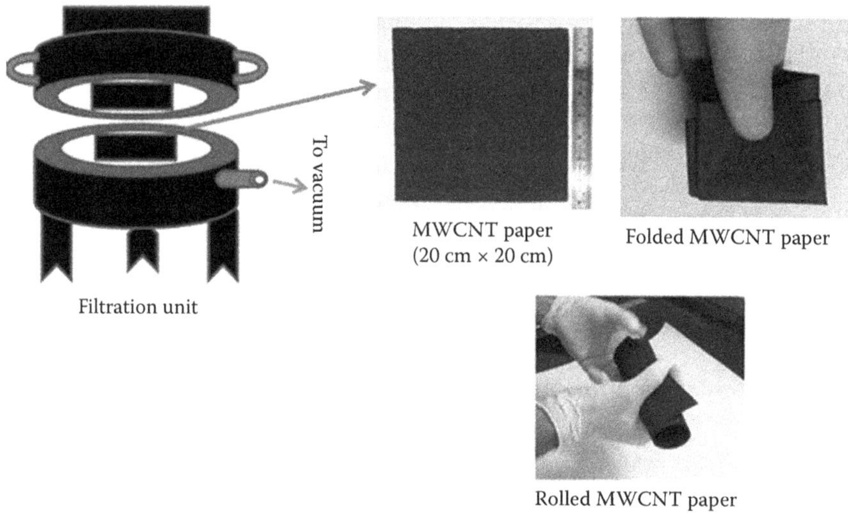

FIGURE 4.15
Flexible MWCNT-based anode for lithium-ion battery prepared at CSIR-NPL, New Delhi, India.

Because of the high aspect ratio (>1000), well-dispersed MWCNTs can form self-supporting films that can be used as anodes for lithium-ion batteries. Therefore, free-standing CNT papers have been tried in LIBs as anodes. Moreover, the high tensile strength, high flexibility, and high aspect ratio of CNTs make them uniquely suited for making free-standing, flexible anode material for lithium ion cell. A photograph of flexible free-standing paper is shown in Figure 4.15. Since the flexible electrode is electrically conducting itself there is no need to support it on cooper foil which is used in the case of other forms of carbon used as anode for LIBs.

4.6.5 Carbon Nanotubes in Fuel Cells

Proton exchange membrane fuel cells (PEMFCs) are now attracting enormous interest for various applications such as low/zero emission vehicles, stationary power generation, and backup power sources. One of the major challenges in the commercialization of fuel cells is the cost due to use of noble metal platinum (Pt) as a catalyst for oxidation of hydrogen and reduction of oxygen at the anode and cathode, respectively. Different carbon materials have been tried as catalyst supports to reduce the amount of Pt. Recently use of high surface area MWCNTs for Pt catalyst support in fuel cells have been reported with great success, which could reduce the amount of Pt catalyst by 50% without sacrificing the fuel cell performance [75]. In addition to the catalyst support, CNTs are also used for the enhancement in the mechanical and electrical properties of bipolar plates used in PEMFCs. These results have been reported by various authors [76].

4.6.6 Water Purification

Various forms of CNTs have been used in water purification or treatment. The unique characteristics of CNTs have recently formed them into nanoporous membrane structure and molecular transport through the tubes have been observed. Vertically aligned CNTs as membranes with an open CNT tip have been recently used for water purification.

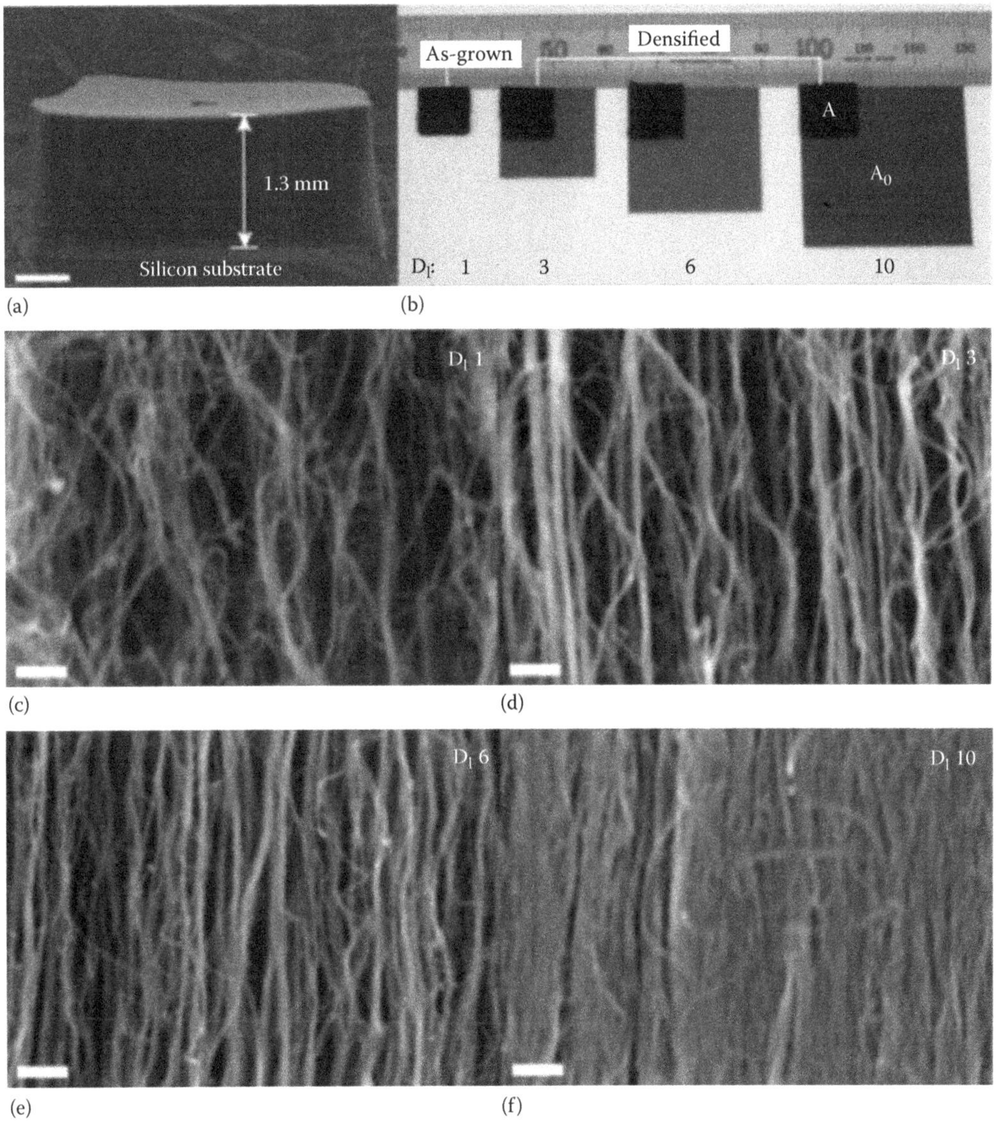

FIGURE 4.16
Densified CNT array and performance of the outer-wall membrane. (a) SEM image of supergrown CNT forest (scale bar 500 μm), (b) optical photograph of densified CNT forests that were used as the outer-wall CNT membranes. The D1 is the ratio of the CNT area before densification to that after densification. The area (A) is 1 cm^2 for all cases, whereas the area occupied by as-grown CNTs (A_0) varies depending on the D_l value. In the case of a D1 of 10, 10 cm^2 of as-grown CNTs were densified into 1 cm^2 of compressed CNTs. (c–f) Cross-sectional SEM images (300,000× magnification, scale bar 100 nm) of densified CNTs. Greater D1 values indicated that the CNTs were more closely packed and more straightened. The figures present the fracture plane images of the CNT array. *(Continued)*

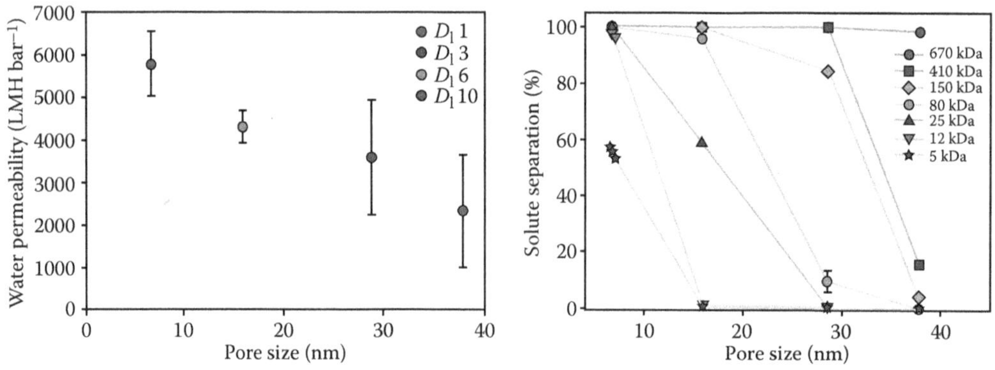

FIGURE 4.16 (CONTINUED)
Densified CNT array and performance of the outer-wall membrane. (g) Water permeability, as affected by the D1 for outer-wall CNT membrane. Data are shown as the mean ±s.d. (h) Rate of rejection by the densified outer-wall CNT membranes obtained with seven types of dextran (5–670 kDa). Dextran solutions with a concentration of 100 mg l^{-1} were filtered by the densified outer-wall CNT membrane under a pressure of 1 bar at 20°C. (Reprinted by permission from Macmillan Publishers Ltd., *Nature Communications*, B. Lee, Y. Baek, M. Lee, D.H. Jeong, H.H. Lee, J. Yoon et al. A carbon nanotube wall membrane for water treatment, 6, copyright 2015.)

Another type of membrane has also been studied recently in which the interstices between nanotubes in a vertical array of CNTs were utilized as membrane pores, which is known as outer-wall CNT membranes [77]. The ability to manipulate the structural characteristics of CNT membranes, such as their pore dimensions, pore density, and tortuosity, is desirable in various applications in addition to the water purification. One efficient approach for manipulating such structural characteristics is through densification of the vertically grown CNT forest. Densification is basically done by two simple techniques: capillary densification and mechanical densification. Water permeability is significantly enhanced by densification. A vertically grown CNT forest and densified CNTs that can be used as outer-wall membranes are shown in Figure 4.16. The water permeability for outer-wall CNT membranes increases with decreasing pore size or increasing densification of the CNT array; the water permeability increases from an average of 2300 ± 1330 Liter/m^2/h (LMH bar^{-1}) for a D_f of 1 (as grown) to an average of 5800 ± 760 (LMH bar^{-1}) for a D_f of 10, which is a twofold increase in permeability by densification [77].

4.6.7 Armor Material

Carbon nanotube is an ideal candidate as armor material due to its unique combination of exceptionally high elastic modulus and high strain to failure (~16%), which are capable of elastically storing an extreme amount of energy that can cause the bullet to bounce off or be deflected. Carbon nanotubes can be utilized to enhance ballistic performance of armor materials by various approaches, such as incorporation of CNTs into polymer matrix composites, metals, or ceramics to enhance their hardness or toughness and erosion resistance. Use of neat or composite fibers of CNTs in the form of woven or nonwoven fabric along with conventional armor-grade fibers like Kevlar, UHMWPE, or PBO will further improve their elastic modulus, energy absorption capacity and exceptional ballistic performance for future armour materials.

4.6.8 CNT as Self-Healing Material

Apart from the aforementioned applications, CNT-based polymer composites can be used in both extrinsic as well as intrinsic self-healing materials. For extrinsic self-healing materials, CNT has been visualized as a nanoreservoir of healing agent for a self-healing polymer system. In fact practically successful intercalation of the healing agent inside the CNT has been reported. It has also been used as a healing agent in structural reinforcing fillers in capsule-based healable systems. A more detailed characterization has to be done to analyze the amount of healing agent inside the CNT and then practically visualizing the crack healing process. Intrinsic self-healing behavior has also been realized by incorporating CNTs along with various kinds of polymers, which has resulted in the fabrication of specific multifunctional smart materials for high-end applications such as shear stiffening self-healing polymer composites, damage-free transparent electrodes, supramolecular healable hydrogels, healable superhydrophobic surfaces, and self-healing synthetic foam [78].

4.6.9 CNT-Based Conducting Cables

Motor and motor-driven systems account for between 43% and 46% of all global electricity consumptions. There is a requirement to develop technologies or materials that can make the motors run more efficiently, which will help in reduction in the energy consumption. Researchers at Rice University in 2011 demonstrated that CNTs braided into wires could outperform copper in conducting electricity, which can improve the efficiency. Building on this concept, a team at the Lappeenranta University of Technology (LUT) in Finland replaced the copper windings used to conduct electricity in electric motors with a woven material made up of CNT yarn and achieved remarkable efficiencies in motors [79]. The replacement of the copper wires with CNT wire could possibly reduce the Joule losses in the winding to half of the present-day machine losses causing an almost twofold increase in the electrical conductivity and hence the efficiency of motors.

References

1. S. Iijima. Helical microtubules of graphitic carbon. *Nature* 354, no. 6348 (1991): 56–58.
2. A. Kukovecz, G. Kozma, and Z. Konya. Multi-walled carbon nanotubes. In *Springer Handbook of Nanomaterials*, edited by R. Vajtai, 147–188. Springer-Verlag, 2013.
3. A comparison of carbon nanotubes and carbon nanofibers, http://pyrografproducts.com/carbon-nanotubes.html.
4. A.K. Arya, B.P. Singh, J. Jyoti, S. Pati, and S. Dhakate. Economic growth of vertically aligned multiwalled carbon nanotubes in nitrogen atmosphere. *Advanced Material Letter* 6, no. 12 (2015): 1094–1097.
5. S. Shukrullah, N.M. Mohamed, M.S. Shaharun, and M.Y. Naz. Mass production of carbon nanotubes using fluidized bed reactor: A short review. *Trends in Applied Sciences Research* 9 (2014): 121–131.
6. X.-L. Xie, Y.-W. Mai, and X.-P. Zhou. Dispersion and alignment of carbon nanotubes in polymer matrix: A review. *Materials Science and Engineering: R: Reports* 49, no. 4 (2005): 89–112.
7. J.N. Coleman, U. Khan, W.J. Blau, and Y.K. Gun'ko. Small but strong: A review of the mechanical properties of carbon nanotube–polymer composites. *Carbon* 44, no. 9 (2006): 1624–1652.

8. K. Balasubramanian and M. Burghard. Chemically functionalized carbon nanotubes. *Small* 1, no. 2 (2005): 180–192.
9. B.P. Singh, V. Choudhary, S. Teotia, T.K. Gupta, V.N. Singh, S.R. Dhakate, and R.B. Mathur. Solvent free, efficient, industrially viable, fast dispersion process based amine modified MWCNT reinforced epoxy composites of superior mechanical properties. *Advanced Materials Letters* 6, no. 2 (2015): 104–113.
10. R.B. Mathur, S. Seth, C. Lal, R. Rao, B.P. Singh, T.L. Dhami et al. Co-synthesis, purification and characterization of single- and multi-walled carbon nanotubes using the electric arc method. *Carbon* 45, no. 1 (2007): 132–140.
11. E.A.H. Marsh and F. Rodriguez-Reinoso. *Introduction to Carbon Technologies*. University of Alicante Press, 1997.
12. R.B. Seymour. *Polymeric Composites*. The Netherlands, 1990.
13. P. Morgan. *Carbon Fibers and Their Composites*. CRC Press/Taylor & Francis Group, 2005.
14. J.A.H. Huit and F.G. Rammerstorer. *Engineering Mechanics of Fibre Reinforced Polymers and Composite Structures*. Springer Verlag, 1994.
15. M. Sittig. *Carbon and Graphite Fibers: Manufacture and Applications*. Noyes Data Corporation, 1980.
16. E. Fitzer and B. Terwiesch. Carbon—Carbon composites unidirectionally reinforced with carbon and graphite fibers. *Carbon* 10, no. 4 (1972): 383–390.
17. K. Friedrich, S. Fakirov, and Z. Zhang. *Polymer Composites: From Nano-to-Macroscale*. Springer, 2005.
18. S. Advani. *Processing and Properties of Nanocomposites*. World Scientific, 2006.
19. M.S. Dresselhaus, G. Dresselhaus, and P. Avouris. *Carbon Nanotubes: Synthesis, Structure, Properties, and Applications*. Springer-Verlag, 2001.
20. P.M. Ajayan, J.-C. Charlier, and A.G., Rinzler. Carbon nanotubes: From macromolecules to nanotechnology. *Proceedings of the National Academy of Sciences* 96, no. 25 (1999): 14199–14200.
21. O. Breuer and U. Sundararaj. Big returns from small fibers: A review of polymer/carbon nanotube composites. *Polymer Composites* 25, no. 6 (2004): 630–645.
22. K.W. Putz, C.A. Mitchell, R. Krishnamoorti, and P.F. Green. Elastic modulus of single-walled carbon nanotube/poly(methyl methacrylate) nanocomposites. *Journal of Polymer Science Part B: Polymer Physics* 2004;42, no. 12 (2004): 2286–2293.
23. J.-H. Du, J. Bai, and H.-M. Cheng. The present status and key problems of carbon nanotube based polymer composites. *eXPRESS Polymer Letters* 1, no. 5 (2007): 253.
24. J.N. Coleman, U. Khan U, and YK Gun'ko. Mechanical reinforcement of polymers using carbon nanotubes. *Advanced Materials* 18, no. 6 (2006): 689–706.
25. E.T. Thostenson, Z. Ren, and T.-W. Chou. Advances in the science and technology of carbon nanotubes and their composites: A review. *Composites Science and Technology* 61, no. 13 (2001): 1899–1912.
26. A. Montazeri and N. Montazeri. Viscoelastic and mechanical properties of multi walled carbon nanotube/epoxy composites with different nanotube content. *Materials & Design* 32, no. 4 (2011): 2301–2307.
27. T.K. Gupta, B.P. Singh, R.B. Mathur, and S.R. Dhakate. Multi-walled carbon nanotube–graphene–polyaniline multiphase nanocomposite with superior electromagnetic shielding effectiveness. *Nanoscale* 6, no. 2 (2014): 842–851.
28. R. Haggenmueller, H.H. Gommans, A.G. Rinzler, J.E. Fischer, and K.I. Winey. Aligned single-wall carbon nanotubes in composites by melt processing methods. *Chemical Physics Letters* 330, no. 3–4 (2000): 219–225.
29. F. Du, J.E. Fischer, and K.I. Winey. Coagulation method for preparing single-walled carbon nanotube/poly(methyl methacrylate) composites and their modulus, electrical conductivity, and thermal stability. *Journal of Polymer Science Part B: Polymer Physics* 41, no. 24 (2003): 3333–3338.
30. Z. Jia, Z. Wang, C. Xu, J. Liang, B. Wei, D. Wu et al. Study on poly(methyl methacrylate)/carbon nanotube composites. *Materials Science and Engineering: A* 271, no. 1–2 (1999): 395–400.

31. R. Mathur, S. Pande, B. Singh, and T. Dhami. Electrical and mechanical properties of multi-walled carbon nanotubes reinforced PMMA and PS composites. *Polymer Composites* 29, no. 7 (2008): 717–727.
32. S. Pande, R. Mathur, B. Singh , and T. Dhami. Synthesis and characterization of multiwalled carbon nanotubes-polymethyl methacrylate composites prepared by in situ polymerization method. *Polymer Composites* 30, no. 9 (2009): 1312–1317.
33. Z. Jin, K.P. Pramoda, G. Xu, and S.H. Goh. Dynamic mechanical behavior of melt-processed multi-walled carbon nanotube/poly(methyl methacrylate) composites. *Chemical Physics Letters* 337, no. 1–3 (2001): 43–47.
34. C. Velasco-Santos, A.L. Martínez-Hernández, F.T. Fisher, R. Ruoff, and V.M. Castaño. Improvement of thermal and mechanical properties of carbon nanotube composites through chemical functionalization. *Chemistry of Materials* 15, no. 23 (2003): 4470–4475.
35. R.E. Gorga and R.E. Cohen. Toughness enhancements in poly(methyl methacrylate) by addition of oriented multiwall carbon nanotubes. *Journal of Polymer Science Part B: Polymer Physics* 42, no. 14 (2004): 2690–2702.
36. D. Qian, E.C. Dickey, R. Andrews, and T. Rantell. Load transfer and deformation mechanisms in carbon nanotube-polystyrene composites. *Applied Physics Letters* 76, no. 20 (2000): 2868–2870.
37. B. Safadi, R. Andrews, and E.A. Grulke. Multiwalled carbon nanotube polymer composites: Synthesis and characterization of thin films. *Journal of Applied Polymer Science* 84, no. 14 (2002): 2660–2669.
38. E.T. Thostenson and T.W. Chou. Aligned multi-walled carbon nanotube-reinforced composites: Processing and mechanical characterization. *Journal of Physics D-Applied Physics* 35, no. 16 (2002): L77–L80.
39. P. Jindal, S. Pande, P. Sharma, V. Mangla, A. Chaudhury, D. Patel et al. High strain rate behavior of multi-walled carbon nanotubes-polycarbonate composites. *Composites Part B: Engineering* 45, no. 1 (2012): 417–422.
40. S. Pande, A. Chaudhary, D. Patel, B.P. Singh, and R.B. Mathur. Mechanical and electrical properties of multiwall carbon nanotube/polycarbonate composites for electrostatic discharge and electromagnetic interference shielding applications. *RSC Advances* 4, no. 27 (2014):13839–13849.
41. S.G. Pardo, L. Arboleda, Ana Ares, X. García, S. Dopico, and M.J. Abad. Toughening strategies of carbon nanotube/polycarbonate composites with electromagnetic interference shielding properties. *Polymer Composites* 34, no. 11 (2013): 1938–1949.
42. R. Mathur, B. Singh, T. Dhami, Y. Kalra, N. Lal, R. Rao et al. Influence of carbon nanotube dispersion on the mechanical properties of phenolic resin composites. *Polymer Composites* 31, no. 2 (2010): 321–327.
43. B. Singh, D. Singh, R. Mathur, and T. Dhami. Influence of surface modified MWCNTs on the mechanical, electrical and thermal properties of polyimide nanocomposites. *Nanoscale Research Letters* 3, no. 11 (2008): 444–453.
44. X. Gong, J. Liu, S. Baskaran, R.D. Voise, and J.S. Young. Surfactant-assisted processing of carbon nanotube/polymer composites. *Chemistry of Materials* 12, no. 4 (2000): 1049–1052.
45. M.J. Biercuk, M.C. Llaguno, M. Radosavljevic, J.K. Hyun, A.T. Johnson, and J.E. Fischer. Carbon nanotube composites for thermal management. *Applied Physics Letters* 80, no. 15 (2002): 2767–2769.
46. A. Allaoui, S. Bai, H.M. Cheng, and J.B. Bai. Mechanical and electrical properties of a MWNT/epoxy composite. *Composites Science and Technology* 62, no. 15 (2002): 1993–1998.
47. Y.S. Song and J.R. Youn. Influence of dispersion states of carbon nanotubes on physical properties of epoxy nanocomposites. *Carbon* 43, no. 7 (2005): 1378–1385.
48. P. Garg, B.P. Singh, G. Kumar, T. Gupta, I. Pandey, R. Seth et al. Effect of dispersion conditions on the mechanical properties of multi-walled carbon nanotubes based epoxy resin composites. *Journal of Polymer Research* 18, no. 6 (2010): 1397–1407.

49. T. Ogasawara, S.-Y. Moon, Y. Inoue Y, and Y. Shimamura. Mechanical properties of aligned multi-walled carbon nanotube/epoxy composites processed using a hot-melt prepreg method. *Composites Science and Technology* 71, no. 16 (2011):1826–1833.
50. N. Hameed, N.V. Salim, T.L. Hanley, M. Sona, B.L. Fox, and Q. Guo. Individual dispersion of carbon nanotubes in epoxy via a novel dispersion-curing approach using ionic liquids. *Physical Chemistry Chemical Physics* 15, no. 28 (2013): 11696–11703.
51. Z. Wang, G. Wei, and G.-L. Zhao. Enhanced electromagnetic wave shielding effectiveness of Fe doped carbon nanotubes/epoxy composites. *Applied Physics Letters* 103, no. 18 (2013).
52. M.F.L. De Volder, S.H. Tawfick, R.H. Baughman, and A.J. Hart. Carbon nanotubes: Present and future commercial applications. *Science* 339, no. 6119 (2013): 535–539.
53. M. Zhang, K.R. Atkinson, and R.H. Baughman. Multifunctional carbon nanotube yarns by downsizing an ancient technology. *Science* 306, no. 5700 (2004): 1358–1361.
54. K. Sears, C. Skourtis, K. Atkinson, N. Finn, and W. Humphires. Focused ion beam milling of carbon nanotube yarns to study the relationship between structure and strength, *Carbon* 48 (2010): 4450–4456.
55. Z. Liu, G. Bai, Y. Huang, Y. Ma, F. Du, F. Li, T. Guo, and Y. Chen. Reflection and absorption contributions to the electromagnetic interference shielding of single-walled carbon nanotube/polyurethane composites. *Carbon* 45, no. 4 (2007): 821–827.
56. D.D.L. Chung. Electromagnetic interference shielding effectiveness of carbon materials. *Carbon* 39, no. 2 (2001): 279–285.
57. X. Luo and D.D.L. Chung. Electromagnetic interference shielding reaching 130 dB using flexible graphite. *Carbon* 34, no. 10 (1996): 1293–1294.
58. S. Pande, B. Singh, R. Mathur, T. Dhami, P. Saini, and S. Dhawan. Improved electromagnetic interference shielding properties of MWCNT–PMMA composites using layered structures. *Nanoscale Research Letters* 4, no. 4 (2009): 327–334.
59. N. Li, Y. Huang, F. Du, X. He, X. Lin, H. Gao et al. Electromagnetic interference (EMI) shielding of single-walled carbon nanotube epoxy composites. *Nano Letters* 6, no. 6 (2006): 1141–1145.
60. Y. Yang, M.C. Gupta, K.L. Dudley, and R.W. Lawrence. Novel carbon nanotube–polystyrene foam composites for electromagnetic interference shielding. *Nano Letters* 5, no. 11 (2005): 2131–2134.
61. B. Gao, A. Kleinhammes, X. Tang, C. Bower, L. Fleming, Y. Wu et al. Electrochemical intercalation of single-walled carbon nanotubes with lithium. *Chemical Physics Letters* 307, no. 3 (1999): 153–157.
62. B. Gao, C. Bower, J. Lorentzen, L. Fleming, A. Kleinhammes, X. Tang et al. Enhanced saturation lithium composition in ball-milled single-walled carbon nanotubes. *Chemical Physics Letters* 327, no. 1 (2000): 69–75.
63. H. Shimoda, B. Gao, X. Tang, A. Kleinhammes, L. Fleming, Y. Wu et al. Lithium intercalation into opened single-wall carbon nanotubes: Storage capacity and electronic properties. *Physical Review Letters* 88, no. 1 (2001): 015502.
64. Z. Yang, Y. Zhou, S. Sang S, Feng Y, Wu H. Lithium insertion into multi-walled raw carbon nanotubes pre-doped with lithium. *Materials Chemistry and Physics* 89, no. 2 (2005): 295–299.
65. G. Maurin, C. Bousquet, F. Henn, P. Bernier, R. Almairac, and B. Simon B. Electrochemical lithium intercalation into multiwall carbon nanotubes: A micro-Raman study. *Solid State Ionics* 136 (2000): 1295–1299.
66. F. Leroux, K. Metenier, S. Gautier, E. Frackowiak, S. Bonnamy, and F. Beguin. Electrochemical insertion of lithium in catalytic multi-walled carbon nanotubes. *Journal of Power Sources.* 81 (1999): 317–322.
67. A.S. Claye, J.E. Fischer, C.B. Huffman, A.G. Rinzler, and R.E. Smalley. Solid-state electrochemistry of the Li single wall carbon nanotube system. *Journal of The Electrochemical Society* 147, no. 8 (2000): 2845–2852.
68. Y. Ando, X. Zhao, H. Shimoyama, G. Sakai, and K. Kaneto. Physical properties of multiwalled carbon nanotubes. *International Journal of Inorganic Materials* 1, no. 1 (1991): 77–82.

69. G. Van Lier, C. Van Alsenoy, V. Van Doren, and P. Geerlings P. Ab initio study of the elastic properties of single-walled carbon nanotubes and graphene. *Chemical Physics Letters* 326, no. 1 (2000): 181–185.
70. M. Treacy, T. Ebbesen, and J. Gibson. Exceptionally high Young's modulus observed for individual carbon nanotubes. *Nature* 381 (1996): 678–680.
71. M.-F. Yu, O. Lourie, M.J. Dyer, K. Moloni, T.F. Kelly, and R.S. Ruoff. Strength and breaking mechanism of multiwalled carbon nanotubes under tensile load. *Science* 287, no. 5453 (2000): 637–640.
72. S. Mukherjee, A. Bates, S.C. Lee, D.-H. Lee, and S. Park. A review of the application of CNTs in PEM fuel cells. *International Journal of Green Energy* 12, no. 8 (2015): 787–809.
73. C. de las Casas and W. Li. A review of application of carbon nanotubes for lithium ion battery anode material. *Journal of Power Sources* 208 (2012): 74–85.
74. D.T. Welna, L. Qu, B.E. Taylor, L. Dai, and M.F. Durstock. Vertically aligned carbon nanotube electrodes for lithium-ion batteries. *Journal of Power Sources* 196, no. 3 (2011): 1455–1460.
75. C. Gupta, P.H. Maheshwari, S. Sasikala, and R. Mathur. Processing of pristine carbon nanotube supported platinum as catalyst for PEM fuel cell. *Materials for Renewable and Sustainable Energy* 3, no. 4 (2014): 1–11.
76. S. Dhakate, S. Sharma, N. Chauhan, R. Seth, and R. Mathur. CNTs nanostructuring effect on the properties of graphite composite bipolar plate. *International Journal of Hydrogen Energy* 35, no. 9 (2010): 4195–4200.
77. B. Lee, Y. Baek, M. Lee, D.H. Jeong, H.H. Lee, J. Yoon et al. A carbon nanotube wall membrane for water treatment. *Nature Commununications* 6 (2015): 1709.
78. S. Pati, B.P. Singh, and S.R. Dhakate. Self-healing polymer nanocomposites based on graphene and CNT. In *Smart Polymer Nanocomposites: Energy Harvesting, Self-Healing and Shape Memory*. Springer, in press.
79. D. Johnson. Carbon nanotube yarns could replace copper windings in electric motors. *IEEE Spectrum*. http://spectrum.ieee.org/nanoclast/semiconductors/nanotechnology/carbon-nanotube-yarns-set-to-replace-copper-windings-in-electric-motors.

5

Carbon Nanofibers and Filaments

5.1 Introduction

Carbon fibers, first used as light bulb filaments in 1879, trace their history in the early 1950s and 60s during the requirement of the aerospace industry for lightweight materials. This led to the development of high strength and high stiffness continuous length of carbon fibers with diameter ranging from 7 to 10 μm as reinforcement in polymer composites. Following this realization there were a number of relatively successful attempts to prepare carbon fibers from different precursors, especially those from Roger Bacon at Union Carbide using viscose rayon in 1957 [1], Shindo in Japan using polyacrylonitrile (PAN) in 1961 [2], and of Ōtani, also in Japan in the year 1965, using an isotropic pitch [3]. In most cases these filaments are derived from organic precursors by spinning them into polymer fibers followed by thermal oxidative stabilization at 200°C–300°C and carbonization at 1000°C in an inert atmosphere. Further heat treatment up to temperatures of ~3000°C helped them attain well-oriented graphitic structure with larger crystallite size and high stiffness in the process called graphitization.

The development of a successful commercial process for carbon fibers was particularly due to William (Bill) Watt and his colleagues at the Royal Aircraft Establishment (RAE) in Farnborough, England, in 1963 using commercial PAN fibers as precursor by heating them at 200°C for several hours in air, followed by pyrolysis to 1000°C and above. Later Watt, together with his long-term colleague William Johnson, began the first experiments with the Courtaulds PAN fibers ("Courtle") and produced carbon fiber with Young's modulus (YM) of 150 GPa with heat treatment temperature (HTT) of 1000°C. Further HTT to 2500°C developed even greater stiffness of 380 GPa. Later, the early batch process was developed into continuous scale, full production lines at Morganite Ltd. and Courtaulds Ltd. UK. In 1964, the first true high-modulus fibers were prepared by the Royal Aircraft Establishment in England from PAN and by the Union Carbide Corporation from rayon [4].

These high-modulus rayon-based fibers were used extensively in the United States as well as France and Germany in the development of epoxy-matrix composites for space structures. They were also used for a new generation of high strength and high stiffness carbon–carbon composites for rocket nozzles, rocket motor casings, and missile nose tips.

Concurrently scientists and engineers at Rolls-Royce Ltd. were also developing a batch process for carbon fiber production and were experimenting with carbon fiber reinforced polymer composites/plastics (CFRP) as the material for the turbine blade of the then-new RB 211 engine. Unfortunately, the well-known bird strike problem and delamination due to poor interlaminar shear strength of the composite eventually ruled out the use of CFRP

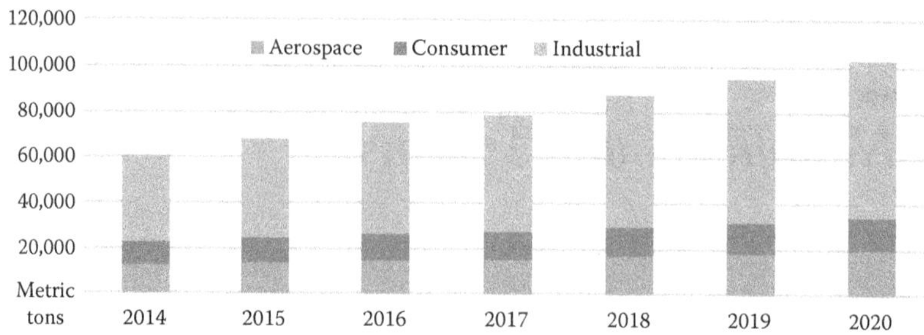

FIGURE 5.1
Carbon fiber production by market 2012–2020. (From Composites Forecast & Consulting LLC, December 2014.)

in turbine blades. And for some time it was a major setback to the carbon fiber industry. Continuous improvement in the precursor characteristics and processing conditions over the years resulted in development of high-performance PAN and pitch-based carbon fibers.

The growing demand for high performance of carbon fiber-reinforced structures in aerospace, sports, and industrial sectors have pushed up the production level of carbon fibers and today it stands at 80,000 metric tons per annum.

The world consumption of continuous carbon fibers for various applications is shown in Figure 5.1.

5.2 Vapor-Grown Carbon Fibers

Another kind of carbon fiber or filament in short lengths have been developed recently via gaseous phase by decomposition of hydrocarbon on a heated surface in the presence of transition metal catalysts (e.g., Fe, Co, Ni, and some of their compounds). The process is known as catalytic chemical vapor decomposition (CCVD). These filaments are not continuous like ex-polymer fibers, but their length can reach several hundred millimeters with diameters as low as 0.3–0.5 mm. It is possible that their production cost will be substantially lower than those of ex-polymer fibers. These have a very high aspect ratio of the order of 100. The so-called vapor-grown carbon fibers or filaments (VGCF) have been the subject of intense research in the last few years because of their distinctive characteristics, such as high aspect ratio, mechanical properties, high electrical and thermal conductivity, and EMI shielding. They are frequently utilized in different material systems like composites due to their exceptional properties and low cost.

5.2.1 Historical Approach

The history of filamentous carbon dates back to 1890 when a number of carbon researchers found fine fibrous carbon by accident in carbonaceous soot and films while studying the thermal decomposition of hydrocarbons, such as benzene and methane, and also through disproportionation reaction of carbon monoxide. The 1889 patent of Hughes and Chambers "Manufacture of carbon filaments," US patent 405,480 [5], which describes the growth of "hair like carbon filaments," utilized the feedstock of hydrogen and methane pyrolyzed in an iron crucible. The fibers were thought to be suitable for electric light bulb filaments, but the lack of modern process control made them noncompetitive. The true appreciation of these fibers, however, came much later when their structure could be analyzed by electron microscopy by Soviet scientist Radushkevich in 1950 [6]. The invention of the electron microscope preceded the discovery of carbon filaments by only a few years.

However, the growth of filamentous carbon is a major concern in many industrial processes involving blast furnaces, steam cracking, fuel pins and boiler tubes of nuclear reactors, and various heterogeneous catalyst systems. In a 1953 study of material deposited on blast furnace brickwork, Davis et al. [7] observed the presence of mold-like twisted carbon filaments about 0.01 μm thick. No doubt, therefore, until recently the motivation for studies relating to the formation of carbon filaments has been the desire to inhibit the growth of this form of carbon deposit.

In the early 1970s, Japanese researchers Koyama and Endo [8] grew good yields of 1 μm diameter and 1 mm long VGCF by thermal decomposition of benzene at about 1200°C. They used catalyst particles for the growth of filamentous carbon. Beginning in 1980, Tibbetts [9] in the United States and Benissad in France described a process for growing the fibers using methane and established the two disjoint steps, lengthening and thickening [10]. He also developed a model for filamentous growth that incorporated both hydrocarbon adsorption on catalyst particle surface and carbon diffusion through the catalyst particle and suggested a model on the poisoning process, that is, the termination of filamentous growth.

The first commercial attempt on the production of VGCF came from the Japanese company Nikasso in 1991, under the trade name "Grasker." Several companies in the world are now involved in the commercial-scale production of VGCF (by M/s Applied Sciences-Grafil, USA) and new engineering applications are forecast.

However, the work of Tibbetts has spurred intense interest concerning the structural properties of this material and its possible application in fiber-reinforced composites.

5.3 Growth Mechanism

Carbon is formed from hydrocarbon in two chemical processes—pyrolysis and incomplete combustion—resulting in two different forms of carbon. The first type is formed by deposition of carbon on the walls of the reactor and is known as pyrocarbon or pyrolytic carbon, and the second type is formed in the bulk of the gas phase. The latter type is characterized by spherical or near-spherical particles, each containing 10^5 to 10^{10} carbon atoms, and is known as soot or carbon black. However, in order to convert this carbon into the filamentous form, a transition metal catalyst is required.

These fibers are believed to grow from the surface of a catalyst particle, with carbon deposited on the particle by decomposition of the hydrocarbon around 1150°C–1200°C and migrating by diffusion through the particle, or over its surface, to the site where the fiber is growing. The fiber size is comparable to the size of the catalytic particle but can thicken if additional pyrolytic carbon is deposited onto the fiber surface.

Several growth models for the formation of VGCF have been proposed.

5.3.1 Growth on Substrate

Figure 5.2 depicts VGCF production equipment. A mixture of hydrogen and hydrocarbon is admitted into the reaction tube in an electrical furnace and the temperature is maintained around 1100°C–1200°C. Carbon fiber forms over a substrate (e.g., silicon, carbon, quartz with catalyst particles, e.g., Fe, Ni, Co, dispersed over it) in the reaction tube, kept inside the constant temperature zone. A reducing atmosphere of hydrogen at this temperature reduces the catalyst particle of a transition metal or cleans its surface, which promotes the polymerization and condensation of the hydrocarbon.

The growth of filaments occurs via a catalytic dehydrogenation reaction of the hydrocarbon in several steps. The fiber first forms over fine catalyst particles of iron or any other transition metal like Co or Ni, dispersed over the substrate. It then grows in the radial direction, resulting in fibers several decimeters long and about 10 micrometer in diameter.

The fibers will continue to grow as long as the particles dispersed on the substrate are active; the fibers cease to grow when the particles surface is covered with carbon layers, oxygen, or other impurities, thus retarding movement of the carbon species. It is therefore vital to control the hydrocarbon partial pressure in the system and to keep the atmosphere free from impurities such as moisture and oxygen with hydrogen flow in order to obtain long fibers or filaments.

Some of the following fundamental questions are important:

- Is the introduction of any catalyst necessary?
- Why does carbon dissolve or at least absorb on a specific location of a catalyst particle and precipitate or grow from another location of the particle?
- What is the cause of the poisoning, which is perhaps associated to the thickening (Figure 5.3), whose occurrence determines the final length of the fiber?

Following are the various parameters that influence the growth mechanism of these fibers.

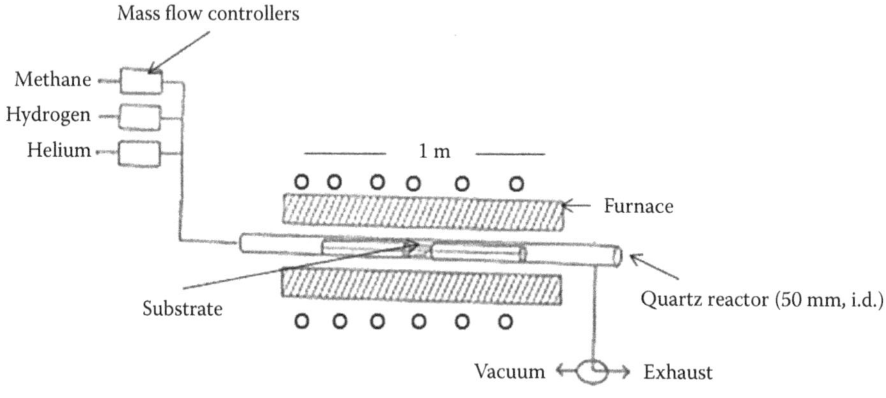

FIGURE 5.2
Schematic of experimental setup for growing VGCF at atmospheric pressure.

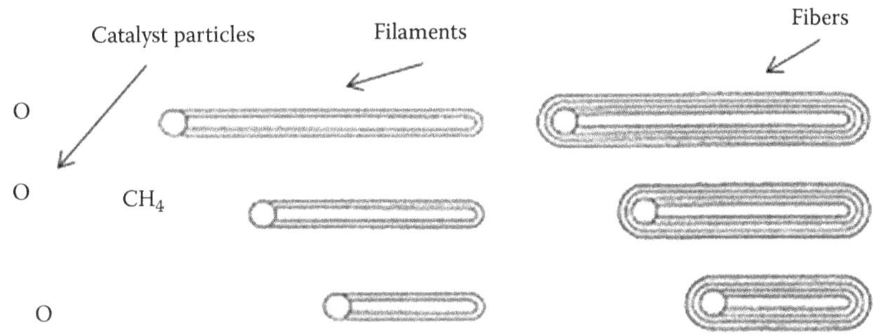

FIGURE 5.3
VGCF growth concept: Schematic showing how catalytic particles exposed to the optimum carburizing atmosphere can produce long filaments.

5.3.1.1 Temperature Gradient

Using the accumulated data obtained from controlled atmosphere electron microscopy (CAEM) and high-resolution electron microscopy studies, a model was developed to account for the growth of filamentous carbon resulting from the interaction of nickel, iron, and cobalt with acetylene. Since the solubility of carbon in a metal is temperature dependent, the key step in the mechanism is believed to be the diffusion of carbon species through the particle from the hotter leading surface, on which hydrocarbon decomposition occurs, to the cooler rear faces, at which the carbon is precipitated from solution. It was suggested that the driving force for carbon diffusion is the temperature gradient created in the particle by the exothermic decomposition of the hydrocarbon at exposed front faces and endothermic deposition of the carbon at the rear faces, which are initially in contact with the support surface. Excess carbon that accumulates at the exposed faces is transported by surface diffusion around the peripheral surfaces of the particle to form the graphitic skin of the filament, as illustrated in Figure 5.4. Eventually growth ceases

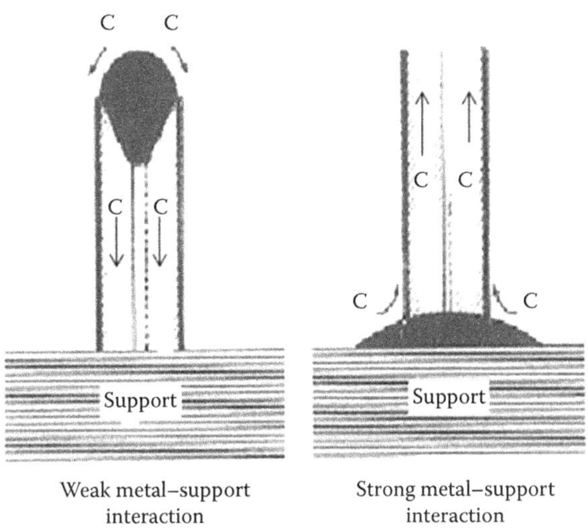

FIGURE 5.4
Influence of the metal–support interaction on the mode of growth of filamentous carbon.

when the leading face is encapsulated by a layer of carbon preventing further hydrocarbon decomposition.

5.3.1.2 Influence of Metal Support Interaction

The interaction between the metal support catalyst particle and supporting medium is another factor that can have an impact on filament growth characteristics. The important role played by metal–support interaction leads to either bottom-up or top-down growth of the filament, as shown in Figure 5.5. In both cases the critical steps in the filament growth are assumed to be carbon solubility in and carbon diffusion through the catalyst particle.

The nature of bonding and crystallographic orientation of metal particles to a support are the two factors [11,12] controlling the extent of interaction of metal particles with a support surface. This can be most easily understood by consideration of the particle configurations depicted in Figure 5.5 and expressed mathematically by Young's equation

$$\gamma_{GS} = \gamma_{MS} + \gamma_{MG} \cos\theta \tag{5.1}$$

where θ is the contact angle, γ is the surface energy, and the subscripts S, M, and G refer to support, metal, and gas, respectively. If γ_{MS} is larger than γ_{GS}, then the contact angle is greater than 90° and the particle will be in a "non-wetting" configuration; if the reverse relationship prevails, then θ will be less than 90° and "wetting" will occur, that is, there will be a significant metal–particle support interaction. The ability of particles to undergo a transformation from a non-wetting to a wetting state and vice versa, suggests that a significant degree of atomic mobility must exist within the particles at temperatures well below that of the bulk melting points of the metals.

5.3.1.3 Diffusion Rate

During the growth process, the metal catalyst is frequently carried away from the support surface and remains at the tip of the growing filament. Diffusion of carbon through the catalyst particle is the rate determining step and is justified from the remarkable agreement between the measured activation energies for filament growth and those for diffusion of carbon through the respective metal [13,14]. The growth rate is also inversely proportional to the diameter of the filament in its initial stage. Only a part of the catalyst particle at the tip of the filament

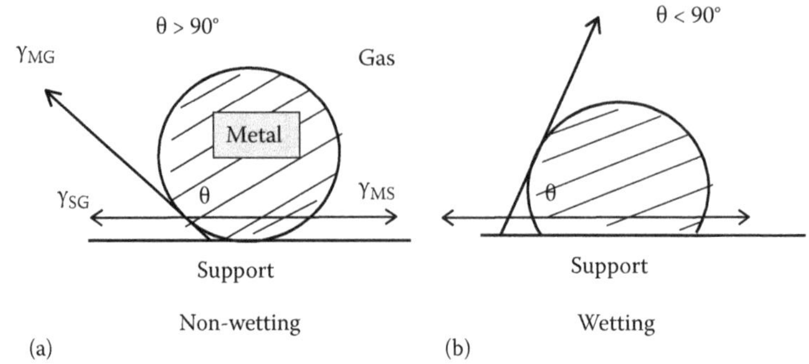

FIGURE 5.5
Schematic representation showing the ramifications on the filament growth mode: (a) a weak interaction being a prerequisite for bottom-up growth and (b) a strong interaction resulting in top-down or growth by the extrusion mode.

is exposed to the hydrocarbon and if the coating of carbon covers the tip completely, then the growth is arrested. Further growth could occur if the surface layer is removed by oxidation.

It may be argued that the driving force for the diffusion of carbon through the metal is probably not due to existence of temperature gradient but rather to a carbon concentration gradient [15]. However, the major disagreement is concerned with the chemical state of the active catalyst.

Looking at the iron phase diagram in Figure 5.6, the allotropic transformation of iron exists in several crystalline forms, and the solubility of carbon is a function of crystal structure and temperature. The figure shows the structural transformation in Fe particles as it cools from the liquid phase to solid phase at around 1539°C. So, the transformation that it goes through first solidifies into a crystalline body-centered cubic (bcc) structure. Next at a little lower temperature (1394°C) this bcc structure gets converted into a face-centered cubic (fcc) structure; again this fcc structure is converted into a bcc structure at temperatures <900°C.

How can carbon precipitated out of carbon be explained on the basis of an iron phase diagram? If we add carbon, this cooling curve will change. And how does it change? This will depend on the extent of solubility of carbon in iron; and this solubility depends on the crystal structure as well as the temperature. In bcc the solubility of carbon is very limited, whereas in an fcc structure the solubility of carbon is much higher. The excess carbon, therefore, precipitates out at around 900°C in the form of filamentous carbon during conversion of Fe from fcc to bcc structure. The concentration of carbon in an Fe catalyst particle is perhaps close to that of eutectic (4.3% carbon) and the activation energy of diffusion 133 kJ/mol is similar to the enthalpy for the growth of filaments in Fe, Ni, and Co catalysts.

The following three important points should be kept in mind:

- The diffusion rate of carbon in liquid iron is much faster than that in solid.
- The growth rate increases with decreasing catalyst particle size.
- The growth rate of VGCF is very quick at first, and then slows with the duration with time. This decrease is caused by an increase of diffusion length for carbon atoms in the catalyst particle.

The initial particle size should be very small; a considerable decrease in activity is noted as the particle diameter exceeds 15 nm. Retaining the small particle size during the warm-up

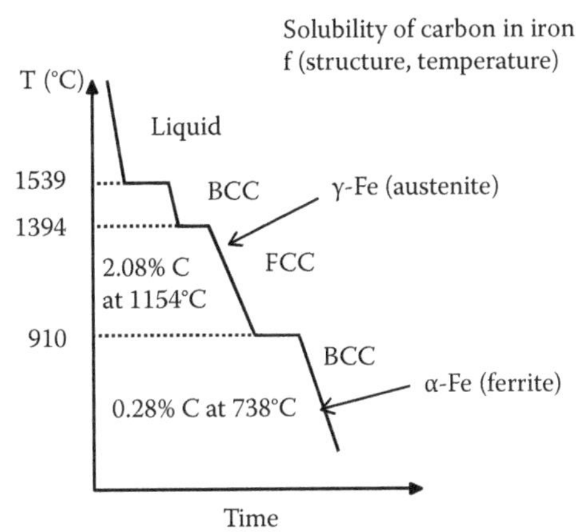

FIGURE 5.6
Carbon diffusion and its precipitation during structural transformation of Fe catalyst with temperature.

phase while avoiding particle coalescence to an ineffective diameter is very important. Sources of methane, hydrogen, and an inert gas can be used at atmospheric pressure for the growth of the fibers ranging up to 30 cm long on nesting cylindrical substrate. Methane, especially natural gas in which methane is mixed with a small fraction of hydrocarbons, inert gas, and oxides of carbon, is a useful and extremely inexpensive feedstock in some parts of the world.

Finding optimal conditions for filament lengthening is a difficult process: around 15% by volume methane in hydrogen, temperature near 1000°C, and gas residence time around 20 sec. The filaments will lengthen to several centimeters in less than 10 min [16].

5.3.1.4 Crystallographic Orientation

In one of the investigations [17] it was experimentally confirmed that the carbon that diffuses through the metal has a graphitic structure and the crystallographic orientations of the metal in contact with the hydrocarbon source can be different than that in the precipitated carbon. This condition has been demonstrated by postreaction selected area electron diffraction studies performed by Yang and Chan [18] on carbon filaments formed on nickel, cobalt, and α-Fe particles from reaction on methane at 700°C. Using this approach they were able to show the existence of (111), (311), and (220) nickel faces at the graphite metal interfaces. They could prove theoretically that the most favored face for graphite precipitation was Ni (111), followed by the order Ni (111) > (311) > (100) > (110). Based on this work it is clear that the orientation of the catalyst particle is the key variable in its ability to not only decompose a hydrocarbon molecule but also on initiating and sustaining carbon filament growth. Indeed it may well be that changes in the crystallographic orientation of the catalyst particle are responsible for the eventual deactivation step.

5.3.1.5 Why the Core Is Hollow

It can be argued that since the surface area of the basal plane is low, the free energy for filament growth must be minimized when the shaft is made up of curved basal planes. Using basic thermodynamics Tibbetts [19] showed that it is energetically more favorable for the filaments to create a hollow core than to precipitate out as highly strained cylindrical planes of narrow diameters. His model established a theoretical framework on why filaments are tubular and accurately predicted the inside diameters of many filaments. If diffusion of the particle is slow enough to limit the rate of the growth process, then eventually the surface of the entire crystallite will be covered with carbon and catalytic activity and filament growth will cease.

To explain the tubular structure of carbon filament, Baird et al. [20] proposed that the filament is formed entirely by a catalytic process involving the surface diffusion of carbon around the metal particle rather than bulk diffusion of carbon through the catalytic particle. Such transport would be easier if the catalyst particle were liquid.

5.4 Large-Scale Production: Gas Phase Process

Since the aforementioned method is basically a batch process, implementing it on the commercial level is less attractive than a comparable continuous process. Moreover, preserving enough catalytic particles of sufficiently small diameter to grow fiber with high yield is also difficult in the batch process.

For all these reasons an alternative method or concept was presented by Koyoma and Endo [21] and developed in more detail by Hatano et al. [22]. In this concept, known as the fluidized bed method, the volume seeding (floating catalyst) particles (5–25 nm diameter) are either incorporated in the feedstock or produced in the reactor (Figure 5.7) by decomposition of an organometallic like ferrocene. The ensuing fiber growth takes place inside the reactor at the same basic temperature (~1200°C) as is used with the substrate method shown in Figure 5.2. The carbon potential may be adjusted to a compromise between lengthening and thickening, allowing fibers a few microns long and a fraction of a micron in diameter to be produced. The growth length of VGCF will increase with duration of time of the catalyst particle floating within the reaction zone, which has been verified by experiment. Fibers move through the reactor with the gas stream and are collected in the reservoir or as they exit. The floating time of the particles coming from the pyrolysis of ferrocene is ~1–2 minutes, which is also an estimate of the fiber growth time. The reaction yield of the floating particle method is high (~80% of the benzene supply is converted into fibers) [23]. Both the hydrocarbon mixture and catalyst particles are fed into the top of the reactor and separated from the gas stream in the colder region of the furnace. Such a method is very flexible and can produce fibers with varying yield, length, diameter, and physical properties. The fiber diameters are uniform but they are not as long as the fibers grown on substrates.

The fibers are collected as a sponge-like deposit [24], consisting of vermicular carbon fibers with diameters between 0.1 and 1.5 microns and typical lengths of about 1 mm (Figure 5.8). At the tip of the fiber small catalytic particles are observed with diameters ~5 nm encapsulated in the carbon deposit of vapor-grown carbon. In the case of fibers grown by floating seed method, the fiber diameter and length can be controlled by the growth conditions.

High-resolution TEM studies show a characteristic central hollow tube along the length of the fiber and the diameter of the hollow tube of the ultrathin fibers is slightly less than the diameter of the catalyst particle (Figure 5.9a–f). The hollow tube is remarkably straight and of constant diameter along the fiber length. The contamination of fibers with catalytic seeds resulting from too high a catalyst concentration interferes with normal fiber growth.

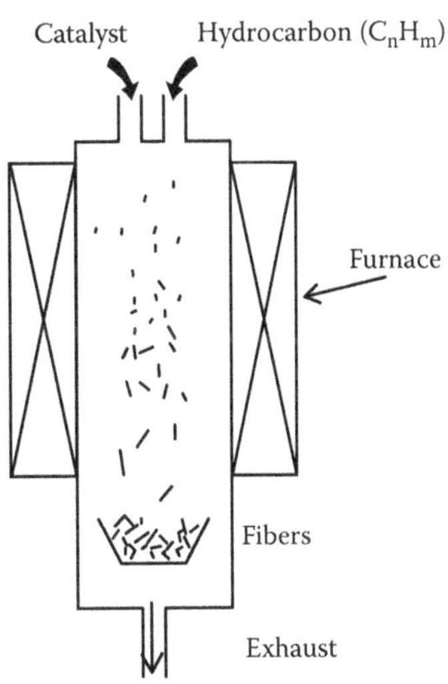

FIGURE 5.7
Schematic representing a continuous process for producing carbon fibers from vapor-phase hydrocarbons. Both the hydrocarbon mixture and catalyst particles are fed into the top of the reactor and separated from the gas stream in the colder region of the furnace.

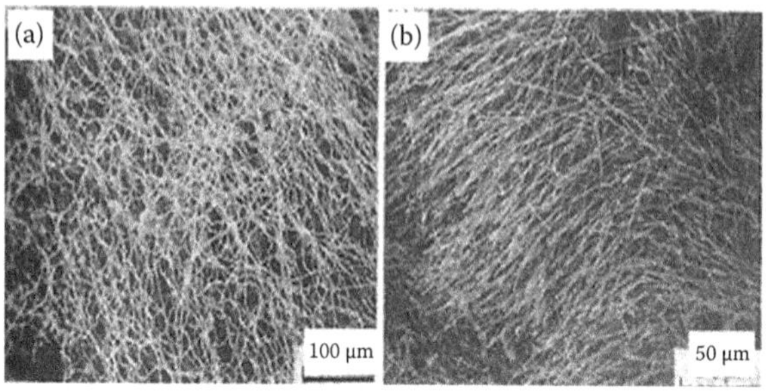

FIGURE 5.8
SEM pictures showing typical aggregation of benzene-derived carbon fibers by the floating catalyst method: (a) twisty and (b) straight fibers. (From Endo M., Extended Abstracts, 18th Biennial Conference on Carbon, July 1987, Worcester Polytechnic Institute, USA, p. 151.)

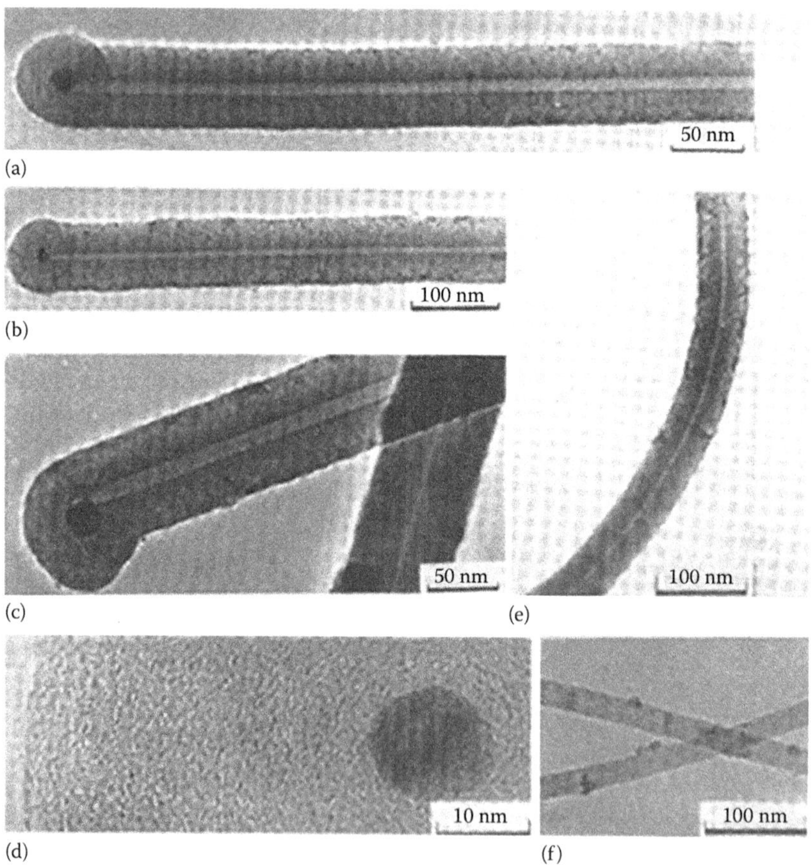

FIGURE 5.9
TEM pictures of the carbon filaments: (a–c) the tip of the fiber and hollow tube, (d) catalyst particle, (e) curved fiber, and (f) contaminated fiber with catalyst seed. (From Endo M., Extended Abstracts, 18th Biennial Conference on Carbon, July 1987, Worcester Polytechnic Institute, USA, p. 151.)

5.5 Morphology

5.5.1 As-Grown Vapor-Grown Carbon Fiber (VGCF)

Morphology is essential to understanding the properties of VGCF. A SEM image of the transverse cross section of a typical fiber is shown in Figure 5.10. The fiber's structure resembles that of a tree trunk, with concentric annular rings. The hollow core lies along the center of symmetry of the filament axis. Extensive electron diffraction studies along with the Debye-Scherrer x-ray diffraction experiments show that these layers exhibit poor interlayer correlation and are partially aligned like turbostratic carbon. The average basal plane misaligned from the fiber axis by about 10° to 15°. The orientational alignment is greater than those of conventional carbon fibers prepared in the same temperature range, which might account for the higher electrical conductivity and modulus of elasticity for the VGCF.

5.5.2 Heat-Treated VGCF

The catalytic particles, approximately 10 nm in diameter, are completely encased in carbon at the conclusion of the vapor deposition (thickening) step. Hence they are not able to participate in the catalytic graphitization of the microscopic fibers. Absence of metallic impurities and the uniform cylindrical geometry of carbon deposition make these fibers eminently graphitizable. Graphitization of the as-grown VGCF to temperatures of ~3000°C results in almost complete graphitization of the structure with long-range order. VGCF is a typically graphitizing carbon, and unlike PAN-based carbon fibers, it can be easily transformed by heat treatment into highly crystalline graphite. The extent of the graphitization is higher than that of any other known fibrous carbon type. These fibers can be easily produced on commercial scale.

FIGURE 5.10
A cross-sectional view of a vapor-grown fiber exposed to 1350°C showing a tree trunk/annular ring microstructure. (Reprinted from *Composites Part A: Applied Science and Manufacturing* 33, no. 2, R. Patton, C. Pittman, L. Wang, J. Hill, and A. Day. Ablation, mechanical and thermal conductivity properties of vapor grown carbon fiber/phenolic matrix composites, 243–251. Copyright 2002, with permission from Elsevier.)

Interestingly, in contrast to polymer-based carbon fibers, after sufficiently high HTT, faceting of the external surface [25] is observed due to release of the strain energy of the curved surface (Figures 5.11 and 5.12). Together with the increase in the crystallite size a high degree of orientation of the graphitic planes along the fiber axis is observed.

The effect of heat treatment on the filaments on atomic level is shown in Figure 5.13. The top panel compares the diffractometer profile for as-grown fibers with those heat treated to 2600°C. The improvement in layer spacing is obvious, as the interplanar spacing begins to approach the values of graphite at 3.35 Å. The bottom panel compares the lattice images of both materials [26].

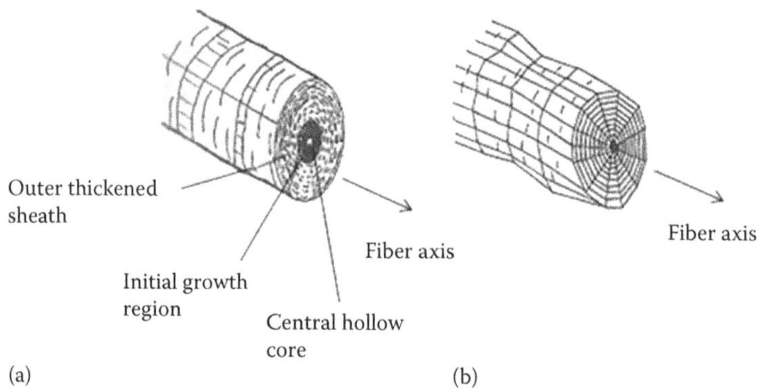

FIGURE 5.11
Sketch illustrating the structure of CCVD filaments (a) as-deposited at 1100°C and (b) after heat treatment to 3000°C. (With kind permission from Springer Science+Business Media: *Graphite Fibres and Filaments*, 1988, M.S. Dresselhaus, G. Dresselhaus, K. Sugihara, I.L. Spain, and H.A. Goldberg.)

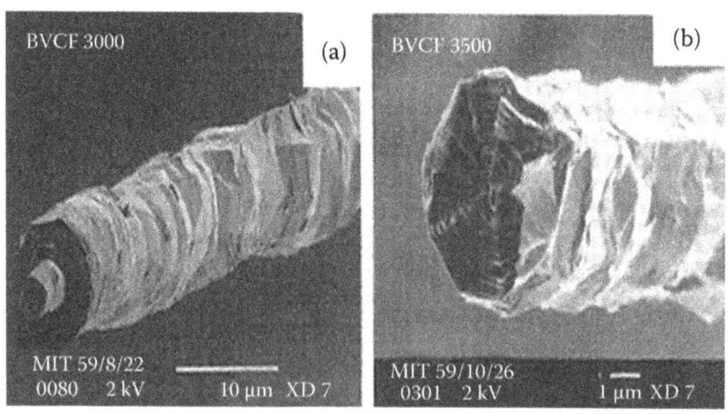

FIGURE 5.12
SEM photographs showing the morphology change on the surface and the cross section of benzene-derived filaments at different heat treatment temperatures: (a) 3000°C and (b) 3500°C. (From Yoshida, A. Hishiyam, Y. Endo, M. Extended Abstract, 17th Biennial Carbon Conference, Lexington, KY, p. 297, 1985.)

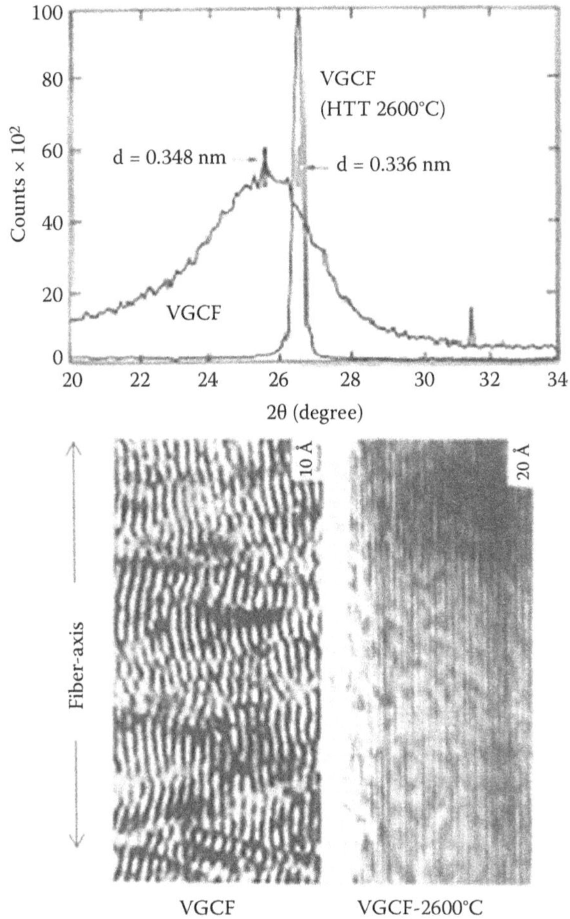

FIGURE 5.13
(Top) X-ray diffraction profiles of the carbon 002 line for as-grown and 2600°C heat-treated VGCF. (Bottom) The high-resolution electron micrographs contrast the 002 lattice images of as-produced and heat-treated fibers. (From Tibbests, G.G. Beetz, Jr. C.P., in *Carbon Fibres, Filaments and Composites,* Edited by Figueirdeo et al., NATO ASI Series E: Applied Sciences, Vol. 177, 1989.)

5.6 Mechanical Properties of VGCF

Carbon fibers customary fail in a brittle fracture. The mode of fracture of the as-produced VGCF is very much like the brittle failure of PAN and pitch based carbon fibers. During this fracture the fiber may even explode in the microtensilometer, and special precautions are necessary to preserve the remainder of the sample in order to examine the mode of fracture. However, the nested basal plane structure of VGCF, particularly when the fibers are graphitized above 1100°C, may slow down the fiber with a "pull out" failure observed in a vapor-grown fiber heat treated to 2200°C [26]. Figure 5.14 shows tensile failure of

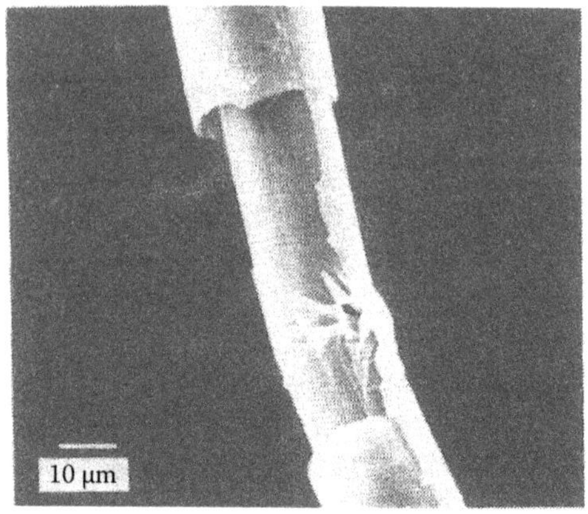

FIGURE 5.14
Fracture behavior of VGCF. Unlike PAN and pitch-based carbon fiber, which has britle failure, the VGCF that has been heat treated to 2200°C instead shows "sword and sheath (pull out)" noncatastrophic failure due to its annular microstructure. (From Tibbests, G.G. Beetz, Jr. C.P., in *Carbon Fibres, Filaments and Composites*, Edited by Figueirdeo et al., NATO ASI Series E: *Applied Sciences*, Vol. 177, 1989.)

the VGCF graphitized to 2200°C. Under the applied tensile load, a surface-initiated crack that is arrested after penetration only a short distance in the fiber may rapidly propagate around the circumference. A series of such events may conclude with thin cylinders of graphite telescoping to increasing length, gradually decreasing the load-bearing capability of the fiber under slowly increasing applied strain. After complete shearing of the fiber, the nesting layers may wedge together as further strain is applied, and continue to support some load [26].

VGCF has desirable mechanical properties, although it is produced at the relatively low temperature of around 1100°C. The product has almost the same properties whether it is made from benzene or methane. Figure 5.15 presents the dependence of the tensile strength and modulus of elasticity of the methane-derived VGCF on fiber diameter. Its

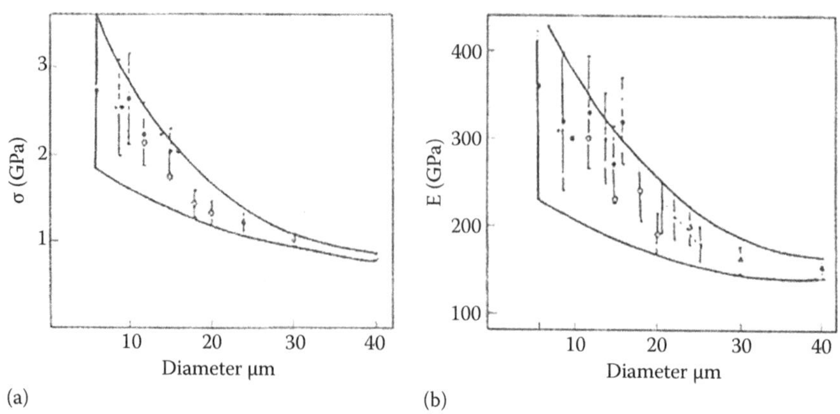

FIGURE 5.15
(a) Dependence of tensile strength (σ) and (b) modulus of elasticity (E) of VGCF on fiber diameter. (From K.N. et al., Extended Abstracts, Int. Carbon Conf. (CARBONE'84), Bordeaux, France, p. 142, 1984.)

tensile strength is comparable to that of commercial carbon fibers, and it can be enhanced to as high as 7 GPa when the fiber is graphitized at 2200°C.

5.7 Transport Properties of VGCF

VGCF has higher electrical and thermal conductivity than any other fiber; its electrical resistivity at room temperature is 10×10^3 μΩ.cm. When it is heat treated at 3000°C, its thermal conductivity at room temperature is about 4 times higher than that of copper [28].

5.7.1 Electrical Conductivity

Figure 5.16 shows measurements of the electrical resistivity of as-produced VGCF and also heat treated at 2000°C and 3000°C, respectively [29]. The resistivity of the 1100°C material as grown, 2×10^{-5} Ω.m, is virtually independent of temperature. It is therefore not surprising that the electrical industry has used vapor-deposited carbon resistors for many years. Heating the fibers at higher temperature results in the expulsion of heavier hydrocarbons, near 2000°C when graphitization begins resulting in an increase in conductivity. At higher temperatures these plots approach the behavior of single-crystal graphite with a 300 K resistivity of about 5×10^{-7} Ω.m. According to Hermans [29] the fibers are semimetal with equal numbers of electrons and holes and a 40 meV overlap between the electron and hole band.

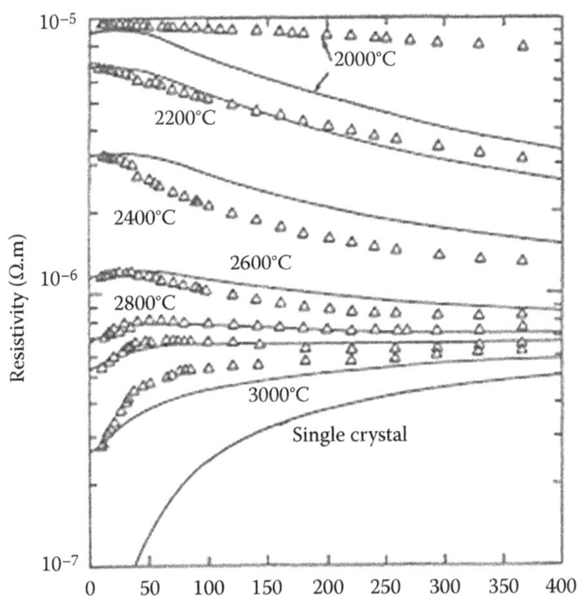

FIGURE 5.16
Model (solid lines) and experiment (points) for temperature-dependent resistivity of carbon fibers HTT at various temperatures. The bottom solid line is the calculated curve fitted to the resistivity of single-crystal graphite. (Reprinted from *Carbon* 23, no. 4, J. Heremans. Electrical conductivity of vapour grown fibres, 432–436. Copyright 1985, with permission from Elsevier.)

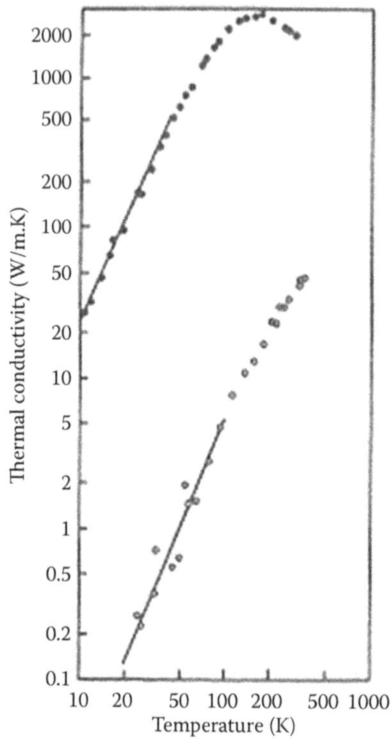

FIGURE 5.17
Temperature dependence of thermal conductivity of as-grown (solid circle) and 3000°C HTT (open circle) VGCF. (Reprinted with permission from J. Heremans and C. Beetz Jr. Thermal conductivity and thermopower of vapor-grown graphite fibers. *Physical Review B* 32, no. 4: 1981. Copyright 1985 by the American Physical Society.)

5.7.2 Thermal Conductivity

The thermal conductivity of VGCF is remarkably high; it may become one of the most compelling reasons for its application. Figure 5.17 shows measurements by Hermans and Beetz [30] of the thermal conductivity of both as-grown fibers and the fibers heat treated to 3000°C. The peak conductivity of the latter curve, 3000 W/m.K, is an excellent value approaching that of single-crystal graphite. Though the as-grown fibers conduct heat only 1/50 as well as the heat-treated fibers, their conductivity peaks at higher temperature.

5.8 Applications of VGCF

Vapor-grown carbon fibers are a promising new technology for the production of strong, stiff, discontinuous carbon fibers that would be useful in applications where cost is an important consideration. In the nuclear energy sector, VGCF has proven to be more resistant to neutron irradiation than other carbon fibers. It can be sufficiently low priced to replace ordinary carbon fibers in discontinuous yarn and thus serve as a useful filler for composites. In that capacity VGCF (thickened at 1100°C) with tensile strength of fibers of ~3 GPa and Young's modulus of ~230 GPa can be bracketed along continuous PAN and pitch-based carbon fibers. Its applications are likely to expand into automobiles and construction materials.

5.8.1 Double-Layer Capacitor

The recent expansion of the electronic industry controlled by microcomputers has created the need for newer, more compact (smaller in size) energy storage devices capable of maintaining the reserve power necessary to protect microcomputer systems from main power shut down. The new storage devices for electronic and space applications were required to be very small in size and with high-charge capacity materials.

In a double-layer capacitor, large storage of charges takes place at the interface, described by Helmoltz [31], and current designs use an electrode consisting of activated carbon. The double layer is created at the interface of the carbon and the electrolyte (usually sulfuric acid). The replacement of activated carbon by other materials that combine a higher surface area with high conductivity could produce a remarkable improvement in the system. Possible use of filamentous carbon can offer a number of advantages over conventional aluminum electrolyte capacitors such as very high capacity in small size, relatively good working voltage, good self discharge, and enhanced electrical conductivity of the electrode.

5.8.2 Intercalation Compounds

VGCF of rather thick diameter (10 micron) can also be used as host material for graphite intercalation compounds (GICs) [32], which can be made as conducting materials and are useful for battery electrodes. VGCF heat treated at 3000°C to form graphite fibers is best suited as the host material for AsF_5 GIC, which has resistivity of the order of 1–3 μΩcm at 300 K and is relatively stable. Therefore it is useful for various lightweight conducting composite materials. Furthermore, fluorine-intercalated VGCF is suitable as an active component for anode of high-performance Li-ion battery.

5.8.3 Carbon Filaments in Reinforcement of Materials

The possibility of mixing carbon filaments with other types of fibrous structures offers the potential of producing a new class of composite materials. The concept of coating carbon filaments with various refractory materials opens up some fascinating applications for new applications. With a carbon core and a refractory exterior, one has the option of either using fiber in their mixed form or selectively removing the carbon core by oxidation treatment to produce ceramic microtubules of nearly atomic dimensions. These structures are expected to exhibit extremely high strength-to-weight ratios and as such will be desirable as ultra lightweight reinforcement in advanced composites. They may be also used as high surface area and high emissivity coatings, novel filtration media, unique catalyst supports, and lightweight insulators.

The notion of growing carbon filaments onto the surfaces of primary carbon fibers by catalytic decomposition of hydrocarbons offers the potential of producing a new generation of composites with unique physical, chemical, and electronic properties.

5.9 Graphite Whiskers

5.9.1 Filaments Prepared from Carbon Arc

Carbon nanocones, whiskers, and larger polygonized nanotubes, called graphite polyhedral crystals, are a whole new class of carbon nanomaterials that, along with vapor-grown carbon fibers, can be placed between graphite and fullerene families of carbon. They are

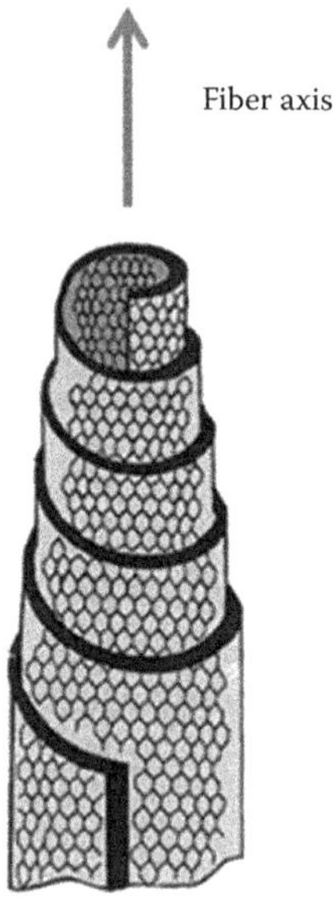

FIGURE 5.18
Sketch illustrating the scroll structure of a carbon whisker. (Courtesy of What-When-How, In Depth Tutorials and Information, http://what-when-how.com/nanoscience-and-nanotechnology/carbon-nanotubes-and-other-carbon-materials-part-1-nanotechnology/.)

elongated and, typically, axially symmetric structures. Graphite whiskers are the first known nonplanar graphitic structures that were obtained through a controlled preparation.

Graphite filaments may be grown from carbon arcs under high pressure of inert gas. Specifically, a positively charged upper electrode strikes an arc against a copper block. Graphite filaments, called "whiskers" by Bacon [33], are embedded inside the solid matrix of graphite, which builds up by diffusion of carbon vapor from the positive to the negative electrode. These can be removed by breaking open the deposit, thereby revealing a forest of filaments, protruding up to 30 mm from the exposed surface. Their diameters generally range from 1 to 5 μm. The maintenance of the arc necessitates the use of an electrode feed mechanism to maintain a constant electrode gap. Growth only occurs at high temperature (~3900 K) and over a pressure range of 90–100 bars. This is close to the triple point (P ~100 bar, and T ~4500 K) of carbon.

Electrical resistivity of these filaments was measured to be close to single-crystal graphite. They consist of one or more concentric tubes, each tube being in the form of a scroll, or rolled-up sheet of graphite layers, wound around the axis (as in rolling up a carpet) and extending continuously along the length of the whisker, with the *c*-axis exactly perpendicular to the whisker axis (Figure 5.18). They exhibit a high degree of flexibility, very high breaking strength, up to 20 GPa and Young's modulus in excess of 700 GPa, which are close to the practical possible limit of highly crystalline graphite. The values of room-temperature resistivity of around 65 μΩcm is approximately the same as the single-crystal value.

5.10 Carbon Nanofibers (CNFs)

Vapor-grown CNFs can be referred to as a smaller size variant of VGCFs. These short-length CNFs have drawn lots of attention to provide solutions to many vexing problems in composite applications. They are electrically conductive and thus are suitable for applications that require the ability to discharge electrostatic potentials, provide sufficient conductivity for electrostatic painting, or even shielding from radiofrequency interference or lightning strike together with excellent thermal and mechanical properties.

Although difficulties in producing these macroscopic vapor-grown carbon fibers with acceptable efficiencies seemed to be formidable, careful studies indicated that the initial nucleation of the predecessor microscopic carbon filaments was quite efficient. In the reactors producing VGCF, nanometer-sized filaments were made in abundance but were being submerged and incorporated in the layer of vapor deposited carbon, which thickened and grew into the longer filaments to macroscopic [34] dimensions. An obvious solution to produce VGCF is therefore to efficiently and continuously produce the carbon filaments while minimizing the inefficient carbon deposition step. Since the filaments themselves are submicron in diameter, these are produced on a continuous basis and then transported them out of the reactor with the gaseous production feedstock. Endo et al. [35] reported such a scheme in 1985. A series of papers from General Motors Research Laboratories described development of a process for continuous production of VGCNF utilizing liquid [36] and gaseous [37] catalysts. A helpful review by Lake and Ting [38] contrasts the properties of the larger diameter VGCF with VGCNF, with particular emphasis on the thermal conductivity of VGCF composites.

5.10.1 Types of Carbon Nanofibers Available

In the early 1980s, while other groups were still attempting to develop a commercially viable macroscopic VGCF, Hyperion Catalysis (fibrils.com) approached the problem of developing carbon nanofibers by first chemically producing dispersed nanometer-sized catalyst particles and then growing "fibrils" from them in reactors using flowing hydrocarbons at high temperatures. These fibrils were relatively long, frequently entangled, and of the coaxial cylinder morphology; they would now be called multiwalled nanotubes. Hyperion developed a large production capacity but did not publish its results due to a very extensive patent policy [39]. In 1991, Applied Sciences Inc. (apsci.com) began marketing a VGCNF (Pyrograf III) developed in collaboration with General Motors Research. Pyrograf III is a very fine, highly graphitic, yet low cost, carbon nanofiber, which is available in diameters ranging from 70 and 200 nanometers and a length in excess of 100 microns. These fibers have the stacked-cup morphology and are manufactured with differing thicknesses and different surface and debulking treatments at prices in the vicinity of US \$200/kg. Because these nanofibers have been available in large quantities and at relatively low prices, many groups have experimented with them and published their results. In Japan, the early work of Endo and collaborators more understanding of the fiber with stacked-cup morphology [40,41] has been developed. Japanese companies such as Sumitomo, Mitsui, Showa Denko, and Nikkiso have apparently developed sizeable production capacities and appear to be working on a host of products. However, few composites papers have been published in the open literature, in contrast to a host of papers on the applications of VGCNF to Li-ion batteries.

5.10.2 Synthesis of Nanofibers

Catalytic chemical vapor deposition (CCVD), assisted with variants like thermal and plasma, remains the dominant commercial technique for the fabrication of VGCF and VGCNF. Here, gas-phase molecules are decomposed at high temperatures and carbon is deposited in the presence of a transition metal catalyst on a substrate where subsequent growth of the fiber around the catalyst particle is realized. In general, this process involves separate stages such as gas decomposition, carbon deposition, fiber growth, fiber thickening, and graphitization, which results in hollow fibers. The nanofiber diameter depends on the catalyst size. The CCVD process for the fabrication of VGCF generally falls into two categories: fixed-catalyst process (batch) and (2) floating-catalyst process (continuous) [42].

In the batch process developed by Tibbetts [43], a mixture of hydrocarbon–hydrogen–helium was passed over a mullite (crystalline aluminum silicate) substrate with fine iron catalyst particle deposits maintained at 1000°C. The hydrocarbon used was methane in the concentration of 15% by volume. Fiber growth in several centimeters was achieved in just 10 minutes with a gas residence time of 20 seconds. In general, fiber length can be controlled by the gas residence time in the reactor. Gravity and direction of the gas flow typically affects the direction of the fiber growth.

The continuous or floating-catalyst process was patented earlier by Koyama and Endo [21] and was later modified by Hatano and coworkers [22]. This process typically yields VGCF with submicrometer diameters and lengths of a few to 100 μm, which accords with the definition of carbon nanofibers. They utilized organometallic compounds dissolved in a volatile solvent like benzene that would yield a mixture of ultrafine catalyst particles (5–25 nm in diameter) in hydrocarbon gas and the temperature allowed to rise to 1100°C. The fiber growth initiates on the surface of the catalyst particles and continues until catalyst poisoning occurs by impurities in the system. In the fiber growth mechanism described by Baker and coworkers [44] only the part of catalyst particle exposed to the gas mixture contributes to the fiber growth and the growth stops as soon as the exposed part is covered, that is, the catalyst is poisoned. The catalyst particle remains buried in the growth tip of the fiber at a final concentration of about a few parts per million. At this stage, fiber thickening takes place.

The most commonly used catalyst is iron, often treated with sulfur and hydrogen sulfide, to lower the melting point and facilitate its penetration into the pores of carbon and hence, to produce more growth sites [45]. Other catalysts used are Fe/Ni, Ni, Co, Mn, Cu, V, Cr, Mo, Pd, MgO, and Al_2O_3 [46,47]. Acetylene, ethylene, methane, natural gas, and benzene are the most commonly used carbonaceous source gases. Often carbon monoxide (CO) is introduced in the gas flow to increase the carbon yield through reduction of possible iron oxides in the system.

The Applied Sciences Inc. process for manufacturing VGCNF is depicted in Figure 5.19. The primary feedstock is natural gas, while the catalytic iron particles are produced by the decomposition of $Fe(CO)_5$. When the catalyst particles are properly dispersed and activated with sulfur, carbon filaments are abundantly produced in a reactor maintained near 1100°C.

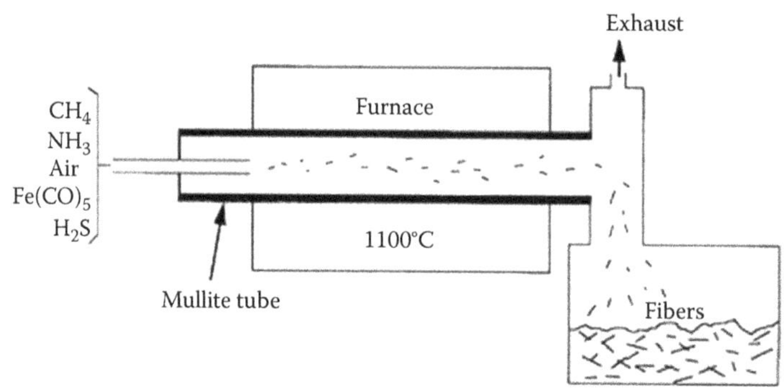

FIGURE 5.19
Apparatus for manufacturing VGCNF.

5.10.3 Growth and Morphology

Figure 5.20 is a TEM micrograph showing the structure of a typical vapor-grown nanofiber. A hollow core is surrounded by a cylindrical fiber grown from a catalyst particle and composed of graphite basal planes stacked at about a 25° angle from the longitudinal axis of the fiber. Note that the diameter of the core of the fiber is more than the thickness of the wall, in contrast to VGCF, wherein the wall thickness is much larger than the diameter.

Tibbets et al. [48] have discussed the lattice continuity conditions that can dictate the formation of nested cone and cone helix morphologies in the core filaments.

Figure 5.21 shows the interior wall of a nanofiber; the nested conical graphene planes are clearly canted with respect to the longitudinal fiber axis. Sometimes the filaments tend to clog the rear of the reactor tube, allowing filaments to spend many seconds within the decomposing methane atmosphere, the layer of carbon deposit thickens the individual fibers and cements them together into large clumps. The exterior layer of vapor-deposited carbon is not as graphitic as the interior cylinder and has graphene planes that are primarily longitudinally oriented. By adjusting the feedstock compositions and furnace operating conditions, different thicknesses of vapor-deposited carbon can be produced.

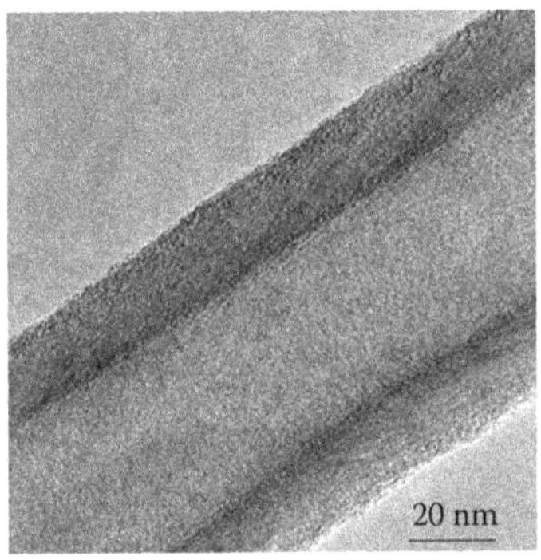

FIGURE 5.20
TEM micrograph showing structure of VGCNF with a cylindrical hollow core at the center. (Reprinted from *Composites Science and Technology* 67, no. 7–8, G.G. Tibbetts, M.L. Lake, K.L. Strong, B.P. Rice. A review of the fabrication and properties of vapor-grown carbon nanofiber/polymer composites, 1709–1718. Copyright 2007, with permission from Elsevier.)

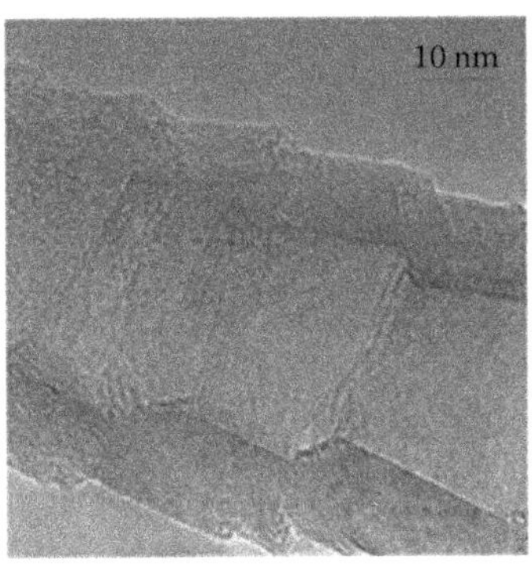

FIGURE 5.21
TEM micrographs of VGCNF graphitized at 2800°C. (Reprinted from *Composites Science and Technology* 67, no. 7–8, G.G. Tibbetts, M.L. Lake, K.L. Strong, B.P. Rice. A review of the fabrication and properties of vapor-grown carbon nanofiber/polymer composites, 1709–1718. Copyright 2007, with permission from Elsevier.)

5.10.4 Heat Treatment

Carbon nanofibers having a filamentary core of conically nested graphene planes preferentially recrystallize into discontinuous conical crystallites after a graphitizing heat treatment above 2800°C. This discontinuous structure improves the crystallinity of the carbon but does not give optimum mechanical or electrical properties to composites in which they are used, or even the lowest resistivity to the fiber itself. This is because the short, nested conical crystallites interface with grain boundaries contributes toward lowering the fiber's mechanical strength, stiffness, and electrical conductivity. The optimum heat treatment temperature for composite mechanical properties and electrical conductivity [49] lies near 1500°C and may vary depending on the specific application. This optimum is achieved when the graphitization of the vapor-deposited carbon exterior layer is maximized in the direction of the fiber's axis (Figure 5.22). Further heat treatment generally recrystallizes this layer on the conical interior planes, sacrificing its desirable longitudinal properties.

5.10.5 Intrinsic Fiber Conductivity

Endo et al. [50] first reported the intrinsic conductivity of large diameter VGCF and the increased conductivity that could be obtained by graphitization at temperatures approaching 3000°C. Heremans [51] measured resistance of VGCF versus temperature as a function of heat treatment temperature and related the improvements to increased electron mobility and higher electronic conduction band occupation. The numbers that define the intrinsic limits for composites fabricated from VGCNF are the intrinsic resistivity at room temperature of VGCF grown near 1100°C, 2×10^{-3} Ω.cm, and the room temperature resistivity of graphitized VGCF, 5×10^{-5} Ω.cm, which is near the resistivity of graphite. These numbers are consistent with resistivities expected at the observed graphitization indices of the VGCNFs [52].

5.10.6 Mechanical Properties of Nanofibers

Unfortunately, direct measurement of the tensile properties of a single VGCNF is not possible experimentally and no data has therefore been reported. However, direct measurements by earlier workers of the macroscopic vapor-grown carbon fibers have given values of 2.9 GPa for the tensile strength and 240 GPa for the tensile modulus [53], values comparable to a medium grade of PAN-based carbon fiber. An indirect method could be the reinforcement of VGCNF in any matrix and measure mechanical properties of the resulting composites. However, to extract intrinsic fiber mechanical properties from composite

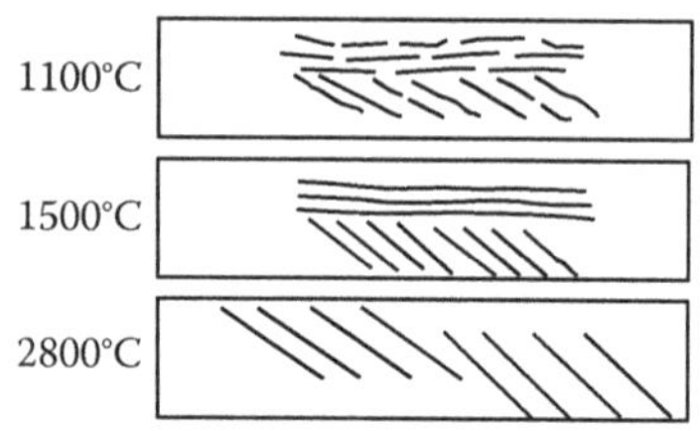

FIGURE 5.22
Schematic of the structure of VGCNF at various stages of graphitization. Each box represents a section of a cylindrical wall. (Reprinted from *Composites Science and Technology* 67, no. 7–8, G.G. Tibbetts, M.L. Lake, K.L. Strong, B.P. Rice. A review of the fabrication and properties of vapor-grown carbon nanofiber/polymer composites, 1709–1718. Copyright 2007, with permission from Elsevier.)

properties would require that fiber position and orientation within the composite, and its coupling with the matrix is controlled optimally.

5.11 Applications of Carbon Nanofiber

Introduction of a new material is typically limited by a number of technical barriers. In the case of VGCNF, these barriers include surface functionalization for bonding to selected matrices, dispersion, and orientation of the VGCNF in the composite. Even though vigorous efforts of numerous researchers have overcome several of these barriers and have established a foundation of rules for composite synthesis needed for the emergence of commercial applications, much work remains before such composite technology is firmed up to standard industrial practice.

5.11.1 Carbon Nanofiber Composites

CNF can be incorporated into commercially available thermoplastics, thermosets, and elastomers and can be used directly in existing high volume molding processes without any significant new manufacturing development. Because of its extraordinary intrinsic properties, particularly its strength and elastic modulus, CNF is expected to enable a reduction in the material required to produce a given strength and/or stiffness, thus providing net weight and cost savings.

The necessity of going beyond conventional compounding techniques has created opportunity for the emergence of a supply chain producing VGCNF compounded materials. Resins are now available that contain 20% by weight loadings of VGCNF that can be utilized in a variety of formulated epoxy systems including pre-pregs, molding compounds, adhesives, and coatings (nanosperse.com). Such formulations are intended for use in conductive adhesives with high-strength characteristics, structural composite panels to replace metal for weight savings and corrosion resistance, and components for medical, aerospace, and electronics applications. Use of VGCNF has been reported for improved mechanical properties of liner-less composite pressure vessels, where performance improvement is attributed to the development of high strain, microcrack-resistant resins, and the inclusion of VGCNF at the ply interfaces. These materials eliminate microcracking as the first failure mode and improve the laminate failure strain to a level that nearly equals that of the reinforcing carbon fiber. Compounded with thermoplastics or thermosets, VGCNF can more than triple the resin's tensile modulus and strength. Compressive strength is generally improved by an even larger margin. Preliminary research gives some hope that a practical method may be found to improve the orientation of VGCNF to achieve even greater improvements.

5.11.2 Conductive Thermoplastic Composites

Since virtually all of the electrical conductivity in carbon fiber–polymer composites is through the network of carbon fibers, it is clear that good fiber dispersion, small fiber diameter, and large aspect ratio will all aid in achieving high composite conductivity. There are really two goals to be sought in fabricating CNF–polymeric composites. The first is a low percolation threshold; it is achieved when a small volume fraction of fibers first establishes this conducting network. The second goal is achieving sufficient conductivity at high fiber loading to meet more ambitious high conductivity goals, such as radio frequency interference shielding and superior mechanical properties of the resulting composites. Recent work by Zhang et al. [54] illustrates the promise of making sensors for organic vapors from VGCNF–polystyrene composites, as the electrical resistivity of such composites rises sharply on exposure to organic vapors. Data from three different types of fibers are plotted and compared to a simple superposition model for graphitized or as-grown fibers in Figure 5.23. In contrast to structural changes induced in carbon black composites, VGCNF composites reproduce cyclability in vapors such as tetrahydrofuran (THF) or benzene.

Finally, the availability of a thermal grease composed of a silicon-free carrier material incorporating VGCNF as the thermal conductor has also been announced (electrovac.com). Here the nanoscale filler enables accurate thin film application and decreases thermal resistance.

5.11.3 Electromagnetic Shielding

The intrinsic conductivity of highly graphitic vapor-grown carbon fiber at room temperature is of the order of 5×10^{-5} Ω.cm, which is near the resistivity of graphite. Due to their high electrical conductivity and high aspect ratio, CNF can impart equivalent electrical conductivity to a composite at lower loadings than conventional conductive fillers. Also, by controlling the loading, one can produce composites with different electrical resistivity values. This is of particular importance for applications that require a resistivity in different ranges such as electrostatic dissipation (ESD) ($10^6 - 10^8$ Ω.cm), electrostatic painting ($10^4 - 10^6$ Ω.cm), EMI shielding ($10^3 - 10^1$ Ω.cm), and lightning strike protection (<10 Ω.cm). Conductivity values for VCGNF/resin composites are sufficiently high to offer significant protection from electromagnetic interference. Donohue and Pittman [55] have measured a 45 db shielding effectiveness at 200 MHz for 15 wt% high temperature heat-treated VGCNF in a vinyl ester matrix. The samples were 1.8 mm thick. Resistivities of below 0.15 Ω.cm

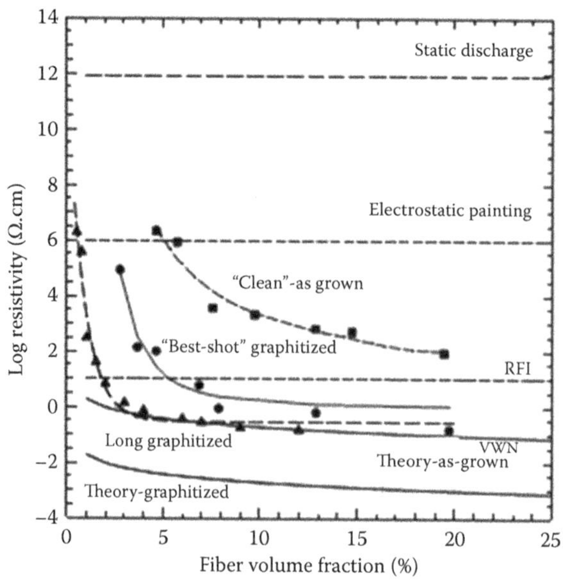

FIGURE 5.23
Resistivities of some polypropylene–VGCNF composites. Approximate resistivity values required for static discharge, electrostatic painting, and radio frequency interference shielding are indicated. (From G.G. Tibbetts, I.C. Finegan, and C. Kwag, *Molecular Crystals and Liquid Crystals* 387 (2002): 129–133.)

may be achieved with a fiber loading near 15 wt%, and a percolation threshold of below 1 wt% can be possible. Optimization of the processing parameters during composite preparation such as dispersion, fiber/matrix interfacial interaction, and the volume fraction of the reinforcement is the key issue in fabricating future shielding materials.

5.11.4 Ablation Resistance

A good example of the multifunctionality of VGCNF is provided by its application in ablation-resistant rocket motor insulation where flame and moisture resistance together with electrical charge dissipation are required in varying degrees. VGCNFs reinforced phenolic resoles and thermoplastic elastomers may make a useful contribution in this field, as they have been able to improve the ablation resistance of the composites [56]. Patton et al. [57] have determined that the low erosion and char rates of VGCNF–phenolic resin composites under a plasma torch at 1650°C are very promising for producing solid rocket motor nozzles. Moreover, the short-length VGCNFs produce a lower thermal conductivity than the competing continuous carbon fibers.

5.12 Carbon Microcoils and Nanocoils

Various types of coiled carbon filaments have been synthesized using chemical vapor deposition and other methods. These carbon filaments exhibit unique electrical and mechanical properties due to their versatile shapes and structures. To form coiled shapes, different types of catalyst compositions and reactive gases have been explored. Generally, coiled

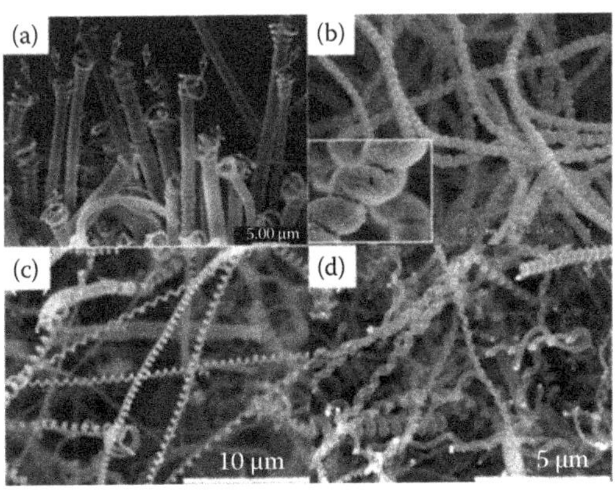

FIGURE 5.24
Different types of coiled carbon filaments: (a) bidirectionally grown double helical carbon filaments (b-DHCFs); (b) bidirectionally grown twisted carbon filaments (b-TCFs); (c) tip grown single helical carbon filaments (t-SHCFs); and (d) tip grown twisted carbon filaments (t-TCFs). ((a) Reprinted with permission from S. Motojima and Q. Chen. Three-dimensional growth mechanism of cosmo-mimetic carbon microcoils obtained by chemical vapor deposition. *Journal of Applied Physics* 85, no. 7: 3919–3921. Copyright 1999 American Institute of Physics. (b) Reprinted from *Diamond and Related Materials* 13, no. 11, S. Yang, X. Chen, S. Motojima, Morphology of the growth tip of carbon microcoils/nanocoils, 2152–2155. Copyright 2004, with permission from Elsevier. (c,d) Reprinted from *Carbon* 48, no. 1, D. Li, L. Pan, J. Qian, and D. Liu, Highly efficient synthesis of carbon nanocoils by catalyst particles prepared by a sol–gel method, 170–175. Copyright 2010, with permission from Elsevier.)

carbon filaments are classified by coil diameter and shape (e.g., microcoil and nanocoil) as shown in Figure 5.24 [58–60].

Carbon nanocoils have many potential applications, including mechanical springs and nanosolanoids. Several energy-related applications of this material, such as catalyst support for polymer electrolyte membrane fuel cells, direct methanol fuel cells, hydrogen storage, and electrodes for supercapacitors are also being looked into. For details of such applications, the reader is advised to refer to the articles by Reddy et al. [61] and Motojima et al. [62].

References

1. R. Bacon. Carbon fibers from rayon precursors. In *Chemistry and Physics of Carbon*, vol. 9, edited by P.L. Walker Jr. and P.A. Thrower, 1–102. Marcel Dekker, 1973.
2. A. Shindo. Studies on Graphite Fiber, Rept. No. 317, Government Industrial Research Institute, Osaka, Japan, 1961.
3. S. Ōtani. On the carbon fiber from the molten pyrolysis products. *Carbon* 3, no. 1 (1965): 31–38.
4. W. Watt, L. Phillips, and W. Johnson. High-strength high-modulus carbon fibres. *Engineer* 221, no. 5757 (1966): 815–816.
5. T.V. Hughes and C.R. Chambers. Manufacture of carbon filaments. US patent 405,48, issued 1889.
6. L. Radushkevich and V. Lukyanovich. About the structure of carbon formed by thermal decomposition of carbon monoxide on iron substrate. *Journal of Physical Chemistry (Moscow)* 26 (1952): 88–95.
7. W. Davis, R. Slawson, and G. Rigby. An unusual form of carbon. *Nature* 171 (1953): 756.
8. T. Koyama and M. Endo. Structure and growth processes of vapour grown carbon fibres. *Oyo Butsuri* 42 (1973): 690.
9. G.G. Tibbetts. Carbon fibers produced by pyrolysis of natural gas in stainless steel tubes. *Applied Physics Letters* 42, no. 8 (1983): 666–668.
10. F. Benissad, P. Gadelle, M. Coulon, and L. Bonnetain. Formation of carbon-fibers from methane. 1. Catalytic growth and pyrolytic thickening. *Carbon* 26 (1988): 61.
11. M. Humenik and W.D. Kingery. Metal-ceramic interactions: III, surface tension and wettability of metal-ceramic systems. *Journal of the American Ceramic Society* 37, no. 1 (1954): 18–23.
12. R. Pilliar and J. Nutting. Solid-solid interfacial energy determinations in metal-ceramic systems. *Philosophical Magazine* 16, no. 139 (1967): 181–188.
13. R.T.K. Baker, M.A. Barber, P.S. Harris, F.S. Feates, and R.J. Waite. Nucleation and growth of carbon deposits from the nickel catalyzed decomposition of acetylene. *Journal of Catalysis* 26, no. 1 (1972): 51–62.
14. A. Oberlin, M. Endo, and T. Koyama. Filamentous growth of carbon through benzene decomposition. *Journal of Crystal Growth* 32, no. 3 (1976): 335–349.
15. J. Rostrup-Nielsen and D.L. Trimm. Mechanisms of carbon formation on nickel-containing catalysts. *Journal of Catalysis* 48, no. 1–3 (1977): 155–165.
16. G.G. Tibbetts and M.G. Devour. Regulation of pyrolysis methane concentration in the manufacture of graphite fibers. US patent 4,565,684, issued 1986.
17. R. Baker, P. Harris, J. Henderson, and R. Thomas. Formation of carbonaceous deposits from the reaction of methane over nickel. *Carbon* 13, no. 1 (1975): 17–22.
18. R.T. Yang and J.P. Chen. Mechanism of carbon filament growth on metal catalysts. *Journal of Catalysis* 115, no. 1 (1989): 52–64.
19. G.G. Tibbetts. Why are carbon filaments tubular? *Journal of Crystal Growth* 66(3) (1984): 632–638.
20. T. Baird, J. Fryer, and B. Grant. Structure of fibrous carbon. *Nature* 233 (1971): 329.

21. T. Koyama and M. Endo. Method for manufacturing carbon fibers by a vapor phase process. Japanese patent 1982-58, 966.
22. M. Hatano, T. Ohsaki, and K. Arakawa. Graphite whiskers by new process and their composites. *Advancing Technology in Materials and Processes* (1985): 1467–1476.
23. M. Endo and M. Sikata. Tanso faiba (carbon fibers). *Oyo Butsuri* 54 (1985): 507.
24. M. Endo. Extended Abstracts, 18th Biennial Conference on Carbon, July 1987, Worcester Polytechnic Institute, USA, p. 151.
25. M.S. Dresselhaus, G. Dresselhaus, K. Sugihara, I.L. Spain, and H.A. Goldberg. *Graphite Fibres and Filaments*, p. 11. Springer-Verlag, 1988.
26. G.G. Tibbests and C.P. Beetz Jr. In Carbon Fibres, Filaments and Composites, Edited by Figueirdeo et al., NATO ASI series E: *Applied Sciences*, Vol. 177, 1989.
27. K.N. et al., Extended Abstracts, Int. Carbon Conf. (CARBONE'84), Bordeaux, France, 1984, p. 142.
28. L. Piraux, B. Nysten, A. Haquenne, J. Issi, M. Dresselhaus, and M. Endo. The temperature variation of the thermal conductivity of benzene-derived carbon fibers. *Solid State Communications* 50, no. 8 (1984): 697–700.
29. J. Heremans. Electrical conductivity of vapour grown fibres. *Carbon* 23, no. 4 (1985): 432–436.
30. J. Heremans and C. Beetz Jr. Thermal conductivity and thermopower of vapor-grown graphite fibers. *Physical Review B* 32, no. 4 (1985): 1981.
31. H.V. Helmholtz, Annals of Physics (Leipzig), 1853, 89, 21.
32. M.S. Dreselhaus and G. Dresselhaus. Intercalation compounds of graphite. *Advances in Physics* 30, no. 2 (1981): 139–326.
33. R. Bacon. Growth, structure, and properties of graphite whiskers. *Journal of Applied Physics* 31, no. 2 (1960): 283–290.
34. G. Tibbetts, M. Devour, and E. Rodda. An adsorption-diffusion isotherm and its application to the growth of carbon filaments on iron catalyst particles. *Carbon* 25, no. 3 (1987): 367–375.
35. M. Endo, M. Shikata, T. Momose, and M. Shiraishi. Vapor grown carbon fibers obtained by fluid ultra fine catalytic particles. In *17th Biennial Carbon Conference Proceedings*, 295–296, American Carbon Society, Lexington, Kentucky, 1985.
36. G.G. Tibbetts, D.W. Gorkiewicz, and R.L. Alig. A new reactor for growing carbon fibers from liquid- and vapor-phase hydrocarbons. *Carbon* 31, no. 5 (1993): 809–814.
37. G.G. Tibbetts, C.A. Bernardo, D.W. Gorkiewicz, and R.L. Alig. Role of sulfur in the production of carbon fibers in the vapor phase. *Carbon* 32, no. 4 (1994): 569–576.
38. M. Lake and J.-M. Ting. Vapor grown carbon fiber composites. In *Carbon Materials for Advanced Technologies*, edited by T.D. Burchell. Pergamon, 1999.
39. H.G. Tennant, J.J. Barber, and R. Hoch. US patent 5,165,909, issued 1992.
40. M. Endo, Y. Kim, T. Hayashi, T. Yanagisawa, H. Muramatsu, M. Ezaka, H. Terrones, M. Terrones, and M. Dresselhaus. Microstructural changes induced in "stacked cup" carbon nanofibers by heat treatment. *Carbon* 41, no. 10 (2003): 1941–1947.
41. Y.-A. Kim, T. Hayashi, S. Naokawa, T. Yanagisawa, and M. Endo. Comparative study of herringbone and stacked-cup carbon nanofibers. *Carbon* 43, no. 14 (2005): 3005–3008.
42. T.D. Burchell. *Carbon Materials for Advanced Technologies*. Elsevier, 1999.
43. G.G. Tibbetts. Lengths of carbon fibers grown from iron catalyst particles in natural gas. *Journal of Crystal Growth* 73, no. 3 (1985): 431–438.
44. R. Baker, M. Barber, P. Harris, F. Feates, and R. Waite. Nucleation and growth of carbon deposits from the nickel catalyzed decomposition of acetylene. *Journal of Catalysis* 26, no. 1 (1972): 51–62.
45. P. Morgan. *Carbon Fibers and Their Composites*. CRC Press, 2005.
46. K.P. De Jong and J.W. Geus. Carbon nanofibers: Catalytic synthesis and applications. *Catalysis Reviews* 42, no. 4 (2000): 481–510.
47. S. Dadvar, H. Tavanai, and M. Morshed. Effect of embedding MgO and Al_2O_3 nanoparticles in the precursor on the pore characteristics of PAN based activated carbon nanofibers. *Journal of Analytical and Applied Pyrolysis* 98 (2012): 98–105.

48. G.G. Tibbetts, M.L. Lake, K.L. Strong, and B.P. Rice. A review of the fabrication and properties of vapor-grown carbon nanofiber/polymer composites. *Composites Science and Technology* 67, no. 7–8 (2007): 1709–1718.
49. J. Howe, G.G. Tibbetts, C. Kwag, and M.L. Lake. Heat treating carbon nanofibers for optimal composite performance. 48th International SAMPE Symposium Proceedings, Long Beach, California, 2003.
50. M. Endo, T. Koyama, and Y. Hishiyama. Structural improvement of carbon fibers prepared from benzene. *Japanese Journal of Applied Physics* 15, no. 11 (1976): 2073.
51. J. Heremans. Electrical conductivity of vapor-grown carbon fibers. *Carbon* 23, no. 4 (1985): 431–436.
52. I.C. Finegan, G.G. Tibbetts, D.G. Glasgow, J.-M. Ting, and M.L. Lake. Surface treatments for improving the mechanical properties of carbon nanofiber/thermoplastic composites. *Journal of Materials Science* 38 (2003): 3485–3490.
53. G.G. Tibbetts, I.C. Finegan, and C. Kwag. Mechanical and electrical properties of vapor-grown carbon fiber thermoplastic composites. *Molecular Crystals and Liquid Crystals* 387 (2002): 129–133.
54. B. Zhang, R. Fu, M. Zhang, X. Dong, L. Wang, and C.U. Pittman. Gas sensitive vapor grown carbon nanofiber/polystyrene sensors. *Materials Research Bulletin* 41, no. 3 (2006): 553–562.
55. J.P. Donohoe. Shielding effectiveness of vapor grown carbon nanofiber/vinyl ester composites. In *International Symposium on Electromagnetic Compatibility*, 313–317, Netherlands, 2004.
56. R. Blanski, J.H. Koo, P. Ruth, H. Nguyen, C. Pittman, and S. Phillips. Polymer nanostructured materials for solid rocket motor insulation–ablation performance. Proceedings of 52nd JANNAF Propulsion Meeting, Las Vegas, Nevada, 2004.
57. R. Patton, C. Pittman, L. Wang, J. Hill, and A. Day. Ablation, mechanical and thermal conductivity properties of vapor grown carbon fiber/phenolic matrix composites. *Composites Part A: Applied Science and Manufacturing* 33, no. 2 (2002): 243–251.
58. S. Motojima and Q. Chen. Three-dimensional growth mechanism of cosmo-mimetic carbon microcoils obtained by chemical vapor deposition. *Journal of Applied Physics* 85, no. 7 (1999): 3919–3921.
59. S. Yang, X. Chen, and S. Motojima. Morphology of the growth tip of carbon microcoils/nanocoils, *Diamond and Related Materials* 13, no. 11 (2004): 2152–2155.
60. D. Li, L. Pan, J. Qian, and D. Liu. Highly efficient synthesis of carbon nanocoils by catalyst particles prepared by a sol–gel method. *Carbon* 48, no. 1 (2010): 170–175.
61. A.L.M. Reddy, R.I. Jafri, N. Jha, S. Ramaprabhu, and P.M. Ajayan. Carbon nanocoils for multifunctional energy applications. *Journal of Materials Chemistry* 21, no. 40 (2011): 16103–16107.
62. S. Motojima, X. Chen, S. Yang, and M. Hasegawa. Properties and potential applications of carbon microcoils/nanocoils. *Diamond and Related Materials* 13, no. 11 (2004): 1989–1992.

6

Graphene and Graphene Oxide

6.1 Introduction

Graphene, a two-dimensional (2D) crystal, is the name given to a single, one-atom thick layer of sp^2 hybridized carbon atoms arranged in a 2D graphitic honeycomb lattice structure [1–4]. Graphene is the basic building block for graphitic materials of all other dimensionalities such as zero-dimensional (0D) fullerenes, one-dimensional (1D) carbon nanotubes, and three-dimensional (3D) graphite [5]. Theoretically, graphene was studied for several decades [6–8] but planar 2D graphene was presumed not to exist in free state, as free-standing graphene was considered thermodynamically unstable [1,4,5]. Graphene, however, was considered a starting point for research of the more complex, 3D graphite [5]. The electronic band structure of graphene was calculated as early as 1947 [6], and graphene was studied purely as an academic material [6–8]. However, this view changed with the experimental discovery and isolation of graphene in 2004 [1,5], the first example of strictly two-dimensional crystals, as free-standing films [9–13]. Konstantin Novoselov and Andre Geim from the University of Manchester along with their team successfully isolated atomically thin films of graphite by peeling off layers from graphite crystal flakes using Scotch tape and created a monolayer graphene on oxidized silicon surface (SiO_2/Si wafer) [1,2,9,12]. Mechanical exfoliation technique allowed isolation of pristine 2D crystals from 3D graphite [11–14]. Resulting single- and few-layer flakes were pinned to the substrate by only van der Waals forces and could be made free-standing by etching away the substrate and were thus stabilized without any supporting substrate [14]. This allowed them to investigate graphene's intrinsic properties [14]. The isolation of monolayer graphene confirmed that graphene is stable, chemically inert, and crystalline under ambient conditions [11,13]. The 2D nature of graphene's honeycomb lattice and its unique electronic band structure give rise to a variety of novel physical, chemical and mechanical properties [9–15]. Much of the initial interest in graphene was due to the electronic properties of graphene [1,2,11,15,16]. Follow-up experiments by Geim and Novoselov showed that electrons in graphene effectively behave like massless relativistic particles, so-called Dirac fermions, that travel at a constant speed (10^6 m/s)—a small but noteworthy fraction of the speed of light [2,5,11,12,15]. The relativistic-like electron transport in graphene is unlike that in conventional solids where electron transport is nonrelativistic [2,11]. This endows graphene with unique electronic properties [9–15]. Graphene is a zero band-gap/gapless semiconductor [11] or a zero-overlap semimetal [12]. Monolayer graphene is transparent, extremely thin (thinnest material ever produced), elastic, light, and an exceptionally strong material [4,12,16,17]. It conducts electricity very well with current carrying capacity much higher than copper [12,17] and conducts heat better than most known materials [16,17]. This discovery has revolutionized the world of carbon nanomaterials. For their "ground breaking experiments regarding the two-dimensional (2D) material graphene," Andre Geim and Konstantin Novoselov were awarded the Nobel Prize in Physics in 2010.

Several methods have been developed over the years for the synthesis of graphene [1,4,18–22]. These include mechanical exfoliation [1], chemical exfoliation [20], chemical vapor deposition (CVD) [21], and thermal decomposition of silicon carbide [22]. One of the economical yet effective routes to produce bulk quantities of graphene is chemical reduction of exfoliated graphite oxide [23–27]. Using inexpensive graphite as the starting material, graphite is first oxidized to graphite oxide by insertion of strong acids and oxidizing agents within the interlayer spacing of graphite [26]. Graphite oxide is then chemically exfoliated into individual layers called graphene oxide [23,26,27]. Just as each layer of graphite is graphene, each layer of graphite oxide is graphene oxide [4], but graphene oxide is nonconducting or has very low electrical conductivity due to disruption of the sp^2 bonding system [19,25]. These graphene oxide layers are precursors for the production of graphene by the removal of oxygen groups, that is, by reduction of graphene oxide [23]. Graphene oxide is easier to process since presence of oxygen-containing groups in the layered material makes it easier to disperse in water and other solvents [19,23,28–31]. Reduction of graphene oxide to reduced graphene oxide restores the structure and properties, particularly electrical conductivity of graphene, though complete removal of oxygen is not achieved [23,26,28]. Graphene oxide and reduced graphene oxide can be chemically modified due to the existence of oxygen-containing functional groups on their surface to obtain chemically modified graphene-based materials [25,26,32,33].

Both graphene and chemically modified graphene (CMG) have generated immense interest in the field of materials science and electronics, and are promising candidates for potential applications in the field of polymer composites, and flexible and transparent electronic devices [12,16,17,28,34]. Polymers reinforced with graphene can result in stronger, tougher, thinner, and lighter composite materials with better thermal and electrical conductivity. Graphene possesses unusual electronic properties and offers potential for faster electronic devices, and graphene-based electronics research has the potential to replace silicon in electronics applications [5].

This chapter presents a review of graphene and graphene oxide; the history and isolation of graphene; and subsequent research on the synthesis, structure, properties, and the potential applications of graphene and graphene oxide.

6.2 History of Graphene and Its Isolation

Graphene belongs to the class of 2D crystals that for several decades were considered to be thermodynamically unstable to exist [5,17,35]. It was believed that planar structure of 2D graphene would be unstable with respect to the formation of curved structures such as soot, fullerenes, and nanotubes, and possibly curl into carbon soot or roll up into scrolls if prepared as single atomic layers [1,5,13,27].

Theoretically, graphene was first discussed by P.R. Wallace in 1947 [6]. He used the two-dimensional honeycomb structure of graphene to obtain information on the properties of graphite and developed the band theory of graphite. For many years, graphene was purely a material of academic interest only [6–8]. The fact that electrons in graphene behave like Dirac fermions was first mentioned by Semenoff in 1984 [8]. The term graphene was first used in 1987 to describe a single layer in graphite intercalation compounds [36]. The name graphene was introduced by Boehm, Setton, and Stumpp in 1994 [3]. According to them the suffix *-ene* is used for fused polycyclic aromatic hydrocarbons, even when the root of the name is of trivial origin (e.g., naphthalene, anthracene, coronene). A single carbon layer of the graphitic structure can be considered as the final member of this series and the term graphene should therefore be used to designate the individual carbon layers in graphite intercalation compounds [3].

The first success in isolation of graphene as a free-standing single layer of carbon came in the year 2004 when Geim and Novoselov isolated few-layer graphene and monolayer graphene on a silicon oxide substrate (SiO_2/Si wafer) by the micromechanical cleavage of bulk graphite (the Scotch tape technique to peel graphite) [1,9]. They repeatedly pulled graphite flakes from originally peeled-off graphite flakes resulting in thinner flakes each time, and several attempts later they obtained flakes of graphite only one atom thick and then they obtained a monolayer graphene, the first instance of physical existence of a two-dimensional material [5,9,11]. This simple process of slicing layered graphite by gently rubbing it against another surface to obtain graphene suggests that graphene is produced every time a pencil is used or from the time graphite was mined and pencil was invented in 1565 [37,38]. The key ingredient for the success of graphene observation by Geim's group was the carefully chosen thickness of SiO_2 (300 nm) on top of a Si wafer that made even the single-layer graphene visible under the optical microscope and they succeeded in discovering the two-dimensional graphene [5]. If not for this, graphene would probably remain undiscovered today [5]. It was experimentally confirmed in the follow-up experiments that the electrons propagating through graphene's honeycomb lattice and interacting with a periodic potential of the honeycomb lattice effectively lose their mass, producing quasiparticles that are described by a 2D analogue of the Dirac equation rather than the Schrödinger equation for spin-½ particles [2]. These quasiparticles are called massless Dirac fermions [6]. Much of the initial interest in graphene was due to this unique nature of graphene's charge carriers and the amazing electronic properties of graphene to explore electronic applications of graphene. The charge carriers in graphene can travel submicrometer distances without scattering, exhibiting high crystal quality of the 2D crystals [5,35]. Mechanical properties of suspended graphene films and free-standing monolayer graphene were subsequently measured [34,39] offering opportunities for structural and mechanical applications [34].

One of the earlier attempts to isolate graphene was based on chemical exfoliation using graphite oxide [40]. However, it was not until the isolation of pristine graphene by micromechanical cleavage and consequent rise in studies on graphene that there was a renewed interest in using graphite oxide as a precursor for graphene. It was realized that chemical exfoliation of graphite oxide to graphene oxide followed by reduction of graphene oxide to reduced graphene-like material called reduced graphene oxide was an economically viable route toward bulk synthesis of graphene [28,30,41,42]. Graphite oxide has a long and established history. The history of graphite oxide can be traced back to 1859 when Brodie oxidized graphite using $KClO_3$ (potassium chlorate) and fuming HNO_3 (nitric acid), which involved repeating the oxidation step several times [43]. After nearly 40 years, Staudenmaier oxidized graphite in a single step using $KClO_3$, fuming HNO_3, and concentrated H_2SO_4 (sulphuric acid) [33]. Both these methods were time consuming and hazardous [44]. Nearly 60 years later Hummers and Offeman developed a rapid and relatively safe method to obtain graphite oxide from graphite by treating graphite with an almost water-free mixture of concentrated H_2SO_4, sodium nitrate ($NaNO_3$), and $KMnO_4$ [44]. The Hummers method and its modified and improved versions have become popular routes toward synthesis of graphite oxide [31,33]. The improved Hummers method, using

$KMnO_4$, H_2SO_4, and H_3PO_4 (phosphoric acid) as the oxidizing agents, avoids the release of toxic NO_x and yields a greater amount of hydrophilic oxidized graphite material compared to the original Hummers method [33,45].

6.3 Preparative Methods for Graphene

A number of approaches have been explored to obtain few-layer or a single-layer graphene [1,4,9,14,18–22,27,46]. One of the approaches to obtain graphene is the top-down approach of mechanical exfoliation or chemical exfoliation to obtain graphene from graphite [27]. In order to exfoliate a single sheet of graphene, van der Waals attractions between the first and second layers have to be overcome without disrupting any subsequent sheets [14,27,45]. The other popular approach, also called the bottom-up approach, is to synthesize graphene by chemical vapor deposition method or by epitaxial growth using SiC substrates [14].

6.3.1 Mechanical Exfoliation

Mechanical exfoliation, also known as the micromechanical cleavage technique or the Scotch tape method, is the technique used by Geim and Novoselov to isolate graphene [1,5,9]. This technique of micromechanical exfoliation of 2D crystals is shown in Figure 6.1 [46]. In this method, a regular Scotch tape is used to peel off a thin layer of graphite from a block of highly oriented pyrolytic graphite (HOPG) [1,5,11,14]. The layer of graphite on the tape is pressed onto other layers of tape to extract thinner layers of graphite. The tape with the thin peeled off layer of graphite is finally pressed onto a ~300 nm smooth SiO_2/silicon substrate [5,14]. The tape is peeled off from the substrate leaving behind a graphene layer, the surface of which can then be carefully scanned in an optical microscope [15].

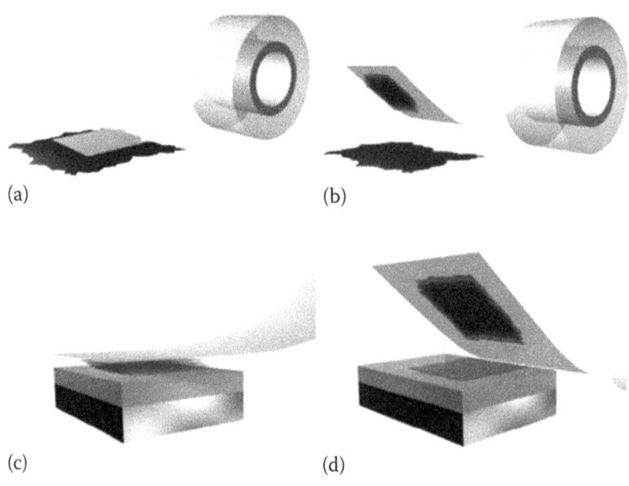

FIGURE 6.1
Micromechanical exfoliation of 2D crystals. (a) Adhesive tape is pressed against a 2D crystal so that (b) the top few layers are attached to the tape. (c) The tape with crystals of layered material is pressed against a surface of choice. (d) Upon peeling off, the bottom layer is left on the substrate. (From K.S. Novoselov and A.C. Neto, *Physica Scripta* 2012, no. T146 (2012): 014006. With permission.)

The thickness of SiO_2 layer is critical for observing graphene [5,37]. A thickness of 300 nm for SiO_2 is widely used for this detection technique [12]. Careful selection of the initial graphite material and the use of freshly cleaved and cleaned surfaces of graphite and SiO_2 make the isolation and detection of graphene crystallites and monolayers successful [5]. Mechanical exfoliation provides pristine graphene with excellent electrical properties. Hence this technique is best suited for fundamental research studies. This technique, however, is not suitable for bulk synthesis of graphene since size, thickness, and location of graphene crystallites cannot be controlled in this technique and the technique is not high yielding [5,14].

6.3.2 Chemical Exfoliation of Graphite/Liquid-Phase Exfoliation of Graphite

Chemical exfoliation or liquid-phase exfoliation method is an important technique and one of the most feasible approach for industrial production of graphene due to its versatility, potential scalability, and low cost [20,32]. This approach is inspired by the theoretical and experimental studies on the solvent dispersion of carbon nanotubes, which showed that nanotubes could be effectively dispersed in solvents whose surface energy matched that of the nanotubes [47,48]. In this method graphite is exfoliated into single graphene layers and individual sheets are stabilized in solution by means of ultrasonication in organic solvents. Ultrasonication is a powerful tool for mixing and obtaining uniform dispersions of solutions [19]. Acoustic cavitation or ultrasound-driven growth and collapse of bubbles accompanied by the generation of shock waves and elevated temperatures and pressures provide useful chemical conditions to induce mechanical effects such as shearing, mixing, and dispersion [19,49]. Hence, ultrasonication is an effective tool for exfoliation of graphite. Graphite powder is dispersed in a suitable solvent and the solution is ultrasonicated to obtain colloidal graphene suspensions. The powder fragments into nanosheets, which are stabilized against aggregation by the solvent [47]. Good solvents for a homogeneous colloidal suspension were classified using a model that included the surface energy of graphene and the cohesive energy of the solvent [50]. The dispersion and exfoliation is possible because the energy required to exfoliate graphene (i.e., to overcome the interacting forces between the graphene layers) is balanced by the solvent–graphene interaction for solvents whose surface energies match that of graphene [19,26,47,50]. The graphitic basal structure is broken and small graphite fragments intercalated by solvent molecules are produced [19]. Highly polar organic solvents, such as dimethylformamide (DMF) [51], N-methyl-2-pyrrolidone (NMP) [50], pyridine, and acetonitrile [52], have been successfully used for this method [19,28,52]. Exfoliation by ultrasonication is generally followed by centrifugation to increase the percentage of the monolayers of graphene. Surfactants can also be employed to stabilize exfoliated graphene in water [47]. Liquid-phase exfoliation yields graphene essentially free of defects and oxides [47]. Liquid-phase exfoliation methods can be used to deposit graphene in a variety of environments and on different substrates and can be employed to produce graphene-based composites or films, which are key components for many applications, such as thin-film transistors, conductive transparent electrodes for indium tin oxide replacement (e.g., in light-emitting diodes, or photovoltaics) [20,47].

In a typical process, crystals of natural graphite were exfoliated by sonication in DMF for over 3 h, and the procedure resulted in a suspension of thin graphitic platelets with a large proportion of monolayer graphene flakes [51]. The solvent DMF also wets the flakes preventing them from conglomerating. The suspension was then centrifuged at 13,000 rpm for 10 min to remove thick flakes. The remaining suspension consisted mostly of graphene and few-layer graphite flakes of submicrometer size. The thickness of the exfoliated flakes was verified by atomic force microscopy and transmission electron microscopy, and was found to have a high proportion of monolayer flakes (up to 50%) [51]. Absence of defects or oxides can also be confirmed by x-ray diffraction, infrared (IR), and Raman spectroscopy [50]. Direct chemical exfoliation of graphite allows production of dispersions of high-quality graphene sheets (almost pristine graphene) and thus allowing exploitation of the excellent conductive properties of graphene in the form of transparent or semitransparent conducting films and conducting composites [26,50,51]. Solution processing of graphene provides a range of potential large-area applications, from device and sensor fabrication to liquid-phase chemistry [50].

Production of high-quality graphene by means of both surfactant-free and surfactant-assisted liquid-phase exfoliation is becoming more popular (Figure 6.2) [20]. Liquid-phase exfoliation has been extensively reviewed by Ciesielski and Samorì [20].

Liquid-phase exfoliation of graphite intercalation compounds (GICs) has also been studied for production of graphene [26,32]. In this method, graphite is first intercalated with a compound followed by expansion of graphite via rapid increase in the vapor pressure of the volatile intercalated substance under microwave or thermal treatment. As nonoxidative agents are applied for intercalation of graphite and microwave or thermal treatment of GIC leads to large expansion of graphite, high-yield production of graphene with high quality can be achieved using this method.

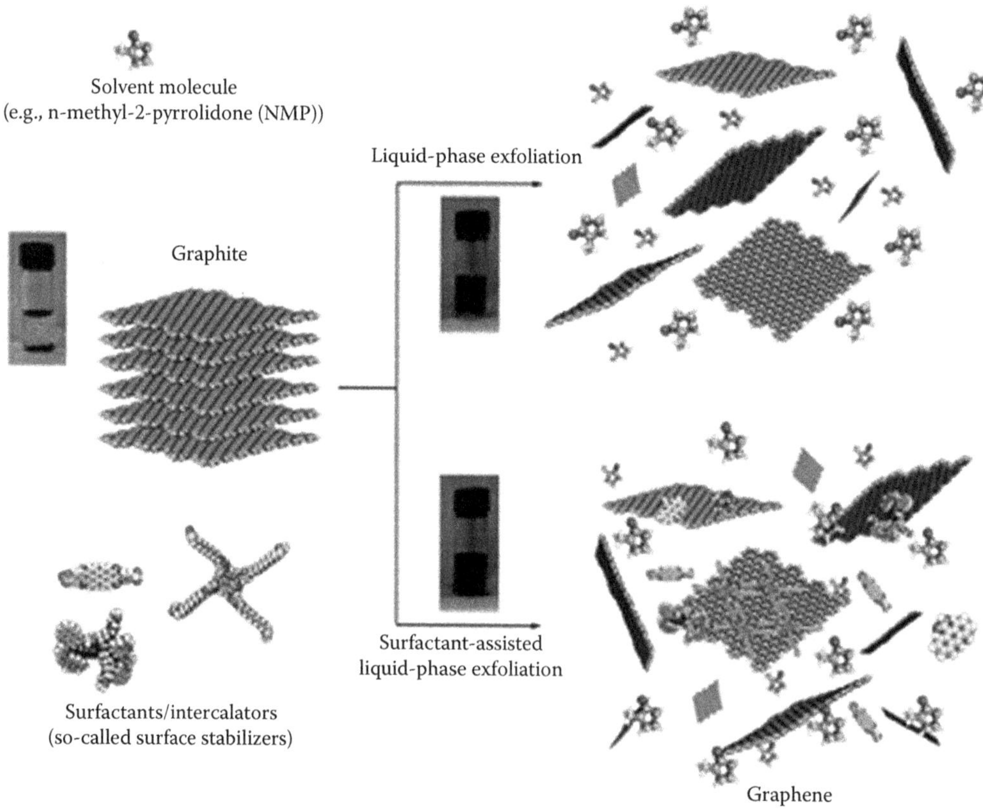

FIGURE 6.2
Schematic representation of liquid-phase exfoliation in the absence (top right) and presence of surfactant (bottom right). (From A. Ciesielski and P. Samorì, *Chemical Society Reviews* 43, no. 1 (2014): 381–398. Reproduced by permission of The Royal Society of Chemistry.)

The solvothermal-assisted exfoliation process in a highly polar organic solvent, acetonitrile, using expanded graphite (EG) as the starting material is an effective method to obtain a high yield of graphene up to 10 to 12 wt% [52]. The dipole-induced dipole interactions between graphene and acetonitrile facilitate the exfoliation and dispersion of graphene to produce high-quality graphene without any significant structural defects, which was confirmed by electron diffraction and Raman spectroscopy [52].

Although the chemical exfoliation technique results in exfoliated layers of graphene, prolonged ultrasonication may reduce the sheet size, introduce defects, and damage the graphitic character of the layers through oxidation of carbon atoms at the edges in the form of epoxy, carbonyl, and carboxyl groups [19]. The elevated temperatures and resultant heat energy can cause chemical reactions to occur in solution if ultrasonication of solution is prolonged.

6.3.3 Chemical Exfoliation and Reduction of Exfoliated Graphite Oxide

Graphite oxide has emerged as a potential precursor for large-scale cost-effective production of graphene-based materials [24], and liquid-phase exfoliation of graphite oxide is one of the most popular methods for preparation of graphene. This method involves oxidation of graphite powder with strong oxidizing agents to produce graphite oxide followed by expansion or exfoliation of oxidized graphite layers via sonication [19]. The reduction of the obtained graphene oxide to graphene is usually conducted by either thermal or chemical approaches [25,29]. The stepwise synthesis of reduced graphene oxide from graphite is shown schematically in Figure 6.3 [29].

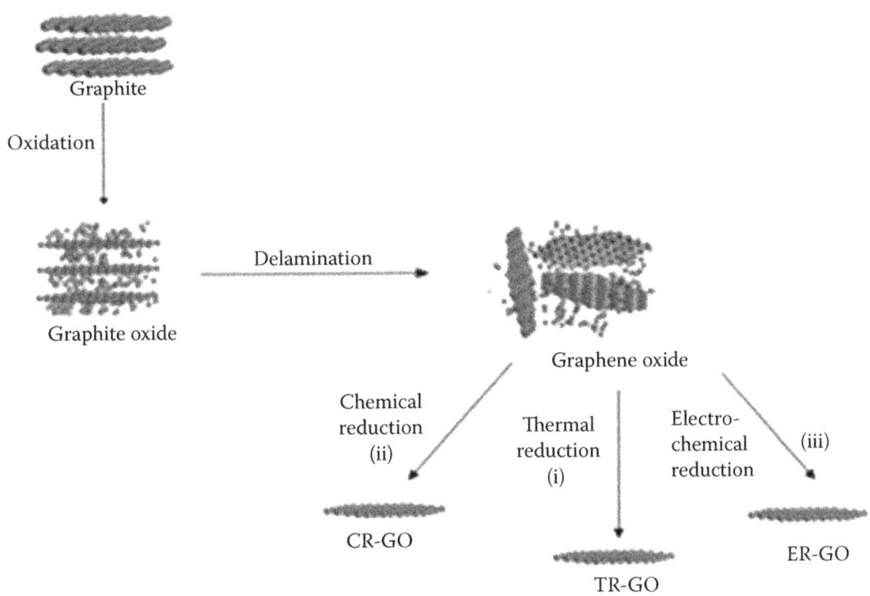

FIGURE 6.3
Schematic representation of the stepwise synthesis of single-/few-layer graphene from graphite. Oxidation of graphite in the first step is followed by chemical exfoliation to graphene oxide. The graphene oxide is then subjected to reduction by chemical, thermal, or electrochemical means. (From R. Garg, N.K. Dutta, and N.R. Choudhury, *Nanomaterials* 4, no. 2 (2014): 267–300.)

The yields are high; up to >50% of graphene is produced [32]. The vigorous oxidation of graphite often leads to incomplete restoration of the sp^2 hybridized carbon bonds after reduction, and the presence of residual oxygen functional groups results in poor electrical conductivity. There is a significant amount of oxygen in the reduced graphene oxide, indicating that reduced graphene oxide is not the same as pristine graphene [26]. However, graphene oxide provides potential for the production of chemically modified graphene on bulk scale through a variety of chemical modifications due to the presence of reactive oxygen functional groups, which provide sites for useful chemical functionalization reactions [25]. Thus graphene oxide undergoes two major reactions: (1) reduction, which is removal of oxygen groups from graphene oxide, and (2) chemical functionalization, which is adding other functionalities to graphene oxide [25].

Such approaches, which add functionality to groups already present on the graphene oxide, render graphene oxide a more versatile precursor for a wide range of applications such as in optoelectronics, polymer composites, and transparent and electrically conducting thin-film applications [25,26].

A large number of studies have been performed on reduction of graphene oxide and chemical functionalization of graphene oxide, which have been reviewed by several authors [19,24–26,28,29].

6.3.3.1 Synthesis of Graphite Oxide

The Hummers method is an easy and popular method for producing large quantities of graphite oxide through the addition of potassium permanganate to a suspension of graphite, sodium nitrate, and sulfuric acid [44]. Oxidation of graphite leads to the formation of anionic groups on graphitic layers, mostly hydroxylates, carboxylates, and epoxy groups, and also disrupts the aromatic character of the graphitic sheets [19]. The increased spacing due to intercalated species and the anionic or polar character of the oxygen groups formed impart a strongly hydrophilic behavior to graphite oxide, which allows water molecules to penetrate between the layers and thereby increase the interlayer distance even farther. Thus graphite oxide becomes highly dispersible in water [25].

6.3.3.2 Chemical Exfoliation of Graphite Oxide

Graphite oxide consists of a layered structure of "graphene oxide" sheets that are strongly hydrophilic, such that intercalation of water molecules between the layers readily occurs and it is easily exfoliated in aqueous media [26,53]. Graphite oxide is exfoliated in water by mechanical stirring or more effectively by ultrasonication to obtain exfoliated graphene oxide sheets [23,25,28]. The exfoliation is accomplished due to the strong interactions between water and the oxygen-containing (epoxide and hydroxyl) functionalities introduced into the basal plane during oxidation [19]. Sonication results in near-complete exfoliation of graphite oxide to graphene oxide to produce stable aqueous colloidal suspensions [25,26]. These graphene oxide layers are precursors for the production of graphene by the removal of the oxygen groups (i.e., by reduction of graphene oxide).

Graphite oxide can also be dispersed directly in several polar solvents such as ethylene glycol, DMF, NMP, and THF, and fully exfoliated into individual, single-layer graphene oxide sheets by sonication [26,54,55] to obtain stable graphene oxide dispersions. Chemical modification of graphene oxide sheets by organic molecules yields homogeneous suspensions in organic solvents [26].

6.3.3.3 Reduction of Graphene Oxide

The main aim of reduction of graphene oxide is to produce electrically conducting graphene-like materials similar to the pristine graphene achieved from direct mechanical exfoliation (i.e., the Scotch tape method) of individual layers of graphite [25].

Reduction of graphene oxide or the deoxygenating process of graphene oxide generally involves low temperature chemical reduction at <100°C using suitable reducing agents or thermal annealing at temperatures >1000°C [19,23,24,26]. It is important that the reducing agent should not react with the solvent used for dispersing graphene oxide [25]. A variety of reducing agents have been studied for chemical reduction of colloidal dispersions of graphene oxide in water, but complete reduction of graphene oxide into graphene has not been achieved [26]. Strong alkaline agents such as hydrazine hydrate [56] and sodium borohydride [57] are effective reducing agents, as well as acidic reducing agents such as hydroiodic acid [58]. The reduction with hydroiodic acid maintains good integrity and flexibility, and even improves the strength and ductility of the original graphene oxide films (Figure 6.4) [58]. Graphene oxide prepared by the improved Hummers method, when reduced with hydrazine gives equivalent electrical conductivity as graphene oxide prepared by Hummers method and similarly reduced with hydrazine [45]. It has been suggested that the improved method might disrupt the basal plane of the graphite less than Hummers method [45]. One of the disadvantages of using chemical methods of reduction, hydrazine in particular, is the introduction of heteroatomic impurities [25]. After removal of the oxygen groups, reduced graphene oxide can be further graphitized by annealing at elevated temperatures. In this process, defects that remain after reduction are rearranged and the aromatic character of the monolayers increases [19].

Thermal exfoliation and reduction of graphene oxide is another method of reduction [25,26,59]. The fundamental prerequisites that enable preparation of large volumes of dry, single graphene sheets are complete oxidation of graphite and extremely rapid heating of graphite oxide. Complete oxidation to graphite oxide was monitored by the total disappearance of the 0.34 nm intergraphene spacing and the appearance of a new one with a 0.65–0.75 nm range depending on the water content of graphite oxide. Thermal exfoliation is achieved through rapid heating (>2000°C/min) of dry graphite oxide under inert gas argon and high temperature up to 1050°C, which splits the graphite oxide into individual thermally reduced sheets through evolution of CO_2 [25,28,59]. Exfoliation takes place when the pressure generated by the carbon dioxide gas evolved due to the decomposition of the epoxy and hydroxyl sites of graphite oxide exceeds van der Waals forces holding the graphene oxide sheets together [28]. Despite the wrinkled sheet defective structure of graphene sheets and some residual functional sites after the reduction step, the graphene produced is electrically conducting [59].

Dai et al. [60] obtained high-quality graphene films by reparative reduction of graphene oxide in one step by reduction of graphene oxide through thermal annealing and simultaneous use of a carbon source to introduce carbon radicals for restoration of the ordered graphene structure.

Solvothermal reduction has also been shown to be an effective reduction method [61]. Solvothermal reduction was carried out in DMF at 180°C, using hydrazine monohydrate as the reducing agent. This method removed oxygen and defects from graphene sheets, increased the size of sp^2 domains, and produced materials that were as conducting as pristine graphene. The improved effectiveness of solvothermal reduction was considered to be due to more thorough removal of oxygen functional groups by hydrazine at high temperature.

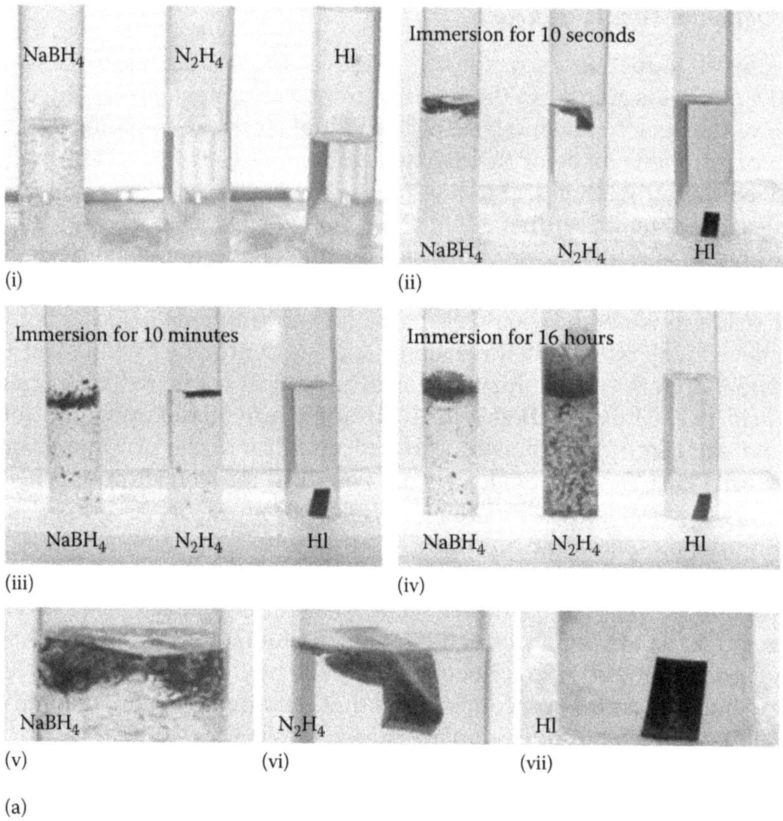

FIGURE 6.4
(a) Optical photographs of the reducing process by immersing a graphene oxide (GO) film into different reducing agents for different times at room temperature. (i) Three liquid reducing agents: 50 mM $NaBH_4$ aqueous solution ($NaBH_4$), 85% $N_2H_4.H_2O$ solution (N_2H_4), and 55% HI acid solution (HI). (ii–iv) The GO films reduced by the three agents for 10 s, 10 min, and 16 h. (v–vii) Enlarged views from (ii) that show that the phenomenon occurred after 10 s immersion of GO films to the three liquid reducing agents. *(Continued)*

Electrochemical reduction of graphene oxide relies on the electrochemical removal of the oxygen functionalities [25]. This method avoids the use of dangerous reductants (e.g., hydrazine) and the need to dispose of the by-products, but scalability is a fundamental issue in this method [25].

6.3.4 Chemical Vapor Deposition (CVD)

The CVD growth of graphene using transition metal (Ni [nickel], Cu [copper], etc.) substrates or thin films as catalysts is a promising, economical, and simple method for single-layer or multilayer graphene production [21,62]. Figure 6.5 [21] schematically shows a common setup for CVD of graphene.

Methane, commonly used as the carbon source, is catalytically decomposed on the catalyst surface at around 1000°C–1200°C in the presence of hydrogen gas. In this method the substrate (usually copper) is first heated in a furnace at low pressure to about 1000°C. This anneals the copper. Methane and hydrogen gases flow through the furnace. Methane decomposes on the copper surface in the presence of hydrogen. The copper substrate

FIGURE 6.4 (CONTINUED)
(b) Optical photographs and mechanical properties of the graphene oxide (GO) films reduced by different chemical agents. (i) as-assembled GO film; (ii) 1 h HI acid-reduced GO film at 100°C. (iii) Stress–strain curve of the GO film and HI acid-reduced GO film. (iv–vi) Hydrazine vapor-, $N_2H_4.H_2O$-, and $NaBH_4$-reduced GO films. The scale bar in (iv–vi) is 5 mm. (Reprinted from *Carbon* 48, no. 15, S. Pei, J. Zhao, J. Du, W. Ren, and H.M. Cheng, 4466–4474. Copyright 2010, with permission from Elsevier.)

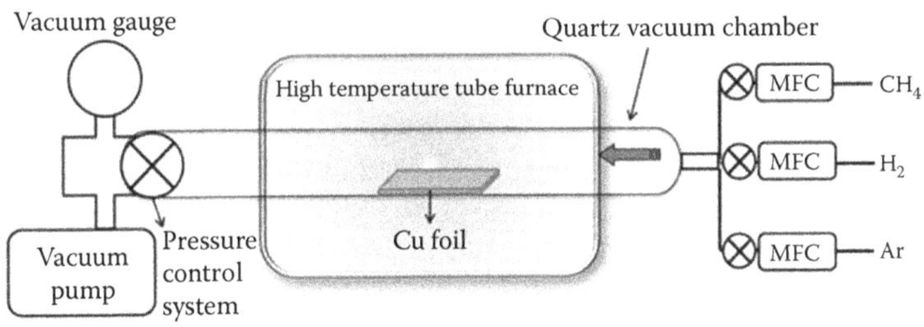

FIGURE 6.5
Schematic of a common setup for chemical vapor deposition of graphene. (From A. Kumar and C.H. Lee, Synthesis and biomedical applications of graphene: Present and future trends, in *Advances in Graphene Science*, InTech, 2013.)

provides nucleating sites for adsorbed and diffusing carbon species, which grow to a continuous graphene sheet. The solubility of carbon in copper is very low, which makes it different from other catalytic surfaces like Fe (iron) and Ni. Single layer graphene can usually be obtained on copper. Growth on copper foils by this approach has produced the largest area of low-defect graphene to date [21]. It was found that graphene CVD growth on copper foil using methane as a carbon source is strongly affected by hydrogen, which appears to serve a dual role: an activator of the surface bound carbon that is necessary for monolayer growth and an etching agent that controls the size and morphology of the graphene domains [63].

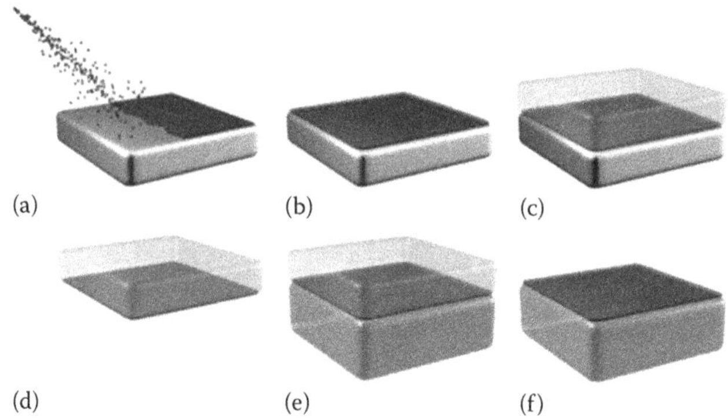

FIGURE 6.6
Transfer of CVD-grown 2D crystals. (a, b) 2D crystals are grown by CVD on a surface of a metal. (c) A sacrificial layer is deposited on top of the 2D crystal. (d) The metal is etched away, leaving 2D crystal stuck on the sacrificial layer. (e) The sacrificial layer, together with the 2D crystal is transferred onto the substrate of choice. (f) The sacrificial layer is removed. (From K.S. Novoselov and A.C. Neto, *Physica Scripta* 2012, no. T146 (2012): 014006. With permission.)

Graphene grown on transition metals must be transferred onto insulating substrates for device fabrication and electronic characterization [21]. This is done by first spin-coating a thin polymeric layer, such as polymethylmethacrylate (PMMA), on top of the as-grown graphene. The polymeric layer provides a supportive framework for graphene before the transfer. Then the underneath Cu substrate is etched away by iron chloride ($FeCl_3$) solution. The floating membrane is scooped and placed on a desired substrate. After drying, the polymeric film is dissolved with acetone or chloroform. The process of transfer of CVD-grown 2D crystals is shown in Figure 6.6 [46].

6.3.5 Thermal Decomposition of Silicon Carbide

Epitaxial growth of monolayer or few-layer graphene on silicon carbide (SiC) by thermal decomposition is attractive for large-scale production and suitable for existing electronic device technology and is a promising route toward epitaxial graphene-based electronics [22,64]. This method involves the conversion of SiC substrate to graphene via sublimation of silicon atoms on the surface at high temperatures of about 1000°C–1600°C in ultrahigh vacuum condition and graphitization of the remaining carbon atoms [19,22,64,65]. When SiC substrates are annealed at high temperatures, Si atoms selectively desorb from the surface and the C atoms left behind naturally form graphene. Figure 6.7 [66] shows the thermal decomposition method of growing graphene epitaxially on silicon carbide.

Riedl et al. [22] have reviewed the controlled growth of epitaxial graphene layers on SiC and the manipulation of their electronic structure. Berger et al. [65] studied the transport

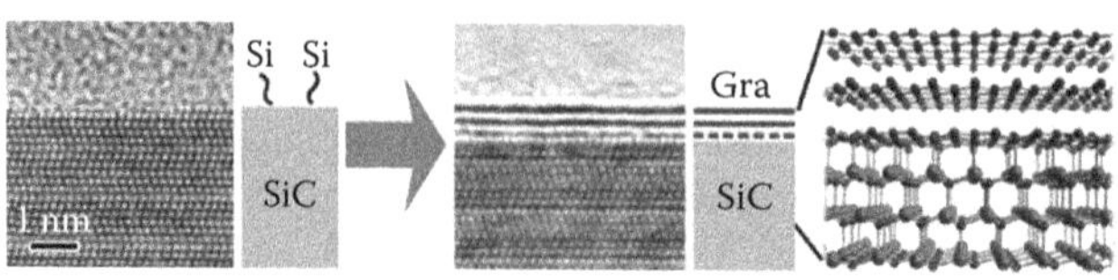

FIGURE 6.7
Thermal decomposition method of growing graphene epitaxially on silicon carbide. (From W. Norimatsu and M. Kusunoki, *Physical Chemistry Chemical Physics* 16 (2014): 3501–3511. Reproduced by permission of The Royal Society of Chemistry.)

and structural properties of graphene layers grown epitaxially on hexagonal SiC and found that the electronic properties of epitaxial graphene are found to be closely related to pristine graphene, revealing high-mobility charge carriers [5,67].

6.4 Characterization of Graphene and Graphene Oxide

Various spectroscopic and microscopic techniques such as optical microscopy, atomic force microscopy (AFM), transmission electron microscopy (TEM), Raman spectroscopy, and x-ray diffraction (XRD) are used to characterize graphene and graphene oxide.

6.4.1 Optical Microscopy

Optical microscopy is a very important characterization technique for detection of graphene. Preliminary identification of few-layer graphene including single-layer graphene amid thicker flakes obtained by micromechanical cleavage was done in an optical microscope, and the critical step that allowed the detection of atomically thin 2D crystallites of graphene was their identification in an optical microscope when they were placed on top of an oxidized Si wafer with a critical thickness of SiO_2 that was 300 nm. Thin flakes were found to attach strongly to SiO_2, presumably due to van der Waals or capillary forces [1]. In fact, 300 nm is the current standard and even a 5% difference in thickness of SiO_2 can make single-layer graphene completely invisible [5,37]. Graphitic films thinner than 50 nm can be seen on the SiO_2 surface because thin flakes are sufficiently transparent to add to the optical path that shifts the interference colors. Even a monolayer adds up sufficiently to the optical path of reflected light so that the interference color changes with respect to the one of an empty substrate [9]. With respect to the color of 300 nm wafer, which is violet-blue, and the shift in the wafer's color due to graphitic films, a color reference was chosen to estimate thickness of graphene films and it was found that at thicknesses of about 1.5 nm, as measured by AFM, graphene films are no longer visible (barely or feebly visible) [1,37]. This provided a natural marker to distinguish between few-layer and multilayer graphene. The few-layer graphene can be studied by AFM.

Blake et al. [37] investigated the problem of visibility of graphene on top of SiO_2/Si wafers (Figure 6.8). They found that by using monochromatic illumination, graphene can be isolated for any SiO_2 thickness except for about 150 nm and below 30 nm. They also suggested the lower thickness of about 90 nm as a substitute for the present benchmark thickness of 300 nm for graphene's visual detection. Their work established a quantitative framework for detecting single, double, and multiple layers of graphene, and they were also able to detect graphene on other insulators such as 50 nm Si_3N_4 using blue light and on 90 nm PMMA using white light.

6.4.2 Atomic Force Microscopy (AFM)

AFM is an important characterization method that allows definitive identification of single-layer crystals. In fact, AFM was used by Novoselov et al. [1] to identify and prove that the graphene films were indeed single-layer graphene. Based on their extensive AFM studies on few-layer graphene films, Novoselov et al. [1] found that the apparent thickness of few-layer graphene as measured by AFM was between 1 and 1.6 nm. The interlayer

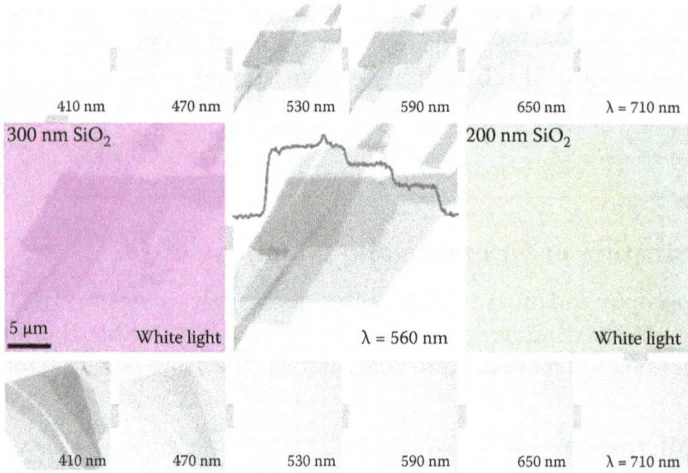

FIGURE 6.8
(a) Graphene crystallites on 300 nm SiO_2 imaged with white light, (b) green light, and (c) another graphene sample on 200 nm SiO_2 imaged with white light. Single-layer graphene is clearly visible on the left image (a), but even three layers are indiscernible on the right (c). Image sizes are 2525 m^2. Top and bottom panels show the same flakes as in (a) and (c), respectively, but illuminated through various narrow bandpass filters with a bandwidth of 10 nm. The flakes were chosen to contain areas of different thickness so that one can see changes in graphene's visibility with increasing numbers of layers. The trace in (b) shows steplike changes in the contrast for 1, 2, and 3-layer trace averaged over 10 pixel lines. This proves that the contrast can also be used as a quantitative tool for defining the number of graphene layers on a given substrate. (Reprinted with permission from P. Blake, E.W. Hill, A.C. Neto, K.S. Novoselov, D. Jiang, R. Yang, T.J. Booth, and A.K. Geim, *Applied Physics Letters* 91, no. 6: 063124. Copyright 2007, American Institute of Physics.)

spacing or interlayer distance in bulk graphite is 0.335 nm, which implies that few-layer graphene is indeed only a few atomic layers thick and contains 1 to 3 layers of graphene. The apparent thickness of single-layer graphene crystals is thus approximately the interlayer distance in the corresponding 3D crystals. AFM images showed folded or pleated areas for single-layer graphene [1].

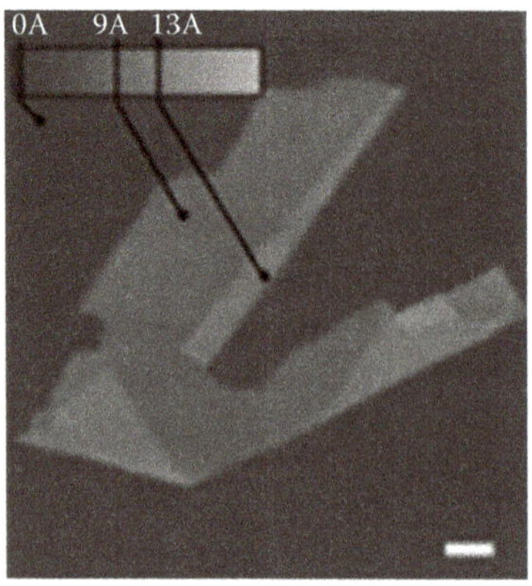

FIGURE 6.9
AFM image of two-dimensional single-layer crystal structure of graphene. (Reprinted from K.S. Novoselov, D. Jiang, F. Schedin, T.J. Booth, V.V. Khotkevich, S.V. Morozov, and A.K. Geim, *Proceedings of the National Academy of Sciences of the United States of America* 102, no. 30: 10451–10453. Copyright 2005, National Academy of Sciences, U.S.A.)

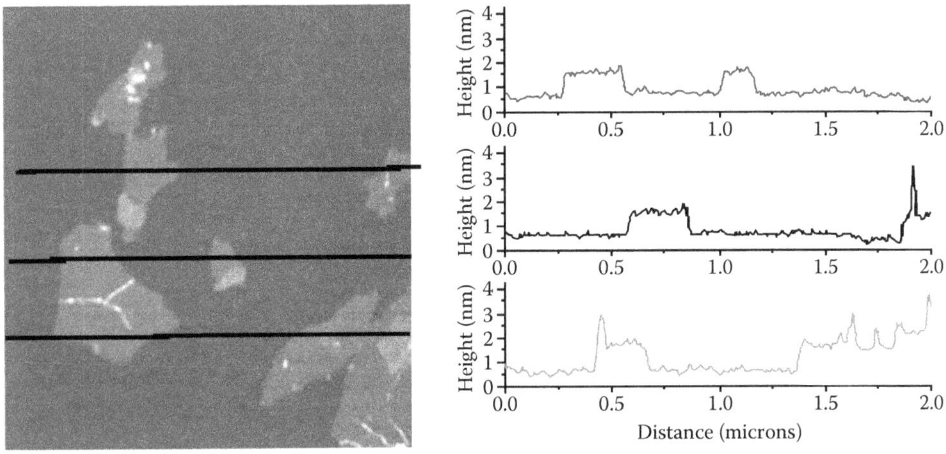

FIGURE 6.10
AFM image of exfoliated graphite oxide sheets with three height profiles at different locations. (Reprinted from *Carbon* 45, no. 7, S. Stankovich, D.A. Dikin, R.D. Piner, K.A. Kohlhaas, A. Kleinhammes, Y. Jia, Y. Wu, S.T. Nguyen, and R.S. Ruoff, 1558–1565. Copyright 2007, with permission from Elsevier.)

In their studies on AFM analysis of 2D single-layer crystals for which the single-layer crystals were selected as those exhibiting an apparent thickness of approximately the interlayer distance in the corresponding 3D crystals, 2D crystallites were often raised by an extra few angstroms above the supporting surface, probably because of a layer of absorbed water [9]. In such cases, the pleated and folded regions seen on many AFM images and having the differential height matching the interlayer distance in the corresponding 3D crystals helped to distinguish between double-layer crystals and true single sheets. Figure 6.9 [9] shows the AFM of 2D single-layer crystal structure of graphene.

AFM has been extensively used for identification of exfoliated graphene oxide sheets to confirm complete exfoliation of graphite oxide to individual graphene oxide sheets and reduced graphene oxide sheets [56]. The Figure 6.10 [56] shows the AFM image of exfoliated graphene oxide sheets; the sheets are ~1 nm thick.

Zhou et al. [32] produced graphene by liquid-phase exfoliation of intercalated graphite and used AFM to determine the thickness of the graphene sheets. An average thickness of about 1 nm calculated from the height difference between the surface of graphene and that of the substrate was consistent with triple-layer graphene [32]. By random sampling of 30 graphene sheets, they estimated that 70% of the graphene sheets were comprised of triple-layer graphene, indicating successful synthesis of few-layer graphene of uniform thickness using their method.

6.4.3 Transmission Electron Microscopy (TEM)

A typical TEM image of single layer graphene shows the transparency of the film [19]. The atomic structure of graphene sheets can be visualized by TEM by advanced imaging at atomic resolution [68,69]. Positions of individual carbon atoms can be distinguished, and defects in the graphene lattice can be identified. Figure 6.11 [69] reveals the honeycomb lattice structure of single-layer graphene under high-resolution TEM.

An important fact that was revealed by TEM studies of freely suspended graphene sheets was that they are not completely flat and the membranes exhibit random microscopic

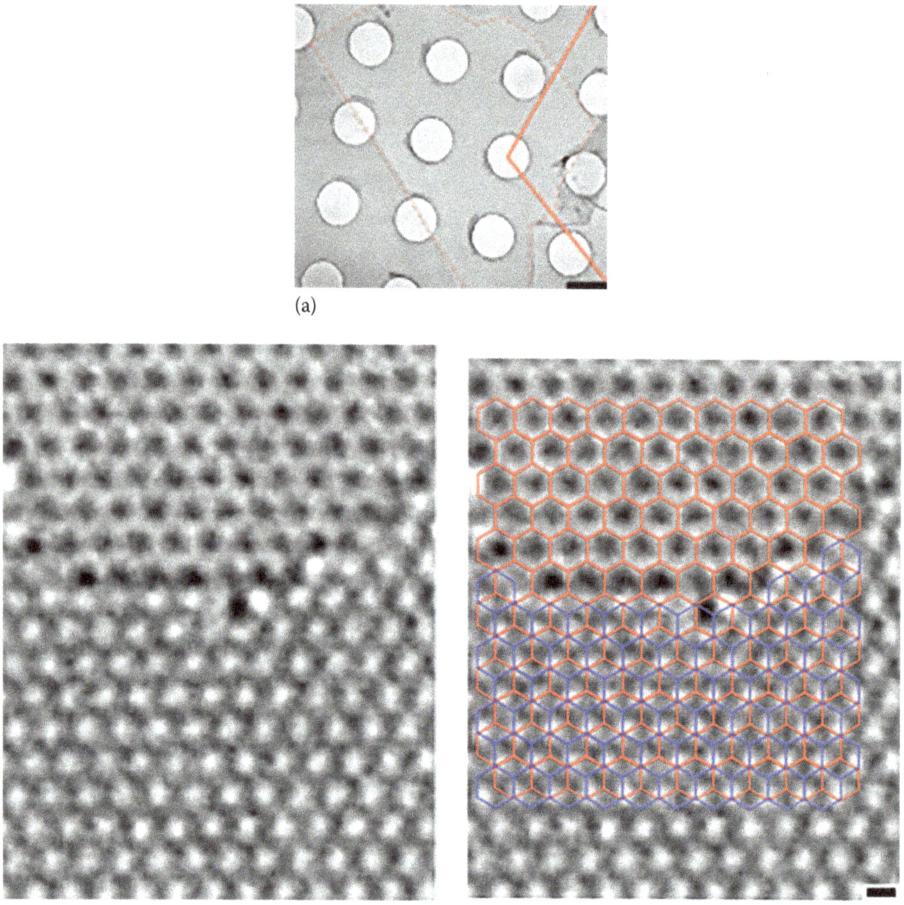

(a) (b) (c)

FIGURE 6.11
TEM images of graphene. (a) Low-magnification TEM image of graphene sheets on the perforated carbon film. A single-layer region is outlined by a dashed line. (b,c) High-resolution images showing atomic structure of graphene step from a monolayer (upper part) to a bilayer (lower part of the image), showing the unique appearance of the monolayer. Panel (c) shows the same image as (b) with an overlay of the graphene lattice (red) and the second layer (blue). (Reprinted with permission from J.C. Meyer et al., *Nano Letters* 8, no. 11 (2008): 3582–3586. Copyright 2008 American Chemical Society.)

curvature or ripples, which are strongest in single-layer membranes and this crumpling in third dimension may possibly lead to intrinsic stability of two-dimensional crystals [5,70–72]. This type of disorder, that is, microscopic corrugations of a graphene sheet, is unavoidable because strictly two-dimensional crystals are extremely soft and flexible and, in fact, on the verge of structural instability [71]. The intrinsic ripples in graphene spontaneously appear owing to thermal fluctuations [72].

6.4.4 Raman Spectroscopy

Raman spectroscopy is a powerful tool to investigate graphene, both single- and few-layer graphene flakes, and to characterize the properties of graphene, both exfoliated and synthesized, and graphene-based related materials, such as graphene oxide [56,73–75]. It is a useful, nondestructive method to obtain information on defects or disorder in the structure and the thickness and, hence, the number of graphene layers [56,73–75]. Raman spectroscopy can clearly distinguish a single layer, from a bilayer from a few (less than five) layers [75]. Ferrari et al. [75] compared Raman spectra of pristine graphene with that of graphite and multilayered graphene nanosheets and showed that for few-layer graphene consisting of one to five layers, the information on the precise number of layers can be extracted from the spectrum (Figure 6.12).

The Raman spectrum of a pristine single graphene layer has two characteristic features at 1580 cm^{-1} (G band) and 2700 cm^{-1} (G′ band). With an increasing number of layers, the G band moves toward the lower wavenumber. When there are sufficient defect sites in graphene, then a band appears at 1350 cm^{-1} called the D band. The presence of D band indicates

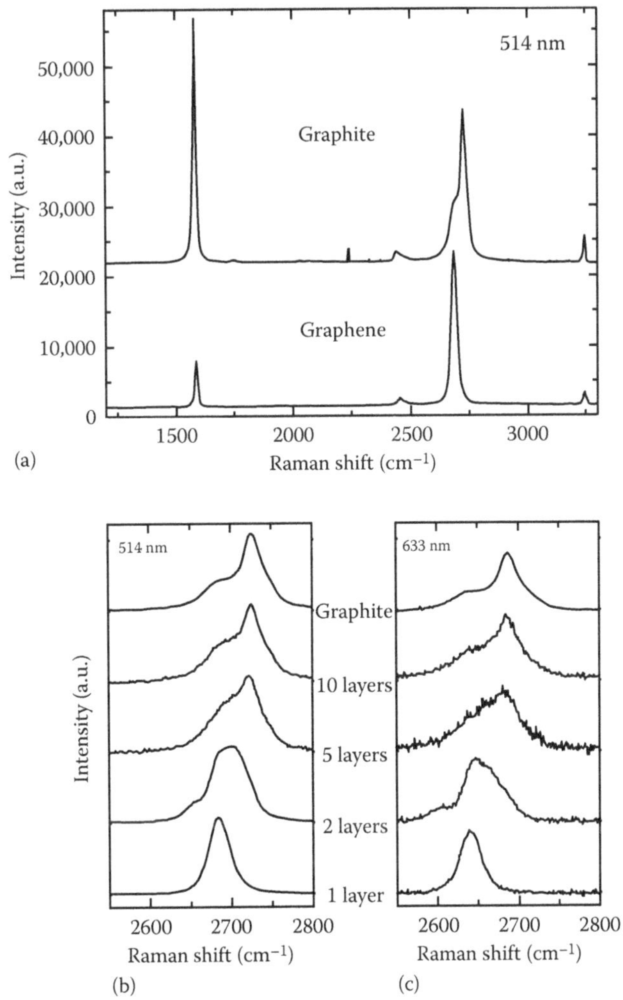

FIGURE 6.12
Comparison of Raman spectra at 514 nm for bulk graphite and graphene. (a) They are scaled to have similar height of the 2D peak at ~2700 cm^{-1}. (b) Evolution of the spectra at 514 nm with the number of layers. (c) Evolution of the Raman spectra at 633 nm with the number of layers. (Reprinted with permission from A.C. Ferrari, J.C. Meyer, V. Scardaci, C. Casiraghi, M. Lazzeri, F. Mauri et al. *Physical Review Letters* 97, no. 18: 187401. Copyright 2006 by the American Physical Society.)

sp^3 carbon atoms on graphene surface and therefore is indicative of the quality of graphene nanosheet. Pristine single-layer graphene obtained by the Scotch tape method is defect-free and does not have a D band. Significant structural changes that occur during chemical processing of pristine graphite to graphene oxide and then to reduced graphene oxide are reflected in their Raman spectra as shown in Figure 6.13 [56]. The pristine graphite (Figure 6.13a) shows only a prominent G peak at 1581 cm^{-1}. Graphene oxide (Figure 6.13b) shows a broader G band and a prominent D band which is indicative of the reduction in size of the in-plane sp^2 domains due to oxidation and a change in hybridization of oxidizing carbon atoms to sp^3 [19,56].

The presence of both D and G bands in reduced graphene oxide spectrum (Figure 6.13c) but with an increased D/G intensity ratio compared to graphene oxide (Figure 6.13b) is due to creation of a large number of smaller-sized new graphitic domains [56]. Raman spectroscopy can quantify the transformation of sp^3-hybridized carbons back to sp^2 on reduction of graphene oxide [28].

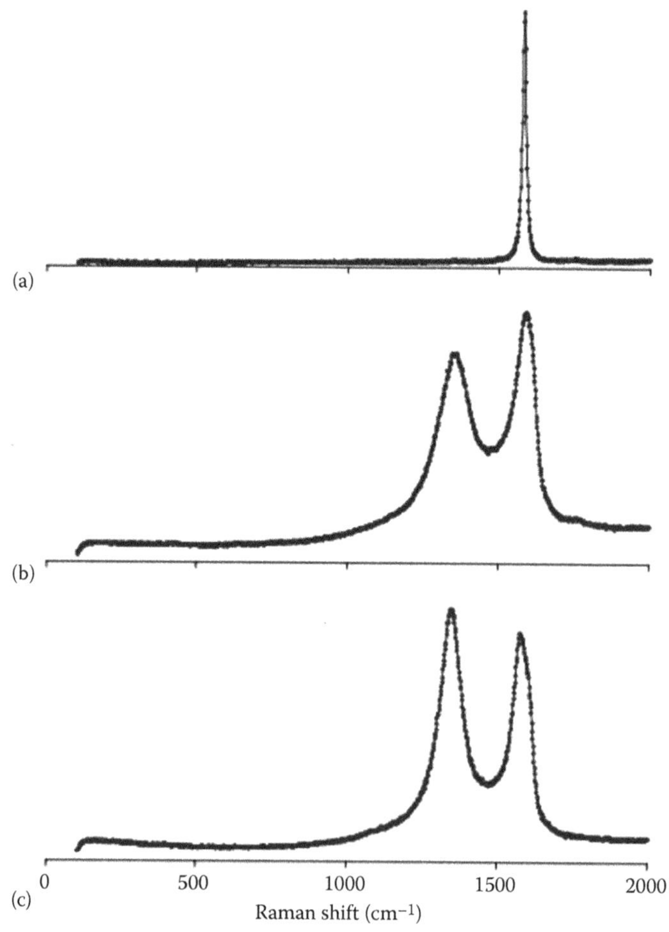

FIGURE 6.13
Raman Spectra of graphite (a), graphene oxide (b), and reduced graphene oxide (c). (Reprinted from *Carbon* 45, no. 7, S. Stankovich, D.A. Dikin, R.D. Piner, K.A. Kohlhaas, A. Kleinhammes, Y. Jia, Y. Wu, S.T. Nguyen, and R.S. Ruoff, 1558–1565. Copyright 2007, with permission from Elsevier.)

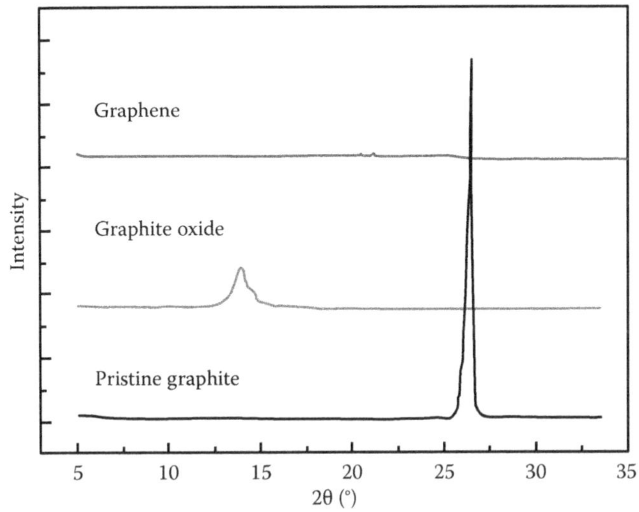

FIGURE 6.14
XRD patterns of pristine graphite, graphite oxide, and graphene. (Reprinted from *Polymer* 51, no. 5, H.B. Zhang, W.G. Zheng, Q. Yan, Y. Yang, J.W. Wang, Z.H. Lu, G.Y. Ji, and Z.Z. Yu, 1191–1196. Copyright 2010, with permission from Elsevier.)

6.4.5 X-Ray Diffraction (XRD)

The XRD is a useful characterization technique since chemical processing of pristine graphite leads to structural changes in the intermediate product graphite oxide and the final product graphene (Figure 6.14) [76].

There are changes in the basal plane reflection peak at 2θ and corresponding d spacing, that is, interlayer distance before and after the oxidation of graphite and before exfoliation of graphite oxide [19]. The basal plane reflection peak at $2\theta = 26.6°$ for graphite, which corresponds to a d spacing of 0.335 nm. The graphite oxide intermediate before exfoliation exhibits a basal reflection (002) peak at 11.2° corresponding to a d spacing of 0.79 nm. The increase in interlayer space in graphite oxide is due to the intercalation of water molecules between the oxidized graphene layers. Graphene oxide sheets are expected to be "thicker" due to the presence of covalently bound oxygen and the displacement of the sp^3-hybridized carbon atoms slightly above and below the original graphene plane [56]. From XRD experiments, the intersheet distance for graphite oxide varies with the amount of absorbed water, with values such as 0.63 nm and 0.61 nm reported for "dry" graphite oxide samples (complete drying of graphite oxide is probably impossible) to 1.2 nm for hydrated graphite oxide [56]. Complete exfoliation of graphite oxide results in disappearance of the diffraction peak at $2\theta = 26.6°$ [19].

6.5 Structure of Graphene, Graphite Oxide, and Graphene Oxide

6.5.1 Structure of Graphene

Graphene is a single layer of sp^2-hybridized carbon atoms hexagonally arranged and tightly packed in a two-dimensional honeycomb lattice structure where each sp^2-hybridized carbon

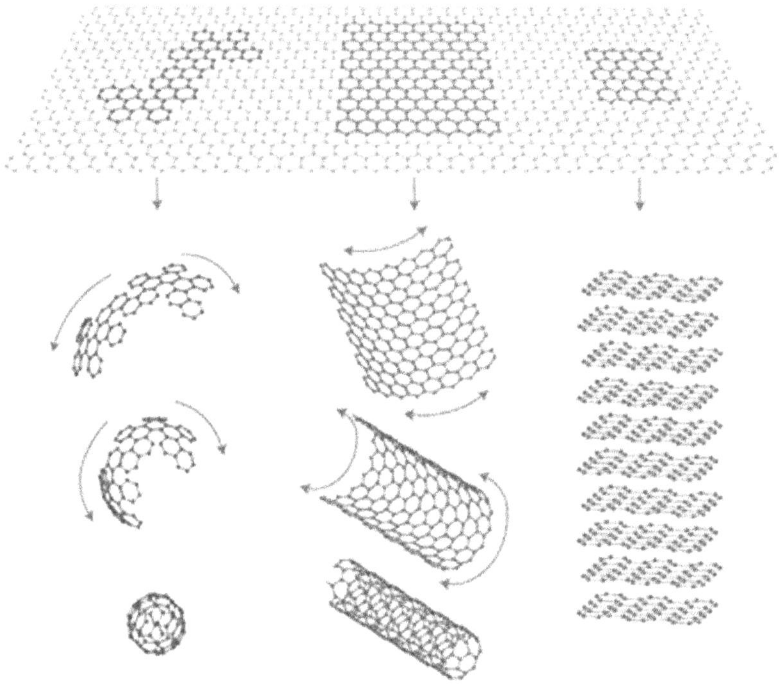

FIGURE 6.15
Graphene is a 2D building material for carbon materials of all other dimensionalities. It can be wrapped up into 0D buckyballs, rolled into 1D nanotubes, or stacked into 3D graphite. (Reprinted by permission from Macmillan Publishers Ltd., *Nature Materials*, A.K. Geim and K.S. Novoselov, 6, no. 3: 183–191, copyright 2007.)

atom is strongly bonded to three other neighboring carbon atoms in the layer by σ bonds [15,40,77]. These strong covalent bonds in the planar honeycomb lattice give strength and rigidity to graphene [13]. The fourth electron in the remaining unhybridized p_z orbital of each carbon atom is involved in forming a delocalized π-bond system with the adjacent carbon atoms, which is responsible for the unique electronic properties and conductivity of graphene [15,24].

Graphene is the basic building block for graphite materials of other dimensions such as graphite, fullerenes, and carbon nanotubes (Figure 6.15) [5]. Graphene can be wrapped into zero-dimensional fullerene. It can be rolled into one-dimensional carbon nanotube. It can be stacked into three-dimensional graphite where interlayer spacing between each graphene layer is about 0.335 nm [1,19,27,56].

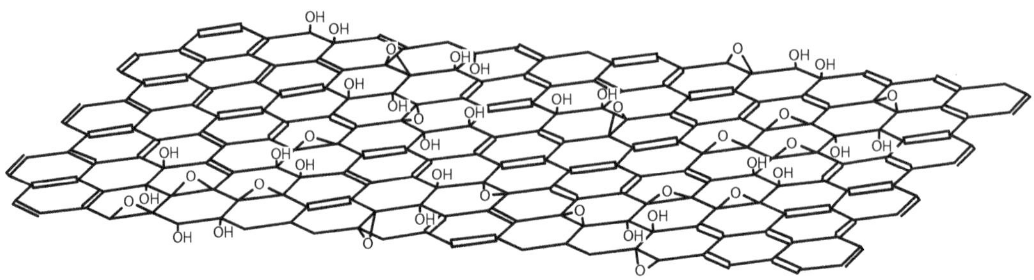

FIGURE 6.16
Structure of graphite oxide. (Reprinted from *Chemical Physics Letters* 287, no. 1, H. He, J. Klinowski, M. Forster, and A. Lerf: 53–56. Copyright 1998, with permission from Elsevier.)

6.5.2 Structure of Graphite Oxide and Graphene Oxide

Graphite oxide is a layered structure like graphite but with structural modifications brought about by oxidation of graphite. The sp^2-bonded carbon network of graphite is disrupted in graphite oxide due to change in hybridization of oxidizing carbon atoms of the graphene planes from sp^2 to sp^3 leading to a myriad of oxygen functionalities on the surfaces and edges on the graphene layers of graphite [19,25,26,56]. Thus graphite oxide consists of oxidized graphene sheets (or graphene oxide sheets) with their basal plane containing predominantly hydroxyl and epoxy groups, while carboxyl or carbonyl groups are located presumably at the edges of the layers in minority [26]. The Lerf-Klinowski model (Figure 6.16) [78] is believed to be the most likely description of graphite oxide structure [19,28,78]. The model describes graphite oxide as built of pristine aromatic "islands" separated from each other by aliphatic regions containing epoxide and hydroxyl groups and double bonds. The out of planar C–O covalent bonds increase the distance between the graphene layers from 0.335 nm in graphite to about 0.68 nm in graphite oxide [19].

The stacked layers of "graphene oxide" sheets in graphite oxide are strongly hydrophilic such that intercalation of water molecules between the layers readily occurs. The interlayer distance between the graphene oxide sheets increases reversibly from about 0.6 nm to about 1.2 nm with increasing relative humidity [26,56]. Graphite oxide can be thought of as a graphite-type intercalation compound with both covalently bound oxygen and noncovalently bound water between the carbon layers [56].

Graphite oxide can be completely exfoliated to produce aqueous colloidal suspensions of graphene oxide sheets by simple sonication or by stirring the water–graphite oxide mixture for a long enough time [26]. Chemically, graphene oxide is similar, if not identical, to graphite oxide, but structurally it is very different. Rather than retaining a stacked structure, the material is exfoliated into monolayers or few-layered stacks [25].

6.6 Properties of Graphene and Graphene Oxide

6.6.1 Electronic Properties of Graphene

Wallace, in 1947, used the two-dimensional honeycomb structure of graphene to obtain information on the properties of graphite and developed the band theory of graphite [6]. In his analyses he assumed that since the spacing of lattice planes of graphite is large (about 0.337 nm) compared with the hexagonal spacing in the layer (0.142 nm), as a first approximation, the interactions between the planes may be neglected and may be supposed that conduction takes place only in the layers [6,40] and this assumption makes subsequent analyses applicable to graphene [40].

Discovery of graphene made it possible to experimentally explore the nature of graphene's conductivity and verify the exotic electrical properties initially predicted—in particular, that its mobile electrons behave as if they were massless, relativistic fermions [13]. Each carbon atom in graphene's honeycomb lattice forms strong covalent bonds with its neighbors, with one unbound electron on each carbon atom left over to interact and move around the carbon atoms across the two-dimensional crystal lattice [5,13]. The electronic π-band structure of graphene derived from its crystal structure governs the behavior of its electrons and describes the energy dependence of that electronic motion, that is, how a particle's energy changes with its momentum along valence and conduction bands. The band structure picture of graphene can be considered as the starting point to understand the amazing electronic properties of graphene.

6.6.1.1 Band Structure of Graphene

The hexagonal crystal structure of graphene consists of two interpenetrating triangular sublattices A and B [5,10,15,77]. Figure 6.17 [15] shows the honeycomb lattice structure of graphene and its lattice vectors, reciprocal lattice vectors, and the first Brillouin zone. The two different but equivalent carbon sublattices in the crystal structure of graphene give graphene its unique electronic band structure and the unusual behavior of charge carriers in graphene [11].

To understand the behavior of electrons in a crystal, it is important to consider the electronic dispersion relation called the E-k relation, or the energy–momentum dispersion relation [13]. Considering only the nearest neighbors in the honeycomb lattice of graphene and using a simple tight-binding approach for electrons in graphene, the energy band structure of graphene is obtained [5,6,15,77]. Figure 6.18 [15] shows the electronic dispersion in graphene or the electronic band structure of graphene.

The upper conduction band (π^* band) and the lower valence band (π band) meet or touch, with a perfect symmetry between the upper band and the lower band, at discrete points in graphene, (at the K or K′ point in the first Brillouin zone) and as the bands approach each other the dispersion of those bands is linear [10,13,15,77]. The energy bands in the vicinity of the K or K′ point in the first Brillouin zone show a linear dispersion relation of energy versus momentum, that is, $E = \hbar v_F k$, where v_F represents the Fermi velocity and is equal to 10^6 m/s [2,12,16]. The linear dispersion curve implies that the electron's effective mass, a parameter that accounts for the interaction of electron with the lattice, vanishes, that is, becomes zero throughout a large range of momentum values in the crystal lattice, and hence the velocity of the electrons confined on graphene remains constant, a transport property similar to the massless particles like photons [13]. The electron behaves more like a photon than a conventional massive particle whose energy–momentum dispersion is parabolic and given by $E = \hbar^2k^2/2m^*$, where m^* is the effective mass of electron [13]. The linear dispersion relation for graphene close to the K or K′ points is similar to the two-dimensional Dirac equation for relativistic massless particles (photons) except that for graphene the Fermi velocity of electrons or holes replaces the speed of light, which appears in the relativistic Dirac equation [2,11]. The K points at the corners of the graphene Brillouin zone where the empty conduction band and the filled valence band meet

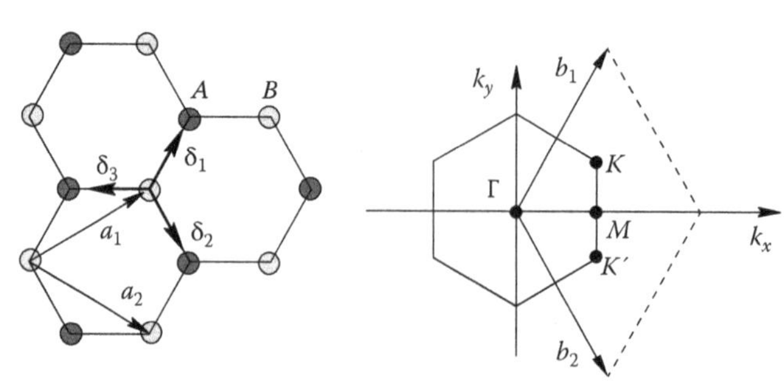

FIGURE 6.17
(a) Lattice structure of graphene made out of two interpenetrating triangular lattices (a_1 and a_2 are the lattice unit vectors and δ_i, i = 1, 2, 3 are the nearest neighbor vectors). (b) Corresponding Brillouin zone. The Dirac cones are located at the K and K′ points. (Reprinted with permission from A.C. Neto, F. Guinea, N.M.R. Peres, K.S. Novoselov, and A.K. Geim, *Reviews of Modern Physics* 81, no. 1: 109. Copyright 2009 by the American Physical Society.)

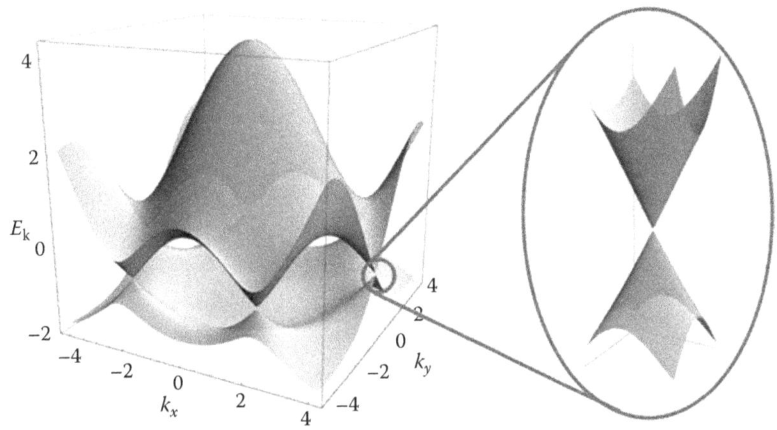

FIGURE 6.18
The electronic band structure of graphene where graphene's valence and conduction bands are represented by two Dirac cones with touching points that cross linearly at the Dirac point. (Reprinted with permission from A.C. Neto, F. Guinea, N.M.R. Peres, K.S. Novoselov, and A.K. Geim, *Reviews of Modern Physics* 81, no. 1: 109. Copyright 2009 by the American Physical Society.)

are called Dirac points [29,77]. There are six Dirac points (two sets of unequal Dirac points K and K′) located at the six corners of the graphene Brillouin zone. The conduction and valence bands are represented by two Dirac cones with touching points that cross linearly at the neutrality point or Dirac point where E(k) = 0 = E_F (Fermi level energy) [10,11,15,77].

6.6.1.2 Electronic Properties of Graphene

Graphene possesses a number of amazing electronic properties as a consequence of its 2D honeycomb crystal lattice [12,14]. Some of the unique electronic properties of graphene are highlighted next.

Measurement of the intrinsic transport properties of graphene have shown that graphene possesses high charge carrier mobilities [1,5,11,14,79]. In their initial studies on few-layer graphene, Novoselov et al. [1] obtained charge mobility values between 3000 and 10,000 $cm^2\ V^{-1}\ s^{-1}$. The mobilities were practically independent of absolute temperature, T, indicating that they were still limited by scattering on defects. Mobilities in excess of 200,000 $cm^2\ V^{-1}\ s^{-1}$ at high electron densities of ~$2 \times 10^{11}\ cm^{-2}$ have been measured for a suspended single-layer mechanically exfoliated graphene [79]. A high quality of 2D crystal of graphene implies an unusually low density of defects, which typically serve as the scattering centers that inhibit charge transport [14]. High carrier mobilities show that charge transport is essentially ballistic on the micrometer-scale at room temperature [5,14].

Another important aspect of charge transport in graphene is the pronounced ambipolar electric field effect shown by graphene [5,14,18]. The charge carriers can be tuned continuously between holes and electrons by using an electric field (Figure 6.19) [5]. This can be done by attaching a gate voltage to the back of the SiO_2 substrate on which graphene sits [13]. This turns the crystals into metallic field-effect transistors (FETs) and the position of the Fermi level, the highest energy level occupied by electrons, can be tuned [13], for example, increasing the gate voltage injects additional electrons into the crystal and thereby increases its carrier concentration and Fermi level shifts to higher energy level. This is possible due to the unique band structure of graphene. The Fermi level coincides with the

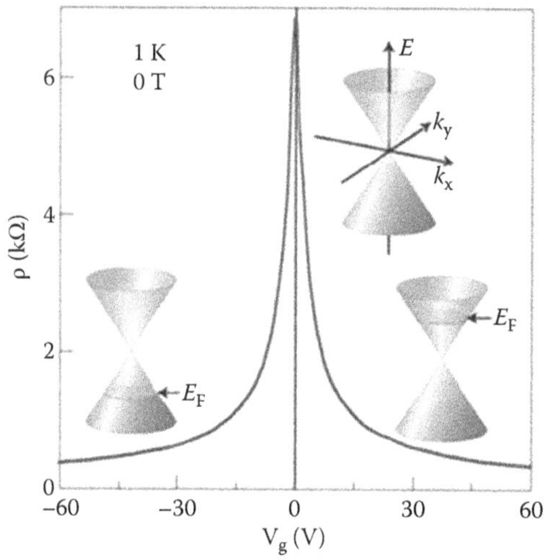

FIGURE 6.19
Electric field effect in graphene. For negative/positive gate voltage (V_g) the charge carriers are holes/electrons. (Reprinted by permission from Macmillan Publishers Ltd., *Nature Materials*, A.K. Geim and K.S. Novoselov, 6, no. 3: 183–191, copyright 2007.)

Dirac point in graphene when there is no electric field, as shown in Figure 6.19 [5]. When a negative gate voltage is applied, the Fermi level drops below the Dirac point, introducing a significant population of holes into the valence band [14]. When a positive gate voltage is applied, the Fermi level rises above the Dirac point, promoting a significant population of electrons into the conduction band. This property is made use of in graphene-based electronic devices such as field-effect transistors and sensors [12,18].

The conductivity of graphene never falls below a certain minimum value even though there is a vanishing density of electrons and holes close to the neutrality point [4,10]. The minimum conductivity of graphene, even as Fermi energy approaches zero, is a result of the unusual Dirac fermion nature of its charge carriers, which are difficult to localize, and therefore even the very last electron or hole provides a minimum conductivity [10]. The inability to localize the massless Dirac fermions enables the relativistic electron to slip through potential barriers that would otherwise trap a nonrelativistic electron [10]. This perfect tunneling through potential energy barriers is the exhibition of the Klein paradox in graphene [5,10,11,40]. The Dirac fermion nature of its charge carriers also endows graphene with another unusual property in the presence of magnetic field, namely, the anomalous quantum Hall effect (QHE), which has been discussed by several authors [4,5,11,13,77]. Graphene is the only material which shows QHE at room temperature.

6.6.2 Mechanical Properties

Mechanical properties, such as intrinsic strength, and elastic properties of free-standing suspended monolayer graphene and suspended stacks of graphene sheets of thickness 2 to 8 nm have been measured by tip-induced deformation (nanoindentation) using AFM [34,39,80]. Graphene possesses outstanding mechanical properties. It is incredibly strong (tensile strength of ~130 GPa, for a defect-free single layer, Young's modulus of 1 TPa), while remaining highly resilient and flexible and very light (0.77 mg m^{-2}) [24,34,80,81], and possesses a very high specific surface area (~2630 m^2 g^{-1}) [24,27,28,30].

Graphene is harder than diamond and about 300 times harder than steel. It is stretchable up to 20% of its initial length [27]. Graphene has been established as the strongest material ever measured, and it has been shown through these measurements of free-standing graphene monolayer membranes that atomically perfect nanoscale materials can be mechanically tested to deformations well beyond the linear regime [34]. Outstanding mechanical properties offer opportunities for structural and mechanical applications of graphene.

The mechanical properties of graphene oxide are expected to be inferior to those of graphene due to disruption of the sp^2-bonded network structure of carbon atoms and change in hybridization of oxidizing carbon atoms of the graphene planes from sp^2 to sp^3 [27]. Reduction of graphene oxide is used to restore the graphene structure in the chemically reduced graphene oxide [25]. The elastic modulus of freely suspended graphene monolayers, obtained via chemical reduction of graphene oxide, was determined through tip-induced deformation experiments [80]. Despite their defect content, the single sheets exhibited an extraordinary stiffness (Young's modulus of 0.25 TPa), approaching that of pristine graphene. The sheets showed high flexibility, which enabled them to bend easily in their elastic regime. The reduced graphene oxide sheets comprised of more than three layers showed a markedly different behavior and displayed 1 order of magnitude lower values of the elastic modulus, as compared to those of the single and double layers [80].

6.6.3 Optical Properties

A single layer of graphene, despite being only one atom thick, is found to absorb a significant (~2.3%) fraction of incident white light [12], a consequence of graphene's unique electronic structure [82]. When pristine graphene sheets are dispersed in an organic solvent, the liquid shows a gray color, which becomes darker as the amount of graphene increases [19]. Dispersions of graphene oxide nanosheets have a brownish tint and the color becomes darker and grayer as graphene oxide is reduced to reduced graphene oxide. Graphene oxide is an insulator and has a much higher transmittance in comparison to pristine graphene or reduced graphene oxide [19].

6.6.4 Thermal Conductivity

Graphene is an ideal thermal conductor showing ballistic and isotropic thermal conductance. Graphene displays high thermal conductivity with a value of ~5000 Wmk^{-1} [24,28,83] measured for a suspended single-layer graphene sheet at room temperature, a value higher than that of graphite, carbon nanotubes, and diamond. The extremely high value of thermal conductivity suggests that graphene can outperform carbon nanotubes in heat conduction. The excellent thermal property of graphene is beneficial for electronic applications and efficient thermal management [83].

6.6.5 Chemical Reactivity of Graphene Oxide

Graphene oxide is a very versatile material. It is not only an important material for preparation of graphene, but the presence of different oxygen functionalities on its surface make graphene oxide an important precursor for a variety of chemically modified graphene-based materials. The chemical conversion of graphene oxide offers a promising route for processing of graphene-based materials for different applications.

Chemical conversion of graphene oxide to graphene via reduction methods is one of the most desirable routes to large quantities of graphene-like materials [25]. The reduction process is among the most important reactions of graphene oxide because of the similarities between reduced graphene oxide and pristine graphene, in terms of their structural and electrical properties. Unlike graphene, graphene oxide is an electrically insulating, hydrophilic material due to the disruption of its sp^2 bonding network and the presence of polar hydroxyl, epoxy, carboxylic acid, and carbonyl groups on its surface. Electrical conductivity of the structure can be recovered by restoring the π-network of graphene by reduction of graphene oxide by removal of oxygen groups. The transformation of sp^3 carbons to sp^2 by reduction of graphene oxide restores electrical conductivity; thus, conductivity is a valuable qualitative measure of the conversion of graphene oxide to graphene [28]. Strong alkaline agents, such as hydrazine hydrate [56] and sodium borohydride [57], are effective reducing agents, as the electrical conductivity of reduced graphene oxide is significantly increased after reduction. While graphene oxide is insulating, reduced graphene oxide films are conducting. The electrical conductivity of reduced graphene oxide (~2×10^2 S/m) increased by about 5 orders of magnitude in comparison to graphene oxide and closely approached that of pristine graphite by reduction with hydrazine [56]. By using an acidic reducing agent, such as hydroiodic acid, electrical conductivity increases to about 298 S/cm [58]. The reduction with hydroiodic acid maintains good integrity and flexibility, and even improves the strength and ductility of the original graphene oxide films.

The first electronic transport studies of chemically reduced graphene oxide sheets, which potentially provide access to large-scale production of graphene monolayers, were performed by Gomez-Navarro et al. [80]. Individual graphene oxide sheets subjected to chemical reduction were electrically characterized as a function of temperature and external electric fields. Before reduction the graphene oxide monolayers were insulating (conductivity about 10^{-3} S/cm). The fully reduced monolayers exhibited conductivities ranging between 0.05 and 2 S/cm and field-effect mobilities of 2–200 cm^2/Vs at room temperature. Thus the conductivity of the reduced graphene oxide monolayers was found to be approximately 3 orders of magnitude higher compared to the starting graphene oxide. However, for fully reduced graphene oxide, the room-temperature conductivity and carrier mobility were found to lag behind those of graphene by 3 and 2 orders of magnitude, respectively, predominantly due to defects present in the structure compared to pristine graphene [80].

Reduction process has a strong effect on solubility and stability [54]. A comparison of the dispersion behavior of graphene oxide and reduced graphene oxide in different polar and nonpolar organic solvents has shown that while graphene oxide has better solubility and stability (long-term stability) in polar solvents, the reduced graphene oxide has better interaction with nonpolar solvents like chloroform, toluene, and chlorobenzene. Similar to graphene oxide, reduced graphene oxide gives very good dispersions in NMP, water, and ethylene glycol, which implies that oxygen-containing functional groups are still present at defect sites in reduced graphene oxide. These studies are important for processing of these materials for different applications when stable aqueous or colloidal dispersions are required.

Covalent chemical functionalization of graphene oxide through the addition of other functional groups to graphene oxide via reactions at the chemically reactive oxygen functionalities, such as carboxylic acid groups, on the edges and epoxy and hydroxyl groups in the basal plane result in chemically modified graphene. Such chemical modifications improve dispersion of these materials in polar aprotic organic solvents [25]. An important reaction of graphene oxide sheets is the isocyanate treatment of graphite oxide where organic isocyanates react with the hydroxyl and carboxyl groups of graphene oxide sheets to form

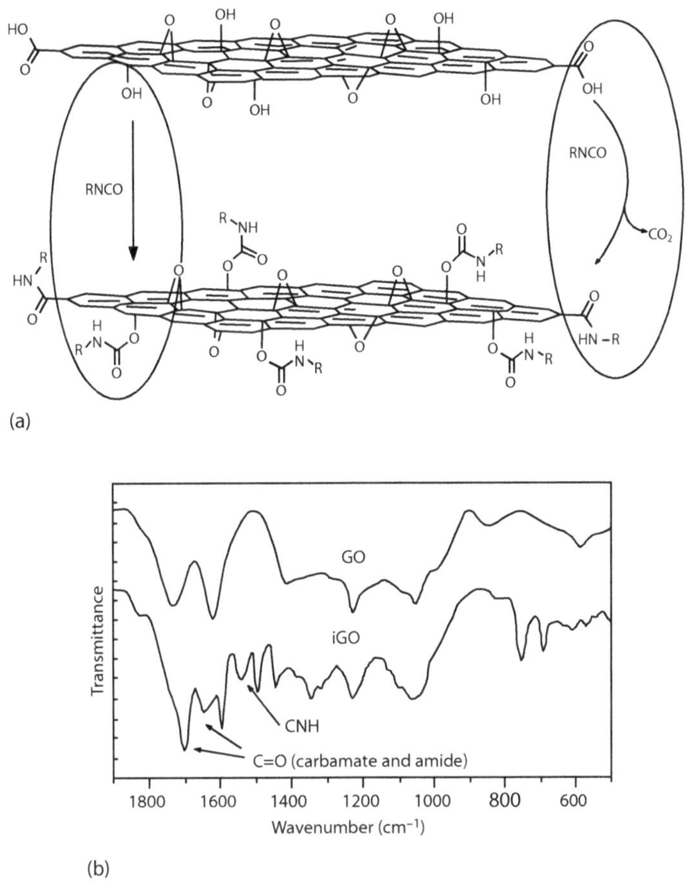

FIGURE 6.20
(a) Proposed reactions during the isocyanate treatment of graphite oxide (GO) where organic isocyanates react with the hydroxyl (left oval) and carboxyl groups (right oval) of graphene oxide sheets to form carbamate and amide functionalities, respectively. (b) FT-IR spectra of GO and phenyl isocyanate-treated GO. (Reprinted from *Carbon* 44, no. 15, S. Stankovich, R.D. Piner, S.T. Nguyen, and R.S. Ruoff, 3342–3347. Copyright 2006, with permission from Elsevier.)

carbamate and amide functionalities, respectively. Well-dispersed modified graphene oxide sheets in polar aprotic solvents can be obtained by isocyanate treatment (Figure 6.20) [84].

In another treatment, covalent derivatization of the acidic functional groups in oxidized graphite with octadecylamine rendered graphite soluble in common organic solvents. Atomic force microscopic characterization of the soluble species supports the idea that the solutions consist of single- and few-layer graphene sheets [85]. Stable dispersions of chemically converted graphene sheets were obtained through epoxy ring opening reactions by covalent functionalization of graphene oxide sheets using amine-terminated ionic liquid [86]. The resulting graphene sheets, without any assistance from polymeric or surfactant stabilizers, can be stably dispersed in water, DMF, and DMSO (dimethyl sulfoxide).

The sp^2 networks in graphene that are not oxidized or engaged in hydrogen bonding in graphene oxide can interact via noncovalent interactions (e.g., via π-π stacking or van der Waals interactions) with other organic moieties [25]. Reduced graphene oxides can interact with organic molecules through noncovalent interactions. Residual functional groups left intact in reduced graphene oxide after incomplete reduction of graphene oxide can undergo covalent functionalization reactions [25]. Park and Ruoff [26] and Dreyer et al. [25] have presented excellent and detailed reviews on the chemistry of graphene oxide and different methods of chemical functionalization to obtain chemically modified graphene.

6.7 Applications of Graphene and Graphene Oxide

6.7.1 Electronic Applications

Graphene and its derivatives offer potential electronic applications. Some of the areas in which electronic properties of graphene can be exploited are field-effect transistors, optoelectronic devices, transparent conducting films, touch screens, flexible electronic devices, solar cells, and sensors [5,6,12,14,24,29].

The ballistic transport has major implications for high-speed graphene-based electronic applications, because it enables, in principle, fabrication of high-speed room-temperature ballistic transistor [5,14,87].

Graphene shows a pronounced ambipolar electric field effect. Both the type (electron or hole) and density of charge carriers in graphene can be easily controlled by using an electric field [12,14]. This property is used in graphene-based electronic devices such as field-effect transistors and sensors [24,87]. In order to exploit the electronic properties of graphene for the majority electronics market, currently dominated by semiconductor silicon, it is desirable to open a band gap in graphene to make graphene behave like a semiconductor and not metallic [5,18]. Opening a sizeable and well-tuned band gap in graphene is a significant challenge for graphene-based electronic devices [87]. Band gap engineering is therefore an important area of research. Band gap opening in graphene or doping is classified into three categories, namely, heteroatom doping [88], chemical modification, and electrostatic field tuning [87]. Whereas heteroatom doping and chemical modification methods can be used to open the band gap and tune the Fermi level of graphene, in the electrostatic field tuning method, the polarity and value of the gate voltage can change the Fermi level of graphene, but the band gap cannot be opened. The boron and nitrogen atoms are the natural candidates for doping in graphene because of their similar atomic size as that of carbon and of their hole acceptor and electron donor characters for substitutional boron- and nitrogen-doping, respectively [87]. Graphene can obtain fine control of electronic properties by the chemical modification without significant degradation in charge carrier mobility. The chemically modified p- or n-typed graphene can be used for high sensitivity single molecule detection. Guo et al. [87] have reviewed the work on graphene doping.

Transparent conducting films and flexible electronics via reduction of graphene oxide is emerging as an important application area for graphene-based electronics. Transparent conducting films are used for optoelectronic devices including displays, LEDs, and solar cells [14,50]. While the current industry standard is indium tin oxide (ITO), carbon nanotubes as well as graphene are a potential alternative [14]. Several studies have been done on obtaining flexible transparent conducting films [23,50,61]. Graphene-based transparent conductive electrodes for solar cells as an alternative to metal oxides were demonstrated by Wang et al. [61] by fabricating graphene films from exfoliated graphite oxide followed by thermal reduction. The films exhibited a high electrical conductivity of 550 S/cm and a transparency of more than 70% over 1000–3000 nm wavelength. Pei et al. [58] have obtained flexible and transparent conducting films by chemical reduction of graphene oxide using HI. The films showed good mechanical and electrical properties and up to 85% transparency at 550 nm wavelength. Semitransparent conducting film was produced by solution processing of graphene by liquid-phase exfoliation in NMP [50]. These films had an electrical conductivity of ~6500 S/m and optical transparencies of ~42%.

The use of graphene nanoribbons in FET is discussed in detail in the Chapter 7 on graphene nanoribbons.

6.7.2 Polymer Composites

The exceptional combination of mechanical, electrical, thermal, and optical properties of graphene has resulted in extensive research on the use of graphene and its derivatives as fillers in polymers to develop graphene-reinforced polymer composites [27,28,30,53,89]. Among the carbon nanomaterials, the most promising materials for use as polymer reinforcement fillers, over the traditional carbon fillers such as carbon black, are carbon nanotubes (CNTs) and graphene [48,89–92]. One aspect of the research takes advantage of their high conductivity, and the other aspect and challenging area involves exploitation of their extraordinary mechanical properties. A unique combination of high flexibility and strength combined with high stiffness and low density offers potential to develop high-strength, lightweight, and high-performance polymer composite materials. As with CNTs, a homogeneous dispersion within the polymer matrix and a strong interfacial adhesion between graphene and the polymer matrix are the main challenges to translate the unique nanoscale properties of graphene into the macroscale properties of polymer composites [27,30,89].

Methods starting from chemically modified graphene precursors, particularly graphite oxide followed by chemical or thermal reduction are better suited for large-scale production of reinforcements for the polymer composites [28,30]. Hence, a majority of graphene/polymer composites investigated are fabricated using graphite oxide, chemically reduced graphene oxide, or thermally reduced graphene oxide as fillers, although they have inferior physical properties than pristine graphene [89]. Graphite oxide contains a range of reactive oxygen functional groups, which renders it a good candidate for use in polymer composite applications through chemical functionalization, which can provide better interaction, either covalent or noncovalent, with the polymer matrices [25]. Unlike CNTs, graphite oxide and reduced graphene oxide are flat or slightly wrinkled sheets so entangled bundles are not an issue. However, restacking of graphene, especially after chemical reduction compared to thermal reduction, can significantly reduce their effectiveness as fillers for composites, and prevention of the restacking of graphene becomes crucial in the fabrication of graphene–polymer composites [28,89]. Du et al. [89] have also compared the properties of CNT–polymer with graphene–polymer composites.

Graphene nanocomposites have been created with different polymer matrices such as polymethylmethacrylate (PMMA) [93], polystyrene (PS) [53], epoxy [30], polyurethane (PU) [94], and polyethylene terephthalate (PET) [76] using different processing methods (solution processing, melt processing, *in situ* polymerization) depending upon the reinforcing filler [27]. The preparation and properties such as electrical percolation threshold, Young's modulus, tensile strength, and ultimate strain of graphene nanocomposites have been reviewed by many authors [27,28,89]. Some of the results of those studies are highlighted next.

Electrically conductive graphene–polymer nanocomposites were prepared by solution phase mixing of the exfoliated phenyl isocyanate-treated graphite oxide sheets with polystyrene, followed by their chemical reduction [53]. The graphene sheets were well dispersed throughout the polymer matrix. The composite exhibited a percolation threshold of ~0.1 vol% for room-temperature electrical conductivity, the lowest reported value for any carbon-based composite except for those involving carbon nanotubes; at only 1 vol%, this composite has a conductivity of ~0.1 S m^{-1}, sufficient for many electrical applications. Polyurethane-based composites have been prepared by solvent casting techniques using different weight percents (0–5 wt%) of reduced graphene oxide as reinforcement [94]. A nanoindentation study carried out on these composite sheets showed dramatic improvement in nanomechanical properties. A maximum nanoindentation hardness of 140 MPa for 5.0 wt% reduced graphene oxide loading was observed as compared to 58.5 MPa for pure PU (an overall improvement of 139%). The nanoindentation elastic modulus for the 5.0 wt% reduced graphene oxide loaded sample was 881.7 MPa as compared to 385.7 MPa for pure PU (an overall improvement of 129%). Due to their excellent nanomechanical properties, these composites find potential use in structural applications such as the automobile and windmill blade industries. These composites can also be used in hard and scratchless coatings on automotive vehicles. In another study, graphene nanosheets prepared by complete oxidation of pristine graphite followed by thermal exfoliation and reduction were incorporated in a PET matrix by melt compounding [76]. Incorporation of graphene greatly improved the electrical conductivity of PET, resulting in a sharp transition from electrical insulator to semiconductor with a low percolation threshold of 0.47 vol%. The nanocomposites showed a high electrical conductivity of 2.11 S/m with only 3.0 vol% of graphene. The TEM studies revealed a uniform dispersion of graphene nanosheets in PET matrix. The low percolation threshold and superior electrical conductivity were attributed to high aspect ratio, large specific surface area, and uniform dispersion of the graphene nanosheets in PET matrix.

The studies reported so far indicate that the graphene–polymer composites are promising multifunctional materials with significantly improved tensile strength, elastic modulus, and thermal and electrical conductivity [89]. The experimental reinforcement level is less than the expected theoretical reinforcement level except at very low volume fractions [27,89]. Despite challenges of uniform dispersion and strong interfacial interaction, to improve load transfer between graphene and polymer matrix, the 2D graphene has provided an alternative for the production of lightweight, low-cost, and high-performance composite materials for a range of applications in future [89].

References

1. K.S. Novoselov, A.K. Geim, S.V. Morozov, D. Jiang, Y. Zhang, S.A. Dubonos, I.V. Grigorieva, and A.A. Firsov. Electric field effect in atomically thin carbon films. *Science* 306, no. 5696 (2004): 666–669.
2. K.S.A. Novoselov, A.K. Geim, S. Morozov, D. Jiang, M. Katsnelson, I. Grigorieva, S. Dubonos, and A. Firsov. Two-dimensional gas of massless Dirac fermions in graphene. *Nature* 438, no. 7065 (2005): 197–200.
3. H.P. Boehm, R. Setton, and E. Stumpp. Nomenclature and terminology of graphite intercalation compounds (IUPAC Recommendations 1994). *Pure and Applied Chemistry* 66, no. 9 (1994): 1893–1901.
4. A. Vijayaraghavan. Graphene–properties and characterization. In *Springer Handbook of Nanomaterials*, edited by R. Vajtai, 39–82. Springer-Verlag, 2013.
5. A.K. Geim and K.S. Novoselov. The rise of graphene. *Nature Materials* 6, no. 3 (2007): 183–191.

6. P.R. Wallace. The band theory of graphite. *Physical Review* 71, no. 9 (1947): 622.
7. J.C. Slonczewski and P.R. Weiss. Band structure of graphite. *Physical Review* 109, no. 2 (1958): 272.
8. G.W. Semenoff. Condensed-matter simulation of a three-dimensional anomaly. *Physical Review Letters* 53, no. 26 (1984): 2449.
9. K.S. Novoselov, D. Jiang, F. Schedin, T.J. Booth, V.V. Khotkevich, S.V. Morozov, and A.K. Geim. Two-dimensional atomic crystals. *Proceedings of the National Academy of Sciences of the United States of America* 102, no. 30 (2005): 10451–10453.
10. A.K. Geim and A.H. MacDonald. Graphene: Exploring carbon flatland. *Physics Today* 60, no. 8 (2007): 35.
11. M.I. Katsnelson. Graphene: Carbon in two dimensions. *Materials Today* 10, no. 1 (2007): 20–27.
12. K.S. Novoselov. Graphene: Materials in the flatland (Nobel Lecture). *Angewandte Chemie International Edition* 50, no. 31 (2011): 6986–7002.
13. M. Wilson. Electrons in atomically thin carbon sheets behave like massless particles. *Physics Today* 59, no. 1 (2006): 21.
14. M.J. Allen, V.C. Tung, and R.B. Kaner. Honeycomb carbon: A review of graphene. *Chemical Reviews*, 110, no. 1 (2009): 132–145.
15. A.C. Neto, F. Guinea, N.M.R. Peres, K.S. Novoselov, and A.K. Geim. The electronic properties of graphene. *Reviews of Modern Physics* 81, no. 1 (2009): 109.
16. A.K. Geim. Graphene: Status and prospects. *Science* 324, no. 5934 (2009): 1530–1534.
17. O.V. Prezhdo. Graphene—The ultimate surface material. *Surface Science* 605, no. 17 (2001): 1607–1610.
18. V. Singh, D. Joung, L. Zhai, S. Das, S.I. Khondaker, and S. Seal. Graphene based materials: Past, present and future. *Progress in Materials Science* 56, no. 8 (2011): 1178–1271.
19. K. Spyrou and P. Rudolf. An introduction to graphene. *Functionalization of Graphene* (2014): 1–20.
20. A. Ciesielski and P. Samorì. Graphene via sonication assisted liquid-phase exfoliation. *Chemical Society Reviews* 43, no. 1 (2014): 381–398.
21. A. Kumar and C.H. Lee. Synthesis and biomedical applications of graphene: Present and future trends. In *Advances in Graphene Science*. InTech, 2013.
22. C. Riedl, C. Coletti, and U. Starke. Structural and electronic properties of epitaxial graphene on SiC (0 0 0 1): A review of growth, characterization, transfer doping and hydrogen intercalation. *Journal of Physics D: Applied Physics* 43, no. 37 (2010): 374009.
23. S. Pei and H.M. Cheng. The reduction of graphene oxide. *Carbon* 50, no. 9 (2012): 3210–3228.
24. Y. Zhu, S. Murali, W. Cai, X. Li, J.W. Suk, J.R. Potts, and R.S. Ruoff. Graphene and graphene oxide: Synthesis, properties, and applications. *Advanced Materials* 22, no. 35 (2010): 3906–3924.
25. D.R. Dreyer, S. Park, C.W. Bielawski, and R.S. Ruoff. The chemistry of graphene oxide. *Chemical Society Reviews* 39, no. 1 (2010): 228–240.
26. S. Park and R.S. Ruoff. 2009. Chemical methods for the production of graphenes. *Nature Nanotechnology* 4, no. 4 (2009): 217–224.
27. R.J. Young, L. Deng, L. Gong, and I.A. Kinloch. Carbon in polymer. In *Springer Handbook of Nanomaterials*, edited by R. Vajtai, 695–728. Springer-Verlag, 2013.
28. H. Kim, A.A. Abdala, and C.W. Macosko. Graphene/polymer nanocomposites. *Macromolecules* 43, no. 16 (2010): 6515–6530.
29. R. Garg, N.K. Dutta, and N.R. Choudhury. Work function engineering of graphene. *Nanomaterials* 4, no. 2 (2014): 267–300.
30. J. Wei, T. Vo, and F. Inam. Epoxy/graphene nanocomposites–processing and properties: A review. *RSC Advances* 5, no. 90 (2015): 73510–73524.
31. P. Marques, A. Sousa, G. Gonçalves, J. Grácio, M. Singh, N. Almeida, and S. Cruz. *Functionalized Graphene Nanocomposites*. InTech Open Access Publisher, 2001.
32. M. Zhou, T. Tian, X. Li, X. Sun, J. Zhang, P. Cui, J. Tang, and L.C. Qin. Production of graphene by liquid-phase exfoliation of intercalated graphite. *International Journal of Electrochemical Science* 9 (2014): 810–820.

33. J. Li, X. Zeng, T. Ren, and E. van der Heide. The preparation of graphene oxide and its derivatives and their application in bio-tribological systems. *Lubricants* 2, no. 3 (2014): 137–161.
34. C. Lee, X. Wei, J.W. Kysar, and J. Hone. Measurement of the elastic properties and intrinsic strength of monolayer graphene. *Science* 321, no. 5887 (2008): 385–388.
35. J.C. Meyer, A.K. Geim, M.I. Katsnelson, K.S. Novoselov, T.J. Booth, and S. Roth. The structure of suspended graphene sheets. *Nature* 446, no. 7131 (2007): 60–63.
36. A. Kaniyoor and S. Ramaprabhu. A Raman spectroscopic investigation of graphite oxide derived graphene. *AIP Advances* 2, no. 3 (2012): 032183.
37. P. Blake, E.W. Hill, A.C. Neto, K.S. Novoselov, D. Jiang, R. Yang, T.J. Booth, and A.K. Geim. Making graphene visible. *Applied Physics Letters* 91, no. 6 (2007): 063124.
38. H. Petroski. 1989. *The Pencil: A History of Design and Circumstance*, 39–51, 53–66. Alfred A. Knopf, 1989.
39. I.W. Frank, D.M. Tanenbaum, A.M. van der Zande, and P.L. McEuen. Mechanical properties of suspended graphene sheets. *Journal of Vacuum Science & Technology B* 25, no. 6 (2007): 2558–2561.
40. Y. Obeng and P. Srinivasan. Graphene: Is it the future for semiconductors? An overview of the material, devices, and applications. *Interface-Electrochemical Society* 20, no. 1 (2011): 47.
41. B. Dai, L. Fu, L. Liao, N, Liu, K. Yan, Y. Chen, and Z. Liu. High-quality single-layer graphene via reparative reduction of graphene oxide. *Nano Research*, 4, no. 5 (2011): 434–439.
42. C.K. Chua and M. Pumera. Chemical reduction of graphene oxide: A synthetic chemistry viewpoint. *Chemical Society Reviews* 43, no. 1 (2014): 291–312.
43. B.C. Brodie. On the atomic weight of graphite. *Philosophical Transactions of the Royal Society of London* 149 (1859): 249–259.
44. S. William, J.R. Hummers, and R.E. Offeman. Preparation of graphitic oxide. *Journal of the American Chemical Society* 80, no. 6 (1958): 1339.
45. D.C. Marcano, D.V. Kosynkin, J.M. Berlin, A. Sinitskii, Z. Sun, A. Slesarev, L.B. Alemany, W. Lu, and J.M. Tour, 2010. Improved synthesis of graphene oxide. *ACS Nano* 4, no. 8 (2010): 4806–4814.
46. K.S. Novoselov and A.C. Neto. Two-dimensional crystals-based heterostructures: Materials with tailored properties. *Physica Scripta* 2012, no. T146 (2012): 014006.
47. J.N. Coleman. Liquid exfoliation of defect-free graphene. *Accounts of Chemical Research* 46, no. 1 (2012): 14–22.
48. J.N. Coleman, U. Khan, W.J. Blau, and Y.K. Gun'ko. Small but strong: A review of the mechanical properties of carbon nanotube–polymer composites. *Carbon* 44, no. 9 (2006): 1624–1652.
49. T. Leong, M. Ashokkumar, and S. Kentish. The fundamentals of power ultrasound—A review. *Acoustics Australia* 39, no. 2 (2011): 54–63.
50. Y. Hernandez, V. Nicolosi, M. Lotya, F.M. Blighe, Z. Sun, S. De et al. High-yield production of graphene by liquid-phase exfoliation of graphite. *Nature Nanotechnology* 3, no. 9 (2008): 563–568.
51. P. Blake, P.D. Brimicombe, R.R. Nair, T.J. Booth, D. Jiang, F. Schedin et al. Graphene-based liquid crystal device. *Nano Letters* 8, no. 6 (2008): 1704–1708.
52. W. Qian, R. Hao, Y. Hou, Y. Tian, C. Shen, H. Gao, and X. Liang. Solvothermal-assisted exfoliation process to produce graphene with high yield and high quality. *Nano Research* 2, no. 9 (2009): 706–712.
53. S. Stankovich, D.A. Dikin, G.H. Dommett, K.M. Kohlhaas, E.J. Zimney, E.A. Stach, R.D. Piner, S.T. Nguyen, and R.S. Ruoff. Graphene-based composite materials. *Nature* 442, no. 7100 (2006): 282–286.
54. D. Konios, M.M. Stylianakis, E. Stratakis, and E. Kymakis. Dispersion behaviour of graphene oxide and reduced graphene oxide. *Journal of Colloid and Interface Science* 430 (2014): 108–112.
55. J.I. Paredes, S. Villar-Rodil, A. Martinez-Alonso, and J.M.D. Tascon. Graphene oxide dispersions in organic solvents. *Langmuir* 24, no. 19 (2008): 10560–10564.
56. S. Stankovich, D.A. Dikin, R.D. Piner, K.A. Kohlhaas, A. Kleinhammes, Y. Jia, Y. Wu, S.T. Nguyen, and R.S. Ruoff. Synthesis of graphene-based nanosheets via chemical reduction of exfoliated graphite oxide. *Carbon* 45, no. 7 (2007): 1558–1565.

57. H.J. Shin, K.K. Kim, A. Benayad, S.M. Yoon, H.K. Park, I.S. Jung et al. Efficient reduction of graphite oxide by sodium borohydride and its effect on electrical conductance. *Advanced Functional Materials* 19, no. 12 (2009): 1987–1992.
58. S. Pei, J. Zhao, J. Du, W. Ren, and H.M. Cheng. Direct reduction of graphene oxide films into highly conductive and flexible graphene films by hydrohalic acids. *Carbon* 48, no. 15 (2010): 4466–4474.
59. H.C. Schniepp, J.L. Li, M.J. McAllister, H. Sai, M. Herrera-Alonso, D.H. Adamson, R.K. Prud'homme, R. Car, D.A. Saville, and I.A. Aksay. Functionalized single graphene sheets derived from splitting graphite oxide. *The Journal of Physical Chemistry B* 110, no. 17 (2006): 8535–8539.
60. B. Dai, L. Fu, L. Liao, N. Liu, K. Yan, Y. Chen, and Z. Liu. High-quality single-layer graphene via reparative reduction of graphene oxide. *Nano Research* 4, no. 5 (2011): 434–439.
61. H. Wang, J.T. Robinson, X. Li, and H. Dai. Solvothermal reduction of chemically exfoliated graphene sheets. *Journal of the American Chemical Society* 131, no. 29 (2009): 9910–9911.
62. L. Zhao, K.T. Rim, H. Zhou, R. He, T.F. Heinz, A. Pinczuk, G.W. Flynn, and A.N. Pasupathy. Influence of copper crystal surface on the CVD growth of large area monolayer graphene. *Solid State Communications* 151, no. 7 (2011): 509–513.
63. I. Vlassiouk, M. Regmi, P. Fulvio, S. Dai, P. Datskos, G. Eres, and S. Smirnov. Role of hydrogen in chemical vapor deposition growth of large single-crystal graphene. *ACS Nano* 5, no. 7 (2011): 6069–6076.
64. H. Hibino, H. Kageshima, and M. Nagase. Graphene growth on silicon carbide. NTT Basic Research Laboratories, 2010.
65. C. Berger, X. Wu, P.N. First, E.H. Conrad, X. Li, M. Sprinkle et al. Dirac particles in epitaxial graphene films grown on SiC. In *Advances in Solid State Physics*, 145–157. Springer-Verlag, 2008.
66. W. Norimatsu and M. Kusunoki, Epitaxial graphene on SiC{0001}: Advances and perspectives, *Physical Chemistry Chemical Physics* 16 (2014): 3501–3511.
67. C. Berger, Z. Song, X. Li, X. Wu, N. Brown, C. Naud et al. 2006. Electronic confinement and coherence in patterned epitaxial graphene. *Science* 312, no. 5777 (2006): 1191–1196.
68. J. Coleman. A new solution to graphene production. *SPIE Newsroom* 2008, doi: 10.1117/2.1200810.1336.
69. J.C. Meyer, C. Kisielowski, R. Erni, M.D. Rossell, M.F. Crommie, and A. Zettl. Direct imaging of lattice atoms and topological defects in graphene membranes, *Nano Letters* 8, no. 11 (2008): 3582–3586.
70. J.C. Meyer, A.K. Geim, M.I. Katsnelson, K.S. Novoselov, D. Obergfell, S. Roth, C. Girit, and A. Zettl. On the roughness of single-and bi-layer graphene membranes. *Solid State Communications* 143, no. 1 (2007): 101–109.
71. M.I. Katsnelson and A.K. Geim. Electron scattering on microscopic corrugations in graphene. *Philosophical Transactions of the Royal Society of London A: Mathematical, Physical and Engineering Sciences* 366, no. 1863 (2008): 195–204.
72. Fasolino, A., Los, J.H. and M.I. Katsnelson. Intrinsic ripples in graphene. *Nature Materials* 6, no. 11 (2007): 858–861.
73. D. Graf, F. Molitor, K. Ensslin, C. Stampfer, A. Jungen, C. Hierold, and L. Wirtz. Spatially resolved Raman spectroscopy of single- and few-layer graphene. *Nano Letters* 7, no. 2 (2007): 238–242.
74. I. Childres, L.A. Jauregui, W. Park, H. Cao, and Y.P. Chen. Raman spectroscopy of graphene and related materials. In *New Developments in Photon and Materials Research*, edited by J.I. Jang, chap. 19. Nova Science, 2013.
75. A.C. Ferrari, J.C. Meyer, V. Scardaci, C. Casiraghi, M. Lazzeri, F. Mauri et al. Raman spectrum of graphene and graphene layers. *Physical Review Letters* 97, no. 18 (2006): 187401.
76. H.B. Zhang, W.G. Zheng, Q. Yan, Y. Yang, J.W. Wang, Z.H. Lu, G.Y. Ji, and Z.Z. Yu. Electrically conductive polyethylene terephthalate/graphene nanocomposites prepared by melt compounding. *Polymer* 51, no. 5 (2010): 1191–1196.
77. S.D. Sarma, S. Adam, E.H. Hwang, and E. Rossi. Electronic transport in two-dimensional graphene. *Reviews of Modern Physics* 83, no. 2 (2011): 407.

78. H. He, J. Klinowski, M. Forster, and A. Lerf. A new structural model for graphite oxide. *Chemical Physics Letters* 287, no. 1 (1998): 53–56.
79. K.I. Bolotin, K.J. Sikes, Z. Jiang, M. Klima, G. Fudenberg, J. Hone, P. Kim, and H.L. Stormer. Ultrahigh electron mobility in suspended graphene. *Solid State Communications* 146, no. 9 (2008): 351–355.
80. C. Gómez-Navarro, M. Burghard, and K. Kern. Elastic properties of chemically derived single graphene sheets. *Nano Letters* 8, no. 7 (2008): 2045–2049.
81. Scientific Background on the Nobel Prize in Physics 2010. Graphene. Compiled by the Class for Physics of the Royal Swedish Academy of Sciences. *The Royal Swedish Academy of Sciences.* October 5, 2010.
82. R.R. Nair, P. Blake, A.N. Grigorenko, K.S. Novoselov, T.J. Booth, T. Stauber, N.M.R. Peres, and A.K. Geim. Fine structure constant defines visual transparency of graphene. *Science* 320, no. 5881 (2008): 1308.
83. A.A. Balandin, S. Ghosh, W. Bao, I. Calizo, D. Teweldebrhan, F. Miao, and C.N. Lau. Superior thermal conductivity of single-layer graphene. *Nano Letters* 8, no. 3 (2008): 902–907.
84. S. Stankovich, R.D. Piner, S.T. Nguyen, and R.S. Ruoff. Synthesis and exfoliation of isocyanate-treated graphene oxide nanoplatelets. *Carbon* 44, no. 15 (2006): 3342–3347.
85. S. Niyogi, E. Bekyarova, M.E. Itkis, J.L. McWilliams, M.A. Hamon, and R.C. Haddon, Solution properties of graphite and graphene. *Journal of the American Chemical Society* 128, no. 24 (2006): 7720–7721.
86. H. Yang, C. Shan, F. Li, D. Han, Q. Zhang, and L. Niu. Covalent functionalization of polydisperse chemically-converted graphene sheets with amine-terminated ionic liquid. *Chemical Commuications* 26 (2009): 3880–3882.
87. B. Guo, L. Fang, B. Zhang, and J.R. Gong. Graphene doping: A review. *Insciences Journal* 1, no. 2 (2011): 80–89.
88. B. Guo, Q. Liu, E. Chen, H. Zhu, L. Fang, and J.R. Gong. Controllable N-doping of graphene. *Nano Letters* 10, no. 12 (2010): 4975–4980.
89. J. Du and H.M. Cheng. The fabrication, properties, and uses of graphene/polymer composites. *Macromolecular Chemistry and Physics* 213, no. 10–11 (2012): 1060–1077.
90. J.H. Du, J. Bai, and H.M. Cheng. The present status and key problems of carbon nanotube based polymer composites. *Express Polymer Letters* 1, no. 5 (2007): 253–273.
91. X-L. Xie, Y-W. Mai, and X-P. Zhou. Dispersion and alignment of carbon nanotubes in polymer matrix: A review. *Materials Science and Engineering: R*: Reports 49, no. 4 (2005): 89–112.
92. R.B. Mathur, S. Pande, B.P. Singh, and T.L. Dhami. Electrical and mechanical properties of multi-walled carbon nanotubes reinforced PMMA and PS composites. *Polymer Composites* 29, no. 7 (2008): 717–727.
93. J.Y. Jang, M.S. Kim, H.M. Jeong, and C.M. Shin. Graphite oxide/poly (methyl methacrylate) nanocomposites prepared by a novel method utilizing macroazoinitiator. *Composites Science and Technology* 69, no. 2 (2009): 186–191.
94. T.K. Gupta, B.P. Singh, R.K. Tripathi, S.R. Dhakate, V.N. Singh, O.S. Panwar, and R.B. Mathur. Superior nano-mechanical properties of reduced graphene oxide reinforced polyurethane composites. *RSC Advances* 5, no. 22 (2015): 16921–16930.

7

Graphene Nanoribbons

7.1 Introduction

With the rise of graphene research since 2004, the past few years have witnessed rapid progress in exploiting graphene as the material for the next generation of dramatically faster, more energy-efficient electronics [1–3]. A sheet of graphene not only conducts electricity and dissipates heat much more efficiently than silicon, the material most commonly found in today's computer chips, its fascinating optical and mechanical properties have ignited enormous interdisciplinary interest in the field of physics, chemistry, and materials science. However, the fact that graphene is a zero-band gap semimetal poses a major problem for its practical applications in making high-performance field-effect transistors (FETs) [2,3].

In order to overcome this problem, a way to open a gap in the electronic structure of graphene has to be found. A straightforward solution is to pattern the graphene sheet into 1D ultranarrow graphene nanoribbons (GNRs) in which the lateral confinement of charge carriers creates an energy gap near the charge neutrality point. This results due to the quantum mechanical constriction of electronic wave functions in the direction perpendicular to the axis of the ribbon [4,5]. Therefore, graphene can be transformed from a semimetal into a semiconductor if it is confined into nanoribbons, narrower than 10 nm, with controlled crystallographic orientation and well-defined armchair/zigzag edges in which the carbon–carbon bonds are parallel to the length of the ribbon. The ribbon forms of graphene inherit almost all of the attractive properties of the carbon nanotube and graphene.

Their band gap roughly varies inversely with ribbon width and the largest band gaps are expected for ribbons with armchair edge orientation [6]. This dependence offers us the opportunity to tailor the electronic structure of the graphene nanoribbons. However, producing nanoribbons of very high quality or in high volumes still remains a challenge [7,8].

The quasi one-dimensional nature of GNRs results in additional advantages over graphene sheets, known as two-dimensional counterparts to GNRs. For example, the high aspect ratio of GNRs significantly lowers the percolation threshold in conductive films and polymer composites, and renders them suitable for spinning fiber from their liquid crystalline alignment.

7.2 Synthesis

The structure and physical properties of GNRs vary significantly depending on the synthesis method. It is necessary to bring down the width of the GNR to ~10 nm in order to induce a technologically relevant band gap that is substantially greater than K_BT (~25 meV) at room temperature [9]. A sub-10 nm resolution, however, is beyond the limits of conventional optical and electron-beam lithography. At present there are three major approaches for fabrication of GNRs: (1) top-down approach, that is, cutting from graphene using lithography; (2) bottom-up synthesis from polycyclic molecules; and (3) unzipping of carbon nanotubes (CNTs). In fact, GNRs produced using those three methods are very different and have little in common.

7.2.1 Cutting from Graphene Using Lithography: Top-Down Approach

The top-down approach uses diverse methodologies with ease for efficient device integration. Following this approach, GNRs have been achieved by etching the graphene surface with a STM (scanning tunneling microscope) tip applying higher constant bias potential with atomic-level precision. The epitaxially grown few-layer graphene has also been patterned by conventional lithographic techniques to fabricate GNR-based devices.

The use of lithographically produced GNRs is limited to applications where GNRs lie flat on a surface, preferably highly oriented pyrolytic graphite (HOPG). Bulk quantities cannot be produced using lithographic methods. Moreover, electron lithographic methods (e-beam lithography) have difficulties in controlling the crystallographic orientation of the ribbons. While this method could produce highly precise and very narrow GNRs, the resulting uncontrolled uneven edges make it difficult to control the electronic properties of the resulting materials [10].

Though the modification of graphite surface by STM has an almost 20-year-old history [11], it is now realized that by using the same technique it is possible to engineer nanostructures with almost atomically precise structure and predetermined electronic properties.

Cutting of GNRs from a flat graphene sheet is done by applying a constant bias potential (significantly higher than the one used for imaging) and simultaneously moving the STM tip with constant velocity to etch the desired geometry fitted to the crystallographic structure. Figure 7.1a shows a 10 nm wide and 120 nm long graphene nanoribbon etched by STM

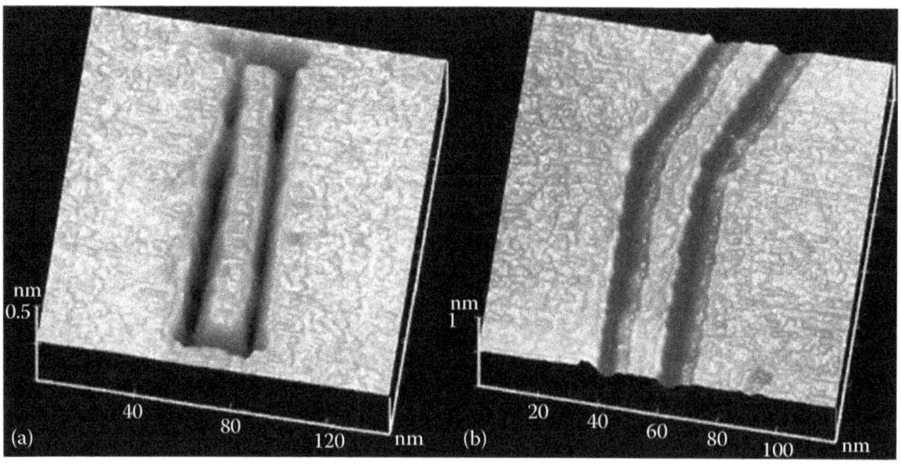

FIGURE 7.1
Graphene nanostructures patterned by STM lithography. (a) 3D STM image of a 10 nm wide and 120 nm long graphene nanoribbon. (b) An 8 nm wide, 30° GNR bent-junction connecting an armchair and a zigzag ribbon. (Reprinted by permission from Macmillan Publishers Ltd., *Nature Nanotechnology*, L. Tapasztó, G. Dobrik, P. Lambin, and L.P. Biró. Tailoring the atomic structure of graphene nanoribbons by scanning tunnelling microscope lithography, 3, no. 7: 397–401, copyright 2008.)

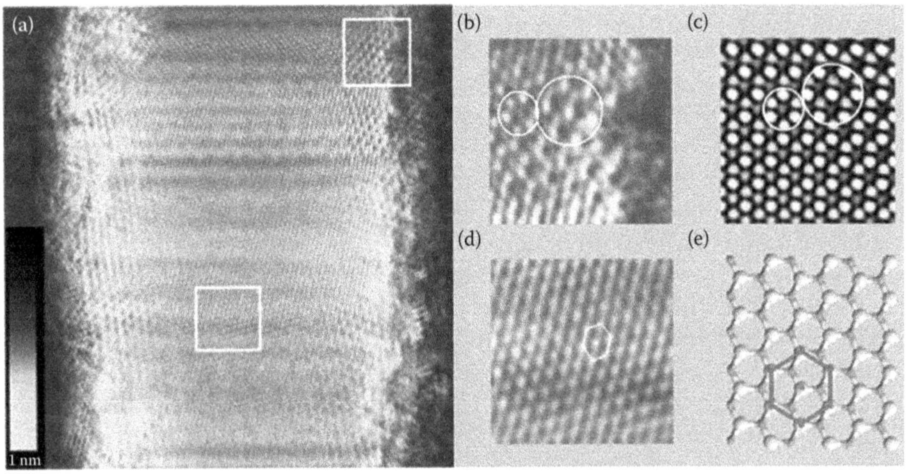

FIGURE 7.2
Atomic structure of graphene nanoribbons. (a) Atomic resolution STM image (20 × 20 nm^2, 1 nA, 200 mV) of a 15 nm wide GNR displaying an atomically flat and defect-free structure. To the right are magnified images of the (b) defect-free lattice taken at the center of the ribbon and (c) position dependent superstructures near the edges. (d) Identification of crystallographic orientation from the triangular lattice observed in atomic resolution STM images of HOPG-supported GNR. (e) Theoretical STM image of the superstructures at the edges of the ribbon. (Reprinted by permission from Macmillan Publishers Ltd., *Nature Nanotechnology*, L. Tapasztó, G. Dobrik, P. Lambin, and L.P. Biró. Tailoring the atomic structure of graphene nanoribbons by scanning tunnelling microscope lithography, 3, no. 7: 397–401, copyright 2008.)

lithography. By setting the optimal lithographic parameters it is possible to cut GNRs with suitably regular edges with reproducibility of GNR-based devices. Furthermore, more complex graphene nanoarchitectures can be tailored by STM lithography. Figure 7.1b shows an 8 nm wide, 30° GNR-bent-junction connecting an armchair and a zigzag ribbon, giving rise to a metal-semiconductor molecular junction [12]. After the patterning process, in situ atomic resolution STM images could be achieved revealing the atomic structure of the GNRs.

Figure 7.2a illustrates a 15 nm wide ribbon with its axis having a crystallographic orientation close to the zigzag direction. For electronic device applications these ribbons have to be deposited on an insulating substrate. The intrinsic characteristics of GNRs can be studied this way.

Using this technique semiconducting GNRs from graphene were produced with predetermined energy gap values up to 0.5 eV, allowing their operation at room temperature.

Thermally activated metallic nanoparticles have also been used for atomically precise etching along crystallographic axes to obtain graphene nanoribbons [13]. It has also been observed recently that the solution dispersion and sonication of exfoliated graphite results in ultrasmooth edges even for sub-10 nm width GNRs [7,14]. GNRs can also be obtained by plasma etching of multiwalled carbon nanotubes partly embedded in a polymer film. This technique has been used to produce single-layer to few-layer GNRs depending on the time period of etching [15].

7.2.2 Nanosphere Lithography

An innovative approach for ultranarrow GNRs fabrication is by utilizing nanosphere lithography (NSL) in combination with low-power O_2 plasma etching. A self-assembled, ordered array of latex nanospheres is used as lithographic masks for deposition of various metal nanostructures as a high-throughput and low-cost nanofabrication process [16].

Single-layered graphene oxide sheets are deposited onto SiO_2/Si substrates. The graphene oxide sheets are reduced to graphene by exposure to hydrazine vapors at elevated temperature followed by thermal annealing. Figure 7.3a shows a representative scanning electron microscopy (SEM) image of the graphene oxide-derived single-layer graphene sheets with lateral dimensions up to several micrometers on a SiO_2/Si substrate. NSL mask was prepared by packing polystyrene (PS) nanospheres with diameter of ~1 μm on the substrate by adopting special techniques [17,18].

Figure 7.3b shows a typical SEM image of a domain of the hexagonal close packed (hcp) nanosphere monolayer. The nanosphere-masked graphene sheets are subsequently subjected to O_2 plasma etching. After etching, the masking PS nanospheres are removed by washing in dichloromethane under mild sonication.

Figure 7.3c shows a typical interconnected, ordered 2D network of GNRs etched through a hcp monolayer NSL mask with an etching duration of a few seconds. An AFM line scan shows its lateral diameter of the order of 12 nm (Figure 7.3d). Even narrower GNRs could

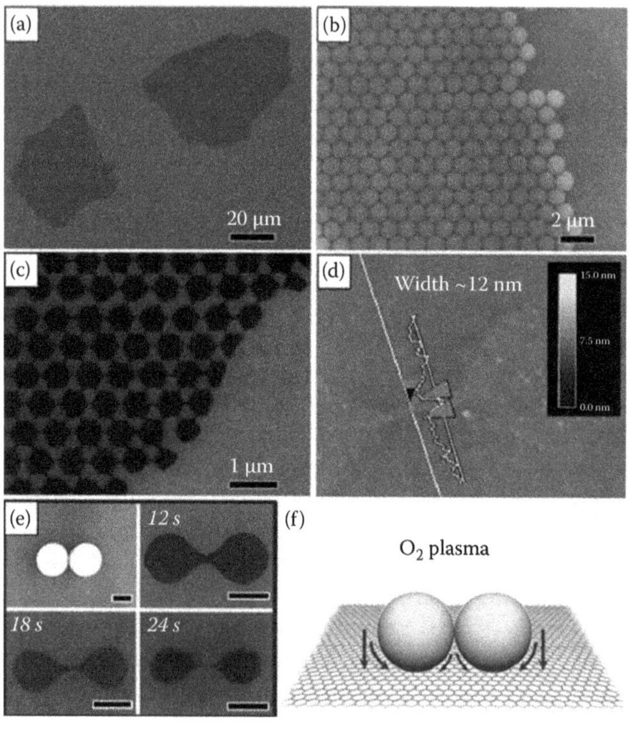

FIGURE 7.3
(a) SEM image of the chemically derived graphene sheets lying on a SiO_2/Si substrate. (b) SEM image showing a domain of the hcp PS nanosphere monolayer that masked the graphene sheets. (c) SEM image of the large area, interconnected 2D GNRs network etched through an hcp monolayer NSL mask. Note that the darker areas in the SEM image are GNRs; the image contrast is due to the local potential differences between graphene and the insulating SiO_2 substrate caused by differential charging. (d) Tapping mode AFM image of a selected typical nanoribbon with a width of ca. 12 nm (measured via line scan profile). (e) SEM images showing the temporal evolution of the individual dumbbell-like GNRs etched through the isolated pairs of packed nanospheres (top panel) with etching durations of 12, 18, and 24 s. Scale bars in (e): 500 nm. (f) Schematic illustration (not to scale) showing the NSL-defined lithographic etching process of graphene sheets. (L. Liu, Y. Zhang, W. Wang, C. Gu, X. Bai, and E. Wang: Nanosphere lithography for the fabrication of ultranarrow graphene nanoribbons and on-chip bandgap tuning of graphene. *Advanced Materials*. 2011. 23. No. 10. 1246–1251. Copyright Wiley-VCH Verlag GmbH & Co. KGaA. Reproduced with permission.)

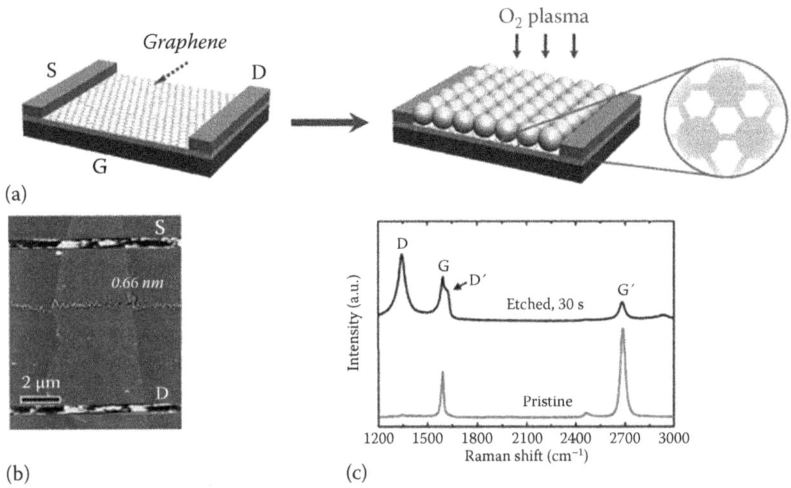

FIGURE 7.4
(a) Schematic illustration of the on-chip band gap tuning of graphene through NSL nanopatterning and subsequent O_2-plasma etching. (b) AFM image of a typical FET device based on the mechanically cleaved single-layer graphene sheet. (c) Micro-Raman spectra of the pristine graphene sheet and the resultant GNRs network produced by 30 s of etching. The spectra were collected at room temperature using 532 nm laser excitation. (L. Liu, Y. Zhang, W. Wang, C. Gu, X. Bai, and E. Wang: Nanosphere lithography for the fabrication of ultranarrow graphene nanoribbons and on-chip bandgap tuning of graphene. *Advanced Materials*. 2011. 23. No. 10. 1246–1251. Copyright Wiley-VCH Verlag GmbH & Co. KGaA. Reproduced with permission.)

be generated with a longer etching time but with lower yield (Figure 7.3e). A key advantage of the NSL nanopatterning approach is the unusual simplicity and ease of fabrication as compared to the other existing methodologies developed for GNR fabrication.

On a practical level, it allows one to directly apply this fabrication approach to graphene sheets that are integrated into electrical devices and it is possible to achieve the straightforward on-chip band gap tuning of graphene, as schematically illustrated in Figure 7.4a. An AFM image of one of the targeted single-layer graphene FET devices that was subjected to NSL masking and a subsequent plasma etching sequence is shown in Figure 7.4b. Raman spectroscopy is used to ascertain the single-layer graphene sheet and is characterized by a single Lorentzian shape of the G′ peak [19] shown in Figure 7.4c.

In light of the simple and low cost of the NSL technique, it is expected that the present GNR fabrication approach may open a new avenue for the development of graphene-based nanoeletronics.

7.2.3 Graphene Nanoribbon (GNR) Fabrication: Bottom-Up Approach

The bottom-up approach follows the strategy of attaching small molecular building blocks to grow giant graphene structures using elegant synthetic chemistry routes. This approach can provide desired widths with full control over their edge geometries, which govern the GNR properties. Most important, these structures with smooth edges can be efficiently isolated in single layers and can be used in device fabrication.

The bottom-up fabrication of GNRs involves multistep organic synthesis based on cyclization of presynthesized polymer chains. Until recently, this type of GNR could only be prepared on the surface of a substrate, thus limiting the potential for bulk production.

Different size and edge configurations of GNRs can be obtained by choosing a suitable precursor polymer structure and a substrate with proper orientation of planes. In one of the recent studies, a bottom-up approach with surface-assisted molecular assembly (SAMA) was adopted to produce defect-free zigzag or armchair-edge GNRs of different widths [20–23]. Figure 7.5 shows the (3,1)-GNR fabricated by thermally induced polymerization of 10,10′-dibromo-9,9′-bianthryl precursor monomers (DBPMs) onto $Cu_{\{111\}}$ substrate held at room temperature in an ultrahigh vacuum.

Figure 7.6a shows STM image of $Cu_{\{111\}}$ surface after DBPM deposition and annealed at 200°C. Annealing the sample to 500°C induces polymerization, yielding clusters of linear structures (Figure 7.6b). High-resolution images reveal the characteristic chiral edge configuration of (3,1)-GNRs (Figure 7.6c) indicating molecular flattening upon polymerization.

In fact the DBPM polymerization on $Cu_{\{111\}}$ begins with debromination, followed by intermolecular cyclodehydrogenation (CDH), as shown in Figure 7.7.

The chemistry of SAMAs can be altered by specific properties of the substrate. The ideal substrates should be just reactive enough to lower the activation barriers of chemical reaction predetermined by the precursor molecular structure [24,25]. The surface atomic structure of $Cu_{\{111\}}$ and its tendency to catalyze CDH combine together to direct the DBPMs to polymerize into (3,1)-GNRs.

Although the high precision and narrow size distribution resulting from the bottom-up synthesis of GNRs may yield future benefits, it is difficult to see their actual applications in the near future. There are currently no ready-to-use techniques to further process these tiny structures and it is difficult to envision its actual application in the near future.

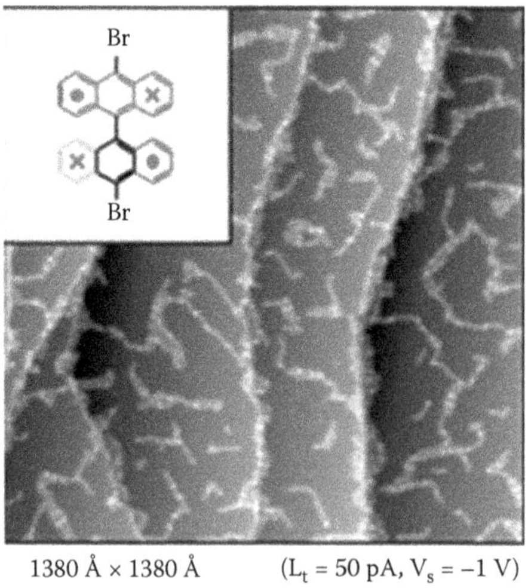

FIGURE 7.5
Self-assembled molecular chains. Scanning tunneling microscope image of a $Cu_{\{111\}}$ surface after precursor deposition at room temperature. The inset shows the structure of 10,10′-dibromo-9,9′-bianthryl (DBPM). (Reprinted with permission from P. Han, K. Akagi, F. Federici Canova, H. Mutoh, S. Shiraki, K. Iwaya et al. Bottom-up graphenenanoribbon fabrication reveals chiral edges and enantioselectivity. *ACS Nano* 8, no. 9: 9181–9187. Copyright 2014 American Chemical Society.)

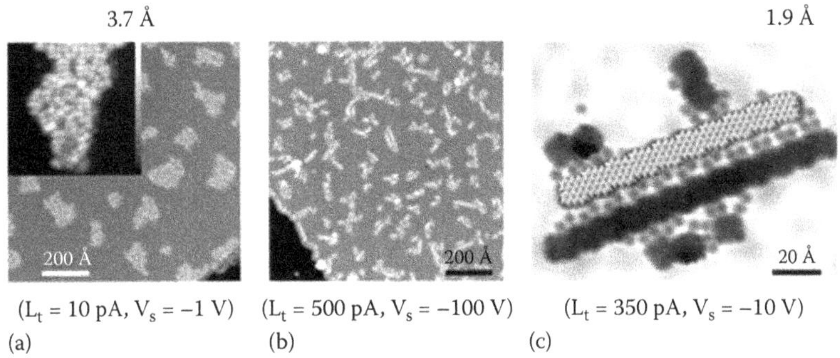

FIGURE 7.6
Thermally induced polymerization. (a) Scanning tunneling microscope image of a $Cu_{\{111\}}$ surface after DBPM deposition at RT, followed by 10 min thermal annealing of the surface at 200°C. The inset shows a high-resolution STM image of a DBPM island. (b) Scanning tunneling microscope image of a $Cu_{\{111\}}$ surface after DBPM deposition at RT, followed by a 10 min thermal annealing of the surface at 500°C. (c) Scanning tunneling microscope image of a chiral-edge GNR cluster. A space-filling model is superimposed over a 10-monomer-long GNR. The circular features in (c) were determined to be Br atoms. (Reprinted with permission from P. Han, K. Akagi, F. Federici Canova, H. Mutoh, S. Shiraki, K. Iwaya et al. Bottom-up graphenenanoribbon fabrication reveals chiral edges and enantioselectivity. *ACS Nano* 8, no. 9: 9181–9187. Copyright 2014 American Chemical Society.)

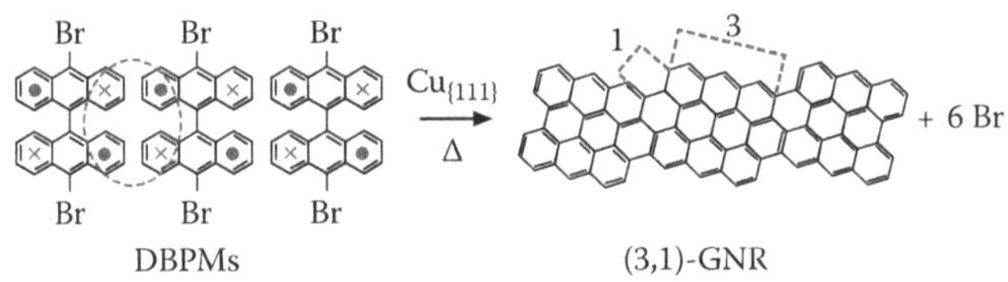

FIGURE 7.7
Reaction scheme for the polymerization of DBPM into (3,1)-GNR. The crosses and filled circles indicate regions of the bianthryl units that are close to and away from, respectively, the $Cu_{\{111\}}$ substrate. The dashed oval highlights a region where cascade CDH reactions occur. The dashed lines demonstrate the (3,1)-chiral edge nomenclature. (Reprinted with permission from P. Han, K. Akagi, F. Federici Canova, H. Mutoh, S. Shiraki, K. Iwaya et al. Bottom-up graphenenanoribbon fabrication reveals chiral edges and enantioselectivity. *ACS Nano* 8, no. 9: 9181–9187. Copyright 2014 American Chemical Society.)

7.2.4 Longitudinal Unzipping of Carbon Nanotubes Using Alkali Metal Vapor

The third approach to the fabrication of GNRs is based on the longitudinal opening, or unzipping, of multiwalled carbon nanotubes (MWCNTs) [26,27]. Two techniques can be employed, the first using potassium (K) vapor and the second is a solution-based approach. The major advantage of this approach over the lithographic approach is the potential of mass production on the kilogram scale. Also, based on the number of recent publications, it is likely that these CNT-derived GNRs will be the first to find actual applications.

The process is based on the hypothesis that thermal motion of a carbon framework in MWCNT sidewalls at elevated temperatures might create large enough transient openings for alkali metal atoms to penetrate and the process is known as intercalation. Intercalation of alkali metals in graphite and carbon nanotubes forms charge transfer compounds. Highly oriented pyrolytic graphite (HOPG) can form well-known layered, stage-I intercalation compounds (C_8K with potassium and C_6Li with lithium) when exposed to all molten alkali metals, except sodium, at moderately elevated temperatures. Similarly MWCNTs with a starting outside diameter of 40–80 nm, and approximately 15–20 inner nanotube layers can also intercalate these alkali metals under appropriate conditions [28,29]. Contrary to the results with lithium and sodium, the change in the nanotube behavior is dramatic when potassium is used. In a typical process, potassium and MWCNTs [30] are sealed in a glass tube and heated in a furnace at 250°C for 14 h. The silvery luster of potassium rapidly changed at 150°C to a uniform appearance of shiny golden-bronze color for C_8K [28]. Potassium atoms expand the interplanar distance in MWCNTs from 0.34 nm to 0.50 nm in C_8K, the saturated stage-I intercalation compound (Figure 7.8). This lattice expansion induces sufficient stress to longitudinally break the nanotube walls. However, the increase in the interstitial distance between the two outermost shells of a MWCNT is too small, 0.373 nm, for stage I lithium intercalation compound and would therefore not provide enough strain in the carbon–carbon bonds to induce breakage at moderately elevated temperatures. Thus experimentally no splitting with lithium is observed.

The carbon atoms at the newly formed edges are reduced to their highly reactive carboanionic form, making them highly susceptible to electrophilic attack. If the intermediate product is quenched with ethanol, followed by aqueous washing, the metal cations on the edges are replaced with protons. This yields H-terminated GNRs. These GNRs, however, are not completely flat due to van der Waals interaction between the nanotube walls.

In order to achieve further exfoliation and flattening of the graphene layers, a superacid such as chlorosulfonic acid is used to protonate the split sidewalls of MWCNTs and thus induce electrostatic repulsive forces, which would facilitate exfoliation of split MWCNTs under sonication [27,31–33]. Figure 7.9 shows fully unzipped MWCNTs after treatment.

The electrical conductivity of the flattened 3.5 to 5 nm thick GNR stacks, comprising 10 to 14 layers, is in the range of 70,000 to 95,000 S/m. These values are comparable to the data reported for other graphitic structures.

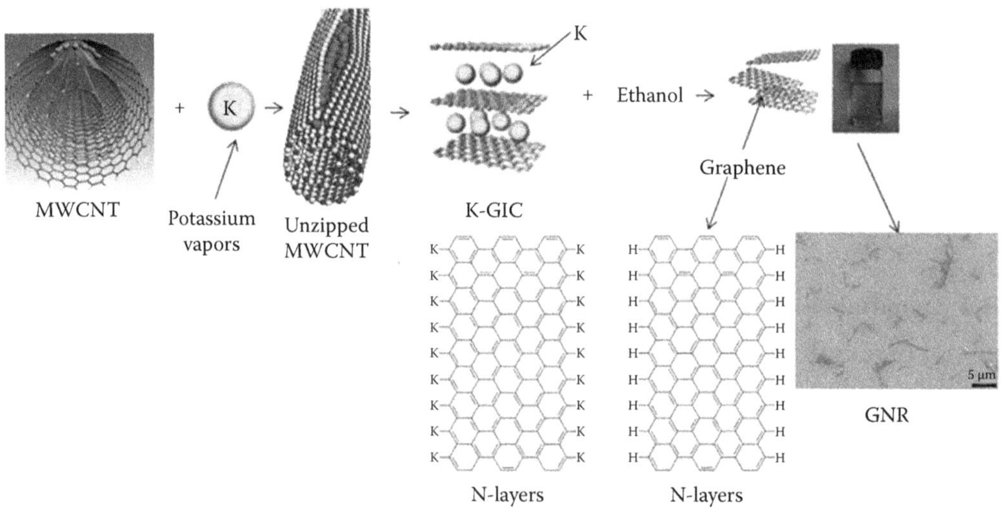

FIGURE 7.8
A schematic of potassium intercalation between the nanotube walls and sequential longitudinal splitting of the walls followed by unraveling to a nanoribbon stack. Ethanol is used to quench the aryl potassium edges; only a single layer is shown for clarity. The actual number of GNR layers correlates with the number of concentric tubes in the MWCNT. (Reprinted with permission from D.V. Kosynkin, W. Lu, A. Sinitskii, G. Pera, Z. Sun, and J.M. Tour. Highly conductive graphene nanoribbons by longitudinal splitting of carbon nanotubes using potassium vapor. *ACS Nano* 5, no. 2: 968–974. Copyright 2011 American Chemical Society.)

FIGURE 7.9
Potassium-treated MWCNTs after bath ultrasonication in chlorosulfonic acid for 24 h. Nearly complete conversion to stacks of straight-edged nanoribbons is evident. (Reprinted with permission from D.V. Kosynkin, W. Lu, A. Sinitskii, G. Pera, Z. Sun, and J.M. Tour. Highly conductive graphene nanoribbons by longitudinal splitting of carbon nanotubes using potassium vapor. *ACS Nano* 5, no. 2: 968–974. Copyright 2011 American Chemical Society.)

The GNRs thus produced could also be an attractive material for reinforcing polymers [34] as the potassium-carrying reactive edges could facilitate the attachment of electrophiles or polymer chains to improve interfacial interaction for load transfer. Consequently, the present procedure opens new directions for preparation of GNRs and subsequent applications.

7.2.5 GNR by Unzipping of Carbon Nanotubes through Chemical Route

A strong oxidizing agent along with sulfuric acid is used to chemically create a hole in the carbon nanotubes. The hole expands along the side of the nanotube, unzipping the tube to form a flat graphene ribbon. The width of the graphene ribbon depends on the diameter of the nanotubes. A simple solution-based oxidative process is used for producing a nearly 100% yield of nanoribbon structures by lengthwise unzipping of MWCNT sidewalls. Although the oxidative process is originally described as an unzipping-based mechanism, it could be suggestive of a longitudinal splitting through $KMnO_4$ intercalation, much like the potassium metal-based method. First, graphene oxide nanoribbons (GONRs) are obtained by suspending MWCNTs in concentrated sulfuric acid followed by treatment with 5 weight equivalents $KMnO_4$ for 1 h at room temperature and 1 h at 55°C–70°C, as shown in Figure 7.10. The mechanism for the opening is based upon the chemistry of oxidation of alkenes by permanganate in acid [33].

Once an opening is initiated, its further opening is enhanced relative to an unopened tube or to an uninitiated site on the same tube due to induced bond-angle strain (4 in Figure 7.10b).

Finally, relief of the bond-angle strain when the nanotube opens to the graphene ribbon (5 in Figure 7.10b) slows further dione formation and cutting.

Sometimes, overoxidation of the nanoribbons formed from MWCNT can be prevented by using a second acid such as TFA or H_3PO_4.

Although the preparative route described here can have the advantage of large-scale nanoribbon production, these unzipping-derived nanoribbons, with their residual oxidized defect sites, possess inferior electronic characteristics when compared to alkali metal exfoliation-derived GNRs.

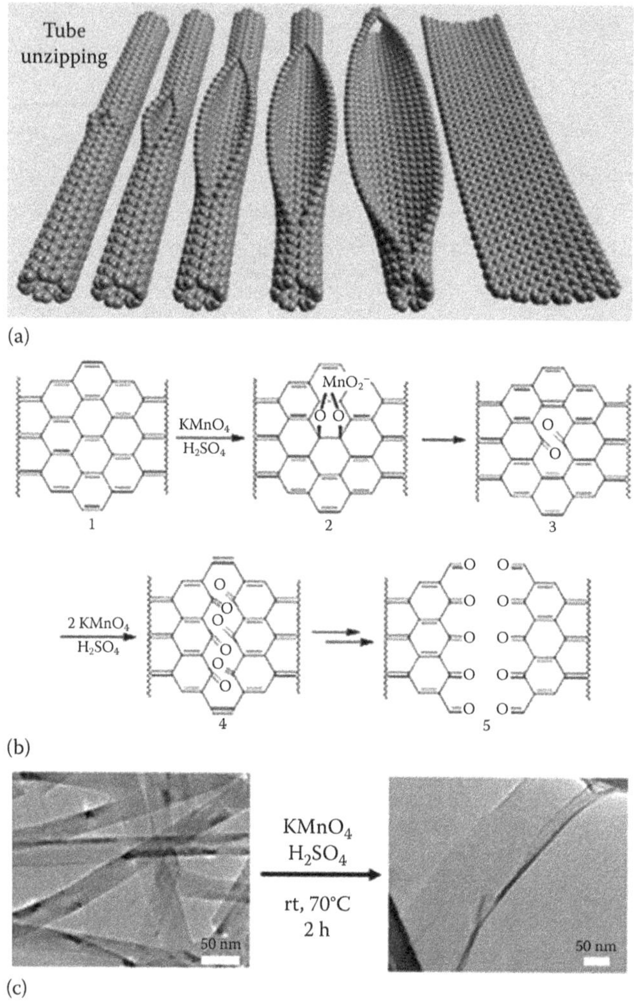

FIGURE 7.10
Nanoribbon formation and imaging. (a) Schematic representation of the gradual unzipping of one wall of a carbon nanotube to form a nanoribbon. Oxygenated sites are not shown. (b) The proposed chemical mechanism of nanotube unzipping. (c) TEM images depicting the transformation of MWCNTs (left) into oxidized nanoribbons (right). The right side of the ribbon is partially folded onto itself. The dark structures are part of the carbon imaging grid. (Reprinted by permission from Macmillan Publishers Ltd., *Nature*, D.V. Kosynkin, A.L. Higginbotham, A. Sinitskii, J.R. Lomeda, A. Dimiev, B.K. Price et al. Longitudinal unzipping of carbon nanotubes to form graphene nanoribbons, 458, no. 7240: 872–876, copyright 2009.)

The oxidative unzipping to form GONRs, with subsequent reduction to the GNRs may introduce damaged planes that might lower their conductivity. However, these could find uses in composites and electronics where high conductance is not essential.

Furthermore, strong intertube intercalation (van der Waals forces) makes it difficult to obtain GNRs from SWCNTs. It is these narrow GNRs that are particularly interesting from the standpoint of easily gatable devices.

Solution-based methodologies are surely the most promising routes toward the production of graphene nanoribbons, where a very stable compromise between quantity and quality of the ribbons can be reached. Dai and coworkers [7,35] have shown a simple method to produce GNRs, starting from expandable graphite. Graphite is exfoliated by rapid heating at elevated temperature (~1000°C) in the atmosphere of forming gas (Ar/H). The expanded graphite flakes are dispersed in a 1,2-dichloroethane (DCE) solution of poly(*m*-phenylenevinylene-co-2,5-dioctoxy-*p*-phenylenevinylene) and then sonicated. Once a homogeneous solution is obtained, centrifugation is used to separate large and heavy graphite pieces from light and narrow graphene small platelets. Ribbons with different widths down to sub-10 nm region can thus be produced. The same group recently developed an unconventional and alternative approach based on unzipping mildly oxidized MWCNTs through mechanical sonication.

7.3 Raman Spectroscopy of GNR

Raman spectroscopy has proven an effective and nondestructive method for the characterization of graphitic carbon nanostructures. A detailed study was carried out by Ryu et al. [34] on the characteristic changes in the Raman spectra of range of GNRs, width ranging from 15 to 100 nm, prepared by e-beam lithography. Figure 7.11 shows the Raman spectra of GNRs excited with laser wavelength λ = 632.8 nm. The spectra show the G band around ~1580 cm^{-1} arising due to optical E_{2g} phonon mode of graphene and two defect-induced bands D (~1330 cm^{-1}) and D′ (~1620 cm^{-1}) originating from scattering from ribbon edges. In fact evolution of the D-band gives valuable information about the nature of as-prepared GNRs. It is observed from Figure 7.11 that the G band energy is almost the same between

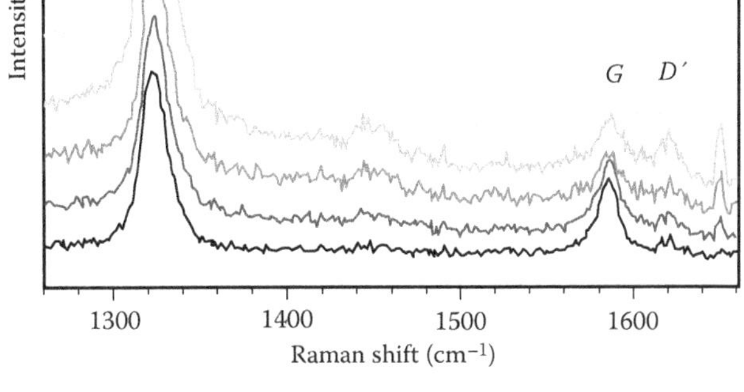

FIGURE 7.11
Raman spectra of as-prepared GNR, excited with laser wavelength λ = 632.8 nm. The spectra were offset for clarity. The bands at 1450 and 1650 cm^{-1} are due to underlying Si and a plasma line of the excitation laser, respectively. (Reprinted with permission from S. Ryu, J. Maultzsch, M.Y. Han, P. Kim, L.E. Brus. Raman spectroscopy of lithographically patterned graphene nanoribbons, *ACS Nano*, 5, 4123–4130. Copyright 2011 American Chemical Society.)

the bulk graphene and GNR up to width down to 25 nm. However, for the width lower than that i.e., 15 nm, the G band upshifts by ~5 cm^{-1} and broadens significantly. The upshift in the G band and enhancement in the intensity of D band of the narrower GNRs can be attributed to the quantum confinement effect or chemical doping (functionalization) of the GNR edge atoms. The ratio I_D/I_G therefore increases with decreasing ribbon width.

7.4 Potential Applications of GNR

Due to their unique structural and electronic properties, GNRs can have broad range of potential applications. The most apparent is the incorporation of GNRs as reinforcement in polymer hosts for the fabrication of novel high-performance composite materials. Though for such applications bulk production of consistent quality of GNRs is required. One of the major producers of GNRs is Sigma Aldrich with H-terminated GNRs (product number 797774) and alkylated GNRs (product number 797766). GNRs have the same high aspect ratio as their parent MWCNTs, but differences in their nanostructure produce unique and unexpected results.

Because of the miniaturization of electronic components, there is a requirement of materials with high permittivity and low loss in the radio and low microwave frequency region. In the high-frequency microwave region, low loss is critical for antennas and other military applications. By varying the type and content of GNRs, the loss and permittivity of composites can be tuned to desirable values over a wide range. The dielectric constant can be tuned from moderate to extremely high (>1000) values, whereas the corresponding loss tangent can be varied from ultralow (<0.02) to high (~1.0). The values are significantly different from that achieved by incorporation of MWCNTs.

Another promising application of GNRs is as an electrode material for batteries and supercapacitors. In one reference [36], a unique hierarchical structure composite of graphene-wrapped MnO_2-GNRs (GMG) was successfully designed and synthesized. In this composite, graphene flakes tightly sandwiched nanosized MnO_2 that grew directly on the GNRs. The synthesis of the MnO_2-GNRs (GMG) composite has been used as an effective electrode to improve the electrochemical stability for lithium-ion batteries.

Electrochemical experiments demonstrate that GMG exhibits enhanced specific capacity and improved cycling stability as anode materials compared to MnO_2–graphene or pure graphene due to synergic effect between the graphene, GNRs, and MnO_2. A stable-specific capacity value of 890 mAh/g could be achieved. Moreover, the coulombic efficiency of GMG was maintained at over 99%, excluding the first several cycles.

Polyaniline (PANI)–GNR composite can be used to fabricate a nanocomposite electrode for high-capacity supercapacitors [37]. As shown in Figure 7.12, PANI nanorods were grown on GNR by the *in situ* polymerization of aniline. In this composite, GNRs not only serve as the substrate to grow the PANI nanorods and improve the electrical conductivity of the composite, but also increase the effective utilization of PANI and enhance the mechanical property of the composite.

The resulting composite has a high specific capacitance of 340 F/g and stable cycling performance with nearly 90% capacitance retention over 4000 cycles. The high performance of the composite is a result of the synergistic combination of electrically conductive GNRs and highly capacitive PANI.

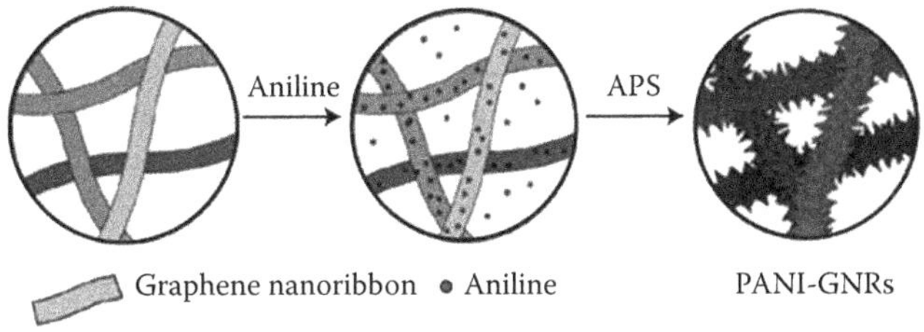

FIGURE 7.12
Illustration of the synthesis of the PANI–GNRs composite with PANI polymerized directly on the GNRs using APS (ammonium persulfate). (Reprinted with permission from L. Li, A-.R.O. Raji, H. Fei, Y. Yang, E.L. Samuel, and J.M. Tour. Nanocomposite of polyaniline nanorods grown on graphene nanoribbons for highly capacitive pseudocapacitors. *ACS Applied Materials & Interfaces* 5, no. 14: 6622–6627. Copyright 2013 American Chemical Society.)

7.4.1 Electronics

Due to their outstanding physical properties, graphene and its allotropes, carbon nanotubes, and graphene nanoribbons are the major candidates to become the silicon of the 21st century and open the era of so-called carbon electronics. Using the property that electrons move like photons could provide the foundation for a new type of electronic device that would capitalize on the ability of graphene to carry electrons with almost no resistance even at room temperature. The property is known as ballistic transport and the electrons in the graphene nanoribbons can move tens or hundreds of micrometers without scattering.

7.4.2 Graphene Transistors

To harness graphene's electronic and magnetic properties, one must first control the fabrication of GNR of desired width and edge configurations. In particular, graphene-based transistors are being developed rapidly and are now considered an option for postsilicon electronics [37]. Such a revolution was started in October 2004 with the discovery of graphene—two-dimensional sheets of carbon atoms—and observation of the electric field effect in their samples. Major chip-makers are now active in graphene research.

Using graphene nanoribbons, it may be possible to make devices with channels that are extremely thin that will allow graphene FETs to be scaled to shorter channel lengths and higher speeds without encountering the adverse short-channel effects that restrict the performance of existing devices. The FET is the most successful device concept in electronics. The FET is a transistor that relies on an electric field to control the conductivity of a channel of one type of charge carrier in a semiconductor material. According to Moore's law, the density of integrated circuits double every 18 months and the trend is being followed over the years. Moreover, this size scaling has enabled the complexity of integrated circuits leading to significant improvements in performance and decreases in price per transistor [38,39]. For decades, making metal-oxide-semiconductor FETs (MOSFETs) smaller in size has been key to the progress in digital electronics. Today, processors containing 2 billion MOSFETs, many with gate lengths of just 30 nm, are in mass production (Figure 7.13).

However, it is well understood now that MOSFET scaling is approaching its limits, and in the long run it will be necessary to introduce new material and device concepts to ensure that performance continues to improve.

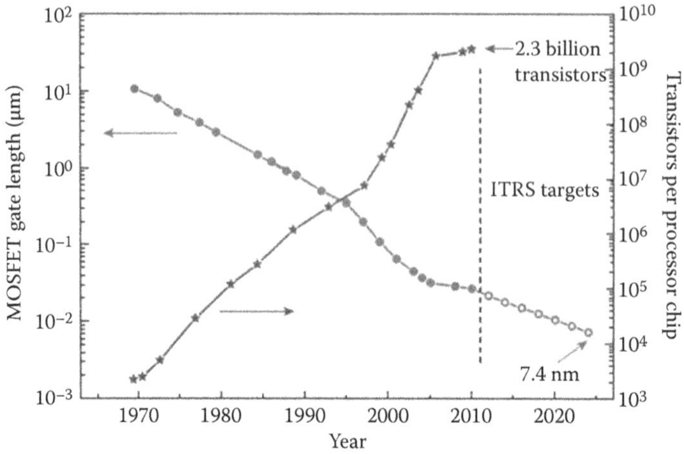

FIGURE 7.13
Trends in digital electronics. Evolution of MOSFET gate length in production-stage integrated circuits (filled circles) and International Technology Roadmap for Semiconductors (ITRS) targets (open circles). As gate lengths have decreased, the number of transistors per processor chip has increased (stars). Maintaining these trends is a significant challenge for the semiconductor industry, which is why new materials such as graphene are being investigated. (Reprinted by permission from Macmillan Publishers Ltd., *Nature Nanotechnology*, F. Schwierz. Graphene transistors, 5, no. 7: 487–496, copyright 2010.)

7.4.3 Field-Effect Transistor Physics

Let us first look into the functioning of an FET. An FET consists of a gate, a channel region connecting source and drain electrodes, and a barrier separating the gate from the channel (Figure 7.14). The operation of a conventional FET relies on the control of the channel conductivity, and thus the drain current, by a voltage, V_{GS}, applied between the gate and source. For high-speed applications, FETs should respond quickly to variations in V_{GS}. This requires short gates and high-speed carriers in the channel. Unfortunately, FETs with short gates frequently suffer from lower performance, known as short-channel effects [40].

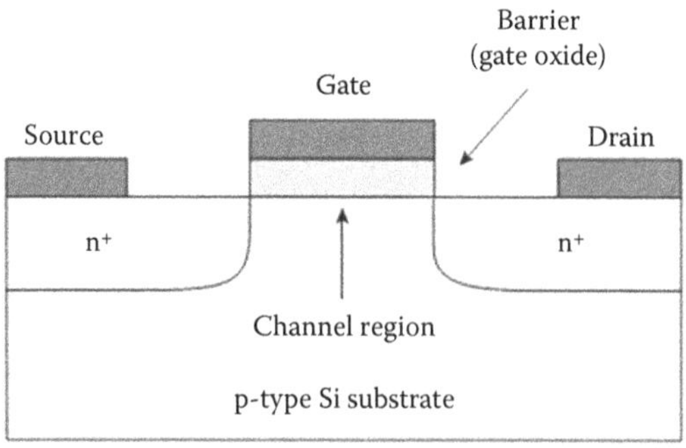

FIGURE 7.14
Conventional FETs: Cross-section of an n-channel Si MOSFET. When the voltage applied between the source and gate electrodes exceeds a threshold voltage, a conducting channel is formed and a drain current flows. The length of the channel is defined by the length of the gate electrode; the thickness of the gate-controlled channel region is the depth to which the electronic properties of the semiconductor (p-doped Si in this case) are influenced by the gate. (Reprinted by permission from Macmillan Publishers Ltd., *Nature Nanotechnology*, F. Schwierz. Graphene transistors, 5, no. 7: 487–496, copyright 2010.)

Scaling theory predicts that a FET with a thin barrier and a thin gate-controlled region will be robust against short-channel effects down to very short gate lengths [41]. The possibility of having channels that are just one atomic layer thick is perhaps the most attractive feature of graphene for use in transistors. The most frequently stated advantage of graphene is its high carrier mobility at room temperature. Mobilities of the order of 10,000–15,000 $cm^2V^{-1}s^{-1}$ with upper limits of 70,000 $cm^2V^{-1}s^{-1}$ have been suggested. By comparison, the channels in high electron mobility transistors are 10–15 nm thick with much lower mobility as compared to graphene.

7.4.4 Band Gap Engineering

Outstanding challenges for graphene transistors include opening a sizeable and well-defined band gap in graphene. Large-area graphene is a semimetal with a zero band gap. Because the band gap is zero, devices with channels made of large-area graphene cannot be switched off and therefore are not suitable for logic applications. However, the band structure of graphene can be modified, and it is possible to open a band gap by constraining large-area graphene in one dimension to form graphene nanoribbons [6]. It has been predicted that both armchair nanoribbons and zigzag nanoribbons (the two ideal types of nanoribbon; Figure 7.15) have a band gap that is, to a good approximation, inversely proportional to the width of the nanoribbon.

As already discussed, the valence and conduction bands in graphene are cone-shaped and meet at the K points of the Brillouin zone (Figure 7.16b(i)). Band gaps in excess of 200 meV for widths below 20 nm can be achieved. However, it should be noted that in reality nanoribbons have rough edges and widths that change along their lengths. Different edge geometries [42] influence the band gap between nanoribbons. To open a band gap useful for conventional field-effect devices, very narrow nanoribbons with well-defined edges are needed. This represents a serious challenge, given the semiconductor processing equipment available at the moment. In general, the larger the band gap that opens in a nanoribbon, the more the valence and conduction bands become parabolic (rather than

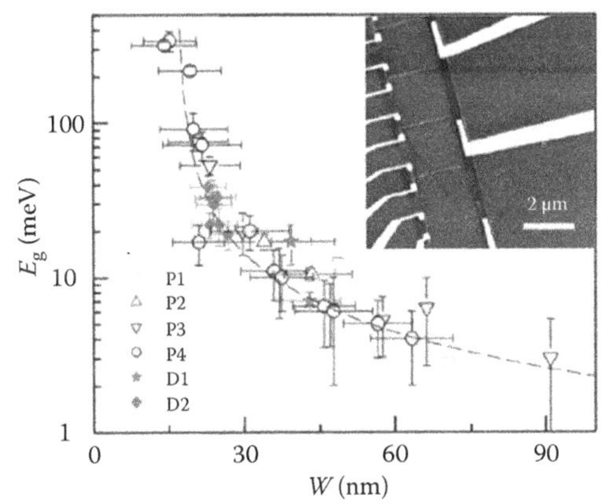

FIGURE 7.15
Energy gap dependence from ribbon width (W) in lithographically fabricated graphene nanoribbons. Dashed lines show the value of as predicted by the empirical scaling of versus W. (Inset) SEM image of a representative graphene nanoribbon-based device. (Reprinted with permission from M.Y. Han, B. Özyilmaz, Y. Zhang, and P. Kim, *Physical Review Letters* 98, no. 20: 206805. Copyright 2007 by the American Physical Society.)

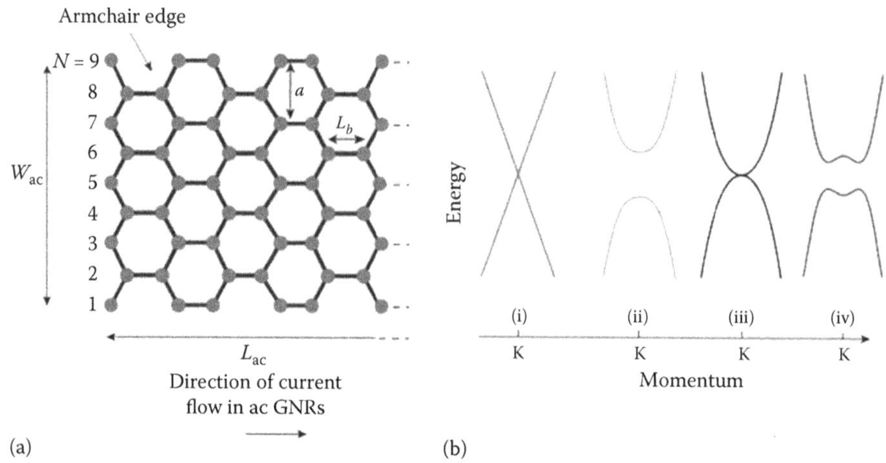

FIGURE 7.16
(a) Schematic of an armchair (ac) graphene nanoribbon (GNR) of length L_{ac} and width W_{ac}. The nanoribbon shown here has $N = 9$ carbon atoms along its width and thus belongs to the $3p$ family, where p is an integer. (b) Band structure around the K point of (i) large-area graphene, (ii) graphene nanoribbons, (iii) unbiased bilayer graphene, and (iv) bilayer graphene with an applied perpendicular field. Large-area graphene and unbiased bilayer graphene do not have a band gap, which makes them less useful for digital electronics. (Reprinted by permission from Macmillan Publishers Ltd., *Nature Nanotechnology*, F. Schwierz. Graphene transistors, 5, no. 7: 487–496, copyright 2010.)

cone-shaped). This decreases the curvature around the K point and increases the effective mass of the charge carriers [43], which is likely to decrease the mobility.

Bilayer graphene is also gapless (Figure 7.16b(iii)), and its valence and conduction bands have a parabolic shape near the K point. If an electric field is applied perpendicular to the bilayer, a band gap opens and the bands near the K point take on the so-called Mexican-hat shape (Figure 7.16b(iv)).

7.4.5 GNR Edges: Where the Action Is

The electronic properties in reduced dimensions are mainly governed by electronic correlations, their size and geometry, and consequent boundary conditions. The quasi one-dimensional ribbon-like structures in GNR offer two different possible edge geometries, namely, zigzag and armchair, as shown in Figure 7.17. They are termed zigzag graphene nanoribbons (ZGNRs) and armchair graphene nanoribbons (AGNRs), respectively, and show very different electronic properties arising from their contrasting boundary conditions. These edges have a 30° difference in their orientation within the graphene sheet [44]. The presence of these edges in GNRs causes the states to localize at the edges of the ribbon, which lead to nonzero magnetic moments on the carbon atoms. The main contribution to the local moments is said to come from dangling bonds on the edges. We assume that all dangling bonds at graphene edges are terminated by hydrogen atoms in order to stabilize the ribbon electronically and structurally, and, thus, do not contribute to the electronic states near the Fermi level. In the following, we will assume the Fermi energy to be zero. However, this level may move to values such as 0.2–0.4 eV, considering the interactions at the GNR with the substrate interface.

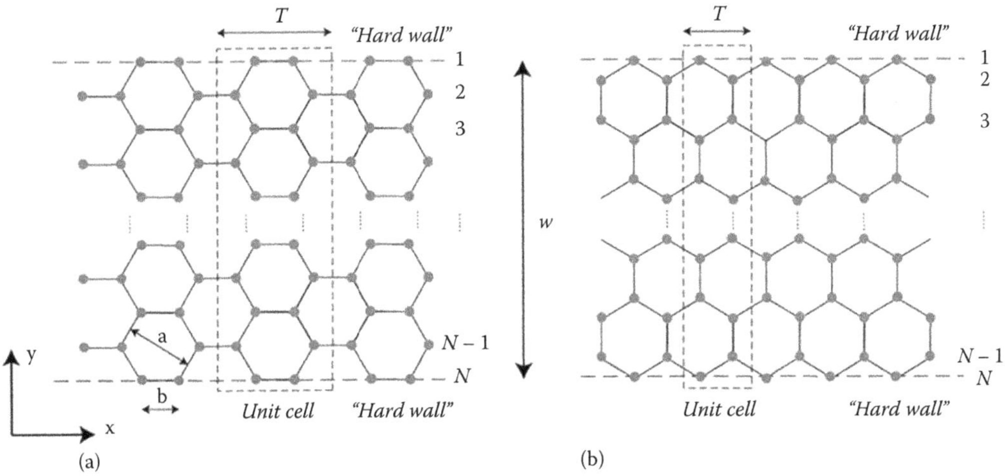

FIGURE 7.17
The schematic representation of cutting the two-dimensional graphene sheet to obtain zigzag and armchair nanoribbons: (a) armchair edges and (b) zigzag edges. (From A. Maffucci and G. Miano, *Applied Sciences* 4, no. 2 (2014): 305–317.)

As shown in Figure 7.17, the width, w, of a graphene nanoribbon is directly related to the integer, N, indicating the number of dimers (two carbon sites) for the armchair nanoribbons and the number of zigzag lines for the zigzag nanoribbons. For armchair GNRs, the unit cell has a length $T = 3a$ and a width $w = Na/2$, whereas for zigzag ones it is $T = a$ and $w = 3Na/2$.

Zigzag and armchair GNRs are metallic or semiconducting electronic properties that depend on the width of the nanoribbon. Zigzag GNRs are metallic and expected to be more conducting. The direct band gap decreases with w increasing and approaches zero in the limit of very large widths.

Note that the atoms along a zigzag edge come from the same sublattice, whereas the atoms from two different sublattices make bonds along the armchair edges. It is now experimentally possible to materialize such systems of varying widths with almost smoothly defined edges.

7.4.6 Band Structure of GNR

Given the number (N) and, thus, the GNR width (w), the energy spectrum of the π-electrons can be obtained by slicing the band structure of graphene. Figure 7.18 shows the band structure for an armchair GNR, assuming $N = 5$ and $N = 6$. For $N = 3p - 1$ ($p = 1, 2, \ldots$), there exists a subband, for which the direct band gap is zero; hence, the GNR behaves as a metal. For other values of p, the armchair GNR is semiconducting. The direct band gap decreases with increasing width and approaches zero in the limit of very large widths, consistent with the behavior of the graphene layer. Therefore, the value of the width (hence, the number, N) determines whether the nanoribbon is metallic or semiconducting.

Figure 7.19 refers to a zigzag GNR. Assuming $N = 5$ and $N = 20$, such nanoribbons have partially flat bands owing to the edge states; hence, their band gap is always zero, and their behavior is always metallic.

The research shows that the graphene nanoribbons act more like optical waveguides or quantum dots, allowing electrons to flow smoothly along the edges of the material. These electrons behave more like light going through an optical fiber. Because of the way

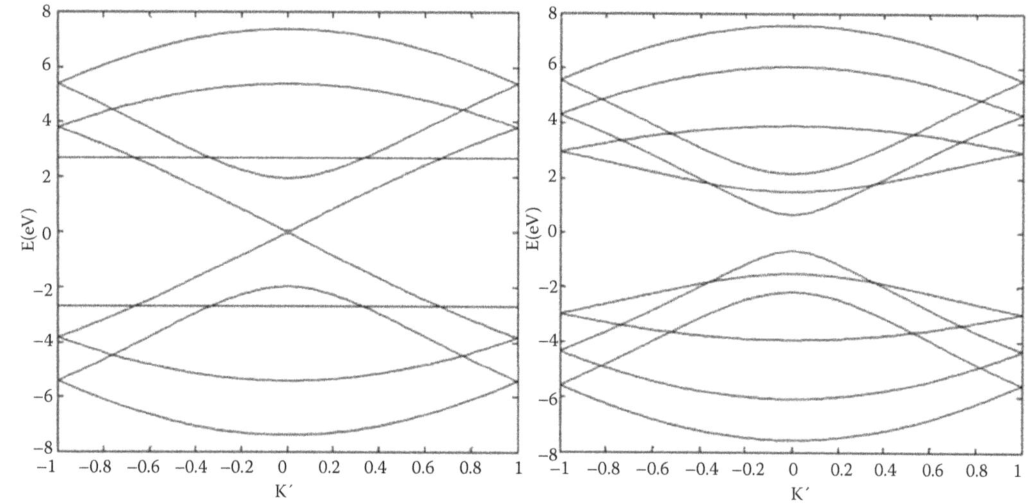

FIGURE 7.18
Energy band structure of an armchair nanoribbon versus the normalized wavenumber K′ = kl/π. (a) N = 5; (b) N = 6. (From A. Maffucci and G. Miano, *Applied Sciences* 4, no. 2 (2014): 305–317.)

the fiber is made, the light transmits without scattering. In ordinary conductors such as copper, resistance increases in proportion to the length as electrons encounter more impurities while moving through the conductor. Electrons in the graphene nanoribbons can move tens or hundreds of microns without scattering. It seems that the current primarily flows on the edges and the other electrons in the bulk portion of the nanoribbons do not interact with the electrons flowing at the edges.

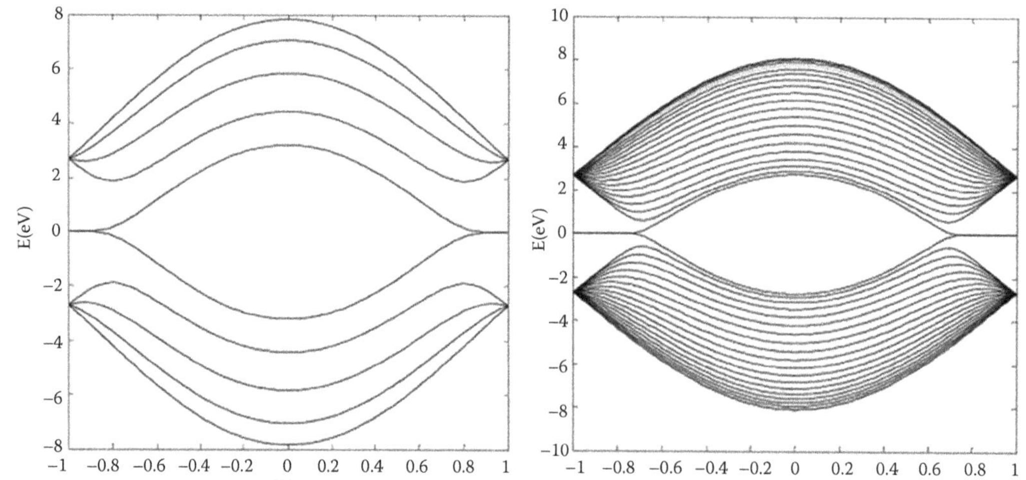

(a) (b)

FIGURE 7.19
The energy band structure of a zigzag nanoribbon versus the normalized wavenumber K′ = kl/π: (a) N = 5 and (b) N = 20. (From A. Maffucci and G. Miano, *Applied Sciences* 4, no. 2 (2014): 305–317.)

7.4.7 GNR Interconnects

In 1965 Gordon Moore observed that silicon transistors were undergoing a continual process of scaling downward. Since his observation, transistor minimum feature sizes have decreased from 10 micrometers to the 22–28 nanometer range in 2011. This exponential increase in the performance of integrated circuits comes from the decrease in dimensions of devices in submicron regime. A quick glance at Figure 7.13 shows that the continued down-scaling of the features on semiconductor chips is expected to reach 7 nm by 2024. For such devices the conventional Cu interconnects might not work because the interconnects need to withstand very high current density as well as efficient heat transfer (high thermal conductivity, k) to dissipate localized heating of the electronics. Graphene, carbon nanotubes, and GNRs are being tried as interconnects in the deep submicron VLSI domain to replace the traditional Cu/low-k interconnects [45]. When compared to Cu/low-k interconnects, graphene nanoribbons hold promising potentials for the following reasons: (a) higher conductivity, with electron mean free path as large as a few hundred nanometers; (b) resistance to electromigration due to the strong inherent carbon–carbon bonds, which accounts for an extraordinary mechanical strength and a very large current conduction capacity of ~10^9 A/cm^2 (as compared to ~10^6 A/cm^2 in Cu); and (c) larger in-plane thermal conductivity of approximately 5000 W/mK, very much comparable to SWNTs and single-crystal graphite and unlike Cu (112–226 W/mK).

7.4.8 GNR for Fuel Cells

Researchers at Rice University have come to realize that graphene's potential as a catalyst doesn't lie along the flat face but along the exposed edges, where molecules prefer to interact. The new material provides an abundance of active sites along the exposed edges for oxygen reduction reactions. Fuel cells turn hydrogen (or sources of hydrogen like methane) into electricity by oxidation of hydrogen at anode and reduction of oxygen at cathode. The by-product of the reaction being water makes the fuel cell an ideal source of a clean energy.

The reactions in most current fuel cells are catalyzed by platinum, but platinum's high cost has prompted the search for alternatives. The Rice team chemically unzipped carbon nanotubes into ribbons and then collapsed them into porous, three-dimensional aerogels (Figure 7.20), simultaneously decorating the ribbons' edges with boron and nitrogen molecules.

The nitrogen- or boron-doped aerogels show promise as a possible alternative to expensive platinum in fuel cells.

7.5 Future Challenges

Carbon-based nanostructures and, in particular, carbon nanotubes and graphene nanoribbons, due to their numerous potential and outstanding properties, have rather quickly attracted the interest of different scientific communities in the quest toward replacement of presently used electronics. However, while moving toward the 20 nm technology node, crucial bottlenecks are now going to be faced. At the nanoscale, where at times only a few hundreds/thousands of atoms are involved, it becomes of crucial importance to

FIGURE 7.20
An illustration shows a three-dimensional aerogel created by researchers at Rice University who combined graphene nanoribbons with boron and nitrogen. (Courtesy of Ajayan Group/Rice University.)

understand the nature of interfaces and possible dynamics taking place at an interface. As a matter of fact, metal contacts and electrodes on nanoscale objects behave differently when compared to micro- and macroscopic devices. Precise control over the nature of defects and their density within GNR is going to play a significant role, since the electronic properties are very sensitive to the presence of any defect and their specificity. There is a real possibility for new types of graphene-based electronics and for that a lot of fundamental physics needs to be done to understand how to make devices.

Even though the field of graphene research is still in its early stages, a blooming and active branch of chemistry and techniques have been developed, with the number of publications growing rapidly in the past few years [46].

Graphene nanoribbons may be producing a new type of electronic transport similar to what is observed in superconductors. Control of the edges and the single-crystal nature of graphene will be critical for future systems. Knowledge of how to do that, however, has yet to be gained.

Controlled growth of high-quality graphene at wafer scale will surely be further needed toward the full implementation of these materials and their architectures in commercially available devices. In fact, while downsizing elements below 100 nm size, interconnects face a number of important challenges. As the lateral dimension of interconnects approaches the mean free path of copper (~40 nm at room temperature), the impact of grain boundary scattering, surface scattering causes a rapid increase in the overall resistivity. A broad overview of such types of materials and architectures that might be suitable to replace current copper/low-k interconnects at various locations onto an integrated circuit is needed. A full implementation of these architectures into commercially available IC in the next few years is expected.

A full integration of CNTs and GNRs in a real microprocessor unit will require the precise and strict control of material properties as well as processes to place those elements at desired locations.

References

1. K.S. Novoselov, A.K. Geim, S. Morozov, D. Jiang, Y. Zhang, S.V. Dubonos, I.V. Grigorieva, and A. A. Firsov. Electric field effect in atomically thin carbon films. *Science* 306, no. 5696 (2004): 666–669.
2. A.K. Geim and K.S. Novoselov. The rise of graphene. *Nature Materials* 6, no. 3 (2007): 183–191.
3. A. Cresti, N. Nemec, B. Biel, G. Niebler, F. Triozon, G. Cuniberti et al. Charge transport in disordered graphene-based low dimensional materials. *Nano Research* 1, no. 5 (2008): 361–394.
4. V. Barone, O. Hod, and G.E. Scuseria. Electronic structure and stability of semiconducting graphene nanoribbons. *Nano Letters* 6, no. 12 (2006): 2748–2754.
5. Y.-W. Son, M.L. Cohen, and S.G. Louie. Energy gaps in graphene nanoribbons. *Physical Review Letters* 97, no. 21 (2006): 216803.
6. L. Yang, C.-H. Park, Y.-W. Son, M.L. Cohen, and S.G. Louie. Quasiparticle energies and band gaps in graphene nanoribbons. *Physical Review Letters* 99, no. 18 (2007): 186801.
7. X. Li, X. Wang, L. Zhang, S. Lee, and H. Dai. Chemically derived, ultrasmooth graphene nanoribbon semiconductors. *Science* 319, no. 5867 (2008): 1229–1232.
8. Z. Chen, Y.-M. Lin, M.J. Rooks, and P. Avouris. Graphene nano-ribbon electronics. *Physica E: Low-Dimensional Systems and Nanostructures* 40, no. 2 (2007): 228–232.
9. M.Y. Han, B. Özyilmaz, Y. Zhang, and P. Kim. Energy band-gap engineering of graphene nanoribbons. *Physical Review Letters* 98, no. 20 (2007): 206805.
10. V. Abramova, A.S. Slesarev, and J.M. Tour. Meniscus-mask lithography for narrow graphene nanoribbons. *ACS Nano* 7, no. 8 (2013): 6894–6898.
11. T. Albrecht, M. Dovek, M. Kirk, C. Lang, C. Quate, and D. Smith. Nanometer scale hole formation on graphite using a scanning tunneling microscope. *Applied Physics Letters* 55, no. 17 (1989): 1727–1729.
12. L. Tapasztó, G. Dobrik, P. Lambin, and L.P. Biró. Tailoring the atomic structure of graphene nanoribbons by scanning tunnelling microscope lithography. *Nature Nanotechnology* 3, no. 7 (2008): 397–401.
13. S.S. Datta, D.R. Strachan, S.M. Khamis, and A.C. Johnson. Crystallographic etching of few-layer graphene. *Nano Letters* 8, no. 7 (2008): 1912–1915.
14. X. Wang, Y. Ouyang, X. Li, H. Wang, J. Guo, and H. Dai. Room-temperature all-semiconducting sub-10-nm graphene nanoribbon field-effect transistors. *Physical Review Letters* 100, no. 20 (2008): 206803.
15. L. Jiao, L. Zhang, X. Wang, G. Diankov, and H. Dai. Narrow graphene nanoribbons from carbon nanotubes. *Nature* 458, no. 7240 (2009): 877–880.
16. X. Zhang, A.V. Whitney, J. Zhao, E.M. Hicks, and R.P. Van Duyne. Advances in contemporary nanosphere lithographic techniques. *Journal of Nanoscience and Nanotechnology* 6, no. 7 (2006): 1920–1934.
17. F. Burmeister, C. Schäfle, T. Matthes, M. Böhmisch, J. Boneberg, and P. Leiderer. Colloid monolayers as versatile lithographic masks. *Langmuir* 13, no. 11 (1997): 2983–2987.
18. L. Liu, Y. Zhang, W. Wang, C. Gu, X. Bai, and E. Wang. Nanosphere lithography for the fabrication of ultranarrow graphene nanoribbons and on-chip bandgap tuning of graphene. *Advanced Materials* 23, no. 10 (2011): 1246–1251.
19. A. Ferrari, J. Meyer, V. Scardaci, C. Casiraghi, M. Lazzeri, F. Mauri et al. Raman spectrum of graphene and graphene layers. *Physical Review Letters* 97, no. 18 (2006): 187401.
20. P. Weiss, M. Kamna, T. Graham, and S. Stranick. Imaging benzene molecules and phenyl radicals on Cu {111}. *Langmuir* 14, no. 6 (1998): 1284–1289.
21. M. Koepf, F. Chérioux, J.A. Wytko, and J. Weiss. 1D and 3D surface-assisted self-organization. *Coordination Chemistry Reviews* 256, no. 23 (2012): 2872–2892.
22. P. Han, K. Akagi, F. Federici Canova, H. Mutoh, S. Shiraki, K. Iwaya et al. Bottom-up graphene-nanoribbon fabrication reveals chiral edges and enantioselectivity. *ACS Nano* 8, no. 9 (2014): 9181–9187.

23. K. Müllen. Evolution of graphene molecules: Structural and functional complexity as driving forces behind nanoscience. *ACS Nano* 8, no. 7 (2014): 6531–6541.
24. M. El Garah, J.M. MacLeod, and F. Rosei. Covalently bonded networks through surface-confined polymerization. *Surface Science* 613 (2013): 6–14.
25. G.S. McCarty and P.S. Weiss. Formation and manipulation of protopolymer chains. *Journal of the American Chemical Society* 126, no. 51 (2004): 16772–16776.
26. B. Genorio, W. Lu, A.M. Dimiev, Y. Zhu, A.-R.O. Raji, B. Novosel et al. In situ intercalation replacement and selective functionalization of graphene nanoribbon stacks. *ACS Nano* 6, no. 5 (2012): 4231–4240.
27. D.V. Kosynkin, W. Lu, A. Sinitskii, G. Pera, Z. Sun, and J.M. Tour. Highly conductive graphene nanoribbons by longitudinal splitting of carbon nanotubes using potassium vapor. *ACS Nano* 5, no. 2 (2011): 968–974.
28. Y.N. Novikov and M.E. Vol'pin. Lamellar compounds of graphite with alkali metals. *Russian Chemical Reviews* 40, no. 9 (1971): 733–746.
29. C. Vallés, C. Drummond, H. Saadaoui, C.A. Furtado, M. He, O. Roubeau et al. Solutions of negatively charged graphene sheets and ribbons. *Journal of the American Chemical Society* 130, no. 47 (2008): 15802–15804.
30. D.V. Kosynkin, A.L. Higginbotham, A. Sinitskii, J.R. Lomeda, A. Dimiev, B.K. Price et al. Longitudinal unzipping of carbon nanotubes to form graphene nanoribbons. *Nature* 458, no. 7240 (2009): 872–876.
31. N. Behabtu, J.R. Lomeda, M.J. Green, A.L. Higginbotham, A. Sinitskii, D.V. Kosynkin et al. Spontaneous high-concentration dispersions and liquid crystals of graphene. *Nature Nanotechnology* 5, no. 6 (2010): 406–411.
32. S. Stankovich, D.A. Dikin, G.H. Dommett, K.M. Kohlhaas, E.J. Zimney, E.A. Stach et al. Graphene-based composite materials. *Nature* 442, no. 7100 (2006): 282–286.
33. S. Wolfe, C.F. Ingold, and R.U. Lemieux. Oxidation of olefins by potassium permanganate. Mechanism of alpha-ketol formation. *Journal of the American Chemical Society* 103, no. 4 (1981): 938–939.
34. S. Ryu, J. Maultzsch, M.Y. Han, P. Kim, L.E. Brus. Raman Spectroscopy of Lithographically Patterned Graphene Nanoribbons. *ACS Nano* 5 (2011): 4123–4130.
35. L. Li, A.-R.O. Raji, and J.M. Tour. Graphene-wrapped MnO_2–graphene nanoribbons as anode materials for high-performance lithium ion batteries. *Advanced Materials* 25, no. 43 (2013): 6298–6302.
36. L. Li, A.-R.O. Raji, H. Fei, Y. Yang, E.L. Samuel, and J.M. Tour. Nanocomposite of polyaniline nanorods grown on graphene nanoribbons for highly capacitive pseudocapacitors. *ACS Applied Materials & Interfaces* 5, no. 14 (2013): 6622–6627.
37. W.M. Arden. The international technology roadmap for semiconductors—Perspectives and challenges for the next 15 years. *Current Opinion in Solid State and Materials Science* 6, no. 5 (2002): 371–377.
38. F. Schwierz. Graphene transistors. *Nature Nanotechnology* 5, no. 7 (2010): 487–496.
39. G.E. Moore. No exponential is forever: But "Forever" can be delayed! [semiconductor industry]. *2003 IEEE International Solid-State Circuits Conference, 2003 Digest of Technical Papers ISSCC* 1 (2003): 20–23.
40. T.H. Ning and Y. Taur. *Fundamentals of Modern VLSI Devices*. Cambridge University Press, 1998.
41. D.J. Frank, Y. Taur, and H.-S.P. Wong. Generalized scale length for two-dimensional effects in MOSFETs. *IEEE Electron Device Letters* 19, no. 10 (1998): 385–387.
42. M. Evaldsson, I.V. Zozoulenko, H. Xu, and T. Heinzel. Edge-disorder-induced Anderson localization and conduction gap in graphene nanoribbons. *Physical Review B* 78, no. 16 (2008): 161407.
43. H. Raza and E.C. Kan. Armchair graphene nanoribbons: Electronic structure and electric-field modulation. *Physical Review B* 77, no. 24 (2008): 245434.
44. A. Maffucci and G. Miano. Electrical properties of graphene for interconnect applications. *Applied Sciences* 4, no. 2 (2014): 305–317.
45. A. Maffucci and G. Miano. Transmission line model of graphene nanoribbon interconnects. *Nanoscience and Nanotechnology Letters* 5, no. 11 (2013): 1207–1216.
46. A.K. Geim. Graphene: Status and prospects. *Science* 324, no. 5934 (2009): 1530–1534.

8

Graphene Quantum Dots

8.1 Introduction

Graphene quantum dots (GQDs) are single-up to five-layer formations of limited size, mostly 3 to 20 nm, fragments or pieces of graphene [1]. The presence of a graphene structure inside the dots not only endows them with some of the unusual properties of graphene but quantum confinement and edge effects resulting from converting two-dimensional graphene to zero-dimensional quantum dots yield novel electronic and optical properties [2–5]. These graphene-based carbon nanomaterials were first synthesized by Pan et al. in 2010 in order to exploit the optical properties that are directly associated with the quantum confinement and edge effects, and they obtained the first fluorescent GQDs about 9.6 nm in diameter [2]. The functionalized GQDs exhibited bright blue photoluminescence, which had never been observed in graphene sheets and graphene nanoribbons owing to their large lateral sizes.

The isolation of monolayer graphene in 2004 led to an explosion in research on graphene to exploit the extraordinary electronic, structural, electrical, and thermal properties possessed by this atomically thin two-dimensional crystal of carbon [6–9]. In particular, graphene emerged as the new candidate material for electronics due to its unusual electronic properties and its unique electronic spectrum, with a relativistic-like nature of charge carriers close to the Fermi level, high charge carrier mobility, ballistic transport, and a pronounced ambipolar effect [10]. However, compared to conventional semiconductors currently used in electronics, graphene does not possess a band gap. In order to exploit the electronic properties of graphene for the majority electronics market, currently dominated by semiconductor silicon, it is important to open a band gap in graphene to make graphene behave like a semiconductor for a controlled switching between a conductive and an insulating state [10,11]. Thus to overcome the limitation of its direct application in nanodevices, opening a band gap in graphene, so-called band gap engineering, has become an active area of research [11]. Heteroatom doping, chemical modification, and electrostatic field tuning are some of the methods to introduce band gap in graphene [11]. An alternative yet remarkable technique to introduce band gaps in graphene is to geometrically confine the size of graphene by producing nanometer-sized pieces of graphene, called graphene quantum dots, with a well-confined shape, which then exhibit pronounced quantum confinement and edge effects and novel electrical and optical properties [2,12], much like the traditional quantum dots (i.e., the semiconductor quantum dots) [4,13–15].

One of the most striking properties of semiconductor nanocrystals is the massive change in optical properties as a function of size. On excitation, the smaller the size of the quantum dot, the higher is the energy and intensity of emitted light [4,14]. The single, intense transition in a quantum dot as a result of quantum confinement can be tuned by changing the size of a quantum dot [14]. This is the most defining behavior of the quantum dot, that is, dependence of its optical behavior on its size, and its light-emitting powers have been put to good use as smart, luminescent materials with applications in biology as biological tracking agents and light-emitting-diode technology [16]. Semiconductor quantum dots have various electronic, electrochemical, sensing, and optoelectronic applications. Their use in biological applications such as bioimaging and biosensing is, however, limited due to their toxic characteristics [3,13].

The GQDs have become the focus of considerable interest in recent times as a new alternative to traditional semiconductor quantum dots. Band gap in GQDs not only makes them semiconducting but also endows them with size-dependent photoluminescence properties that can be utilized in optoelectronic applications [3,5,17]. Additionally, graphene quantum dots have advantages over semiconductor quantum dots in terms of toxicity and biocompatibility [3,18]. These zero-dimensional crystalline carbon nanomaterials, depending on their size, shape, and edges, possess excellent optical properties such as bright excitonic fluorescence, strong excitonic absorption bands, and long-term photostability, and are finding promising applications in bioimaging, biosensing, drug delivery, and photovoltaics [19,20].

8.2 Structure of Graphene Quantum Dots

The history of use of quantum dots starts with the discovery of semiconductor quantum dots. This section therefore starts with an introduction to semiconductor quantum dots.

8.2.1 Semiconductor Quantum Dots

Semiconductor nanocrystals or semiconductor quantum dots are tiny crystalline particles or fragments of semiconductor consisting of hundreds to thousands of atoms that exhibit size-dependent optical and electronic properties [14,15]. With typical dimensions in the range of 1 to 100 nm, these nanocrystals bridge the gap between small molecules and large crystals, displaying discrete electronic transitions reminiscent of isolated atoms and molecules, as well as enabling the exploitation of the useful properties of crystalline materials [15]. The quantum dots exhibit discrete structure in their optical and electrical characteristics, therefore, they are referred to as artificial atoms [14].

Semiconductor quantum dots or the traditional quantum dots were discovered by Ekimov and Onishchenko as a new class of materials exhibiting quantum size effects in the early 1980s [21]. The spectral positions of the exciton absorption lines of CuCl semiconducting microcrystals of varying sizes from tens of angstroms to hundreds of angstroms showed a significant short-wave shift as the size of the crystal decreased. Subsequent work on CdS nanocrystals showed that small nanometer-sized colloidal nanocrystals had a band gap larger than the bulk value shown by larger nanocrystals. Using a three-dimensional quantum confinement model for small nanometer-sized nanocrystals of semiconductors Luis Brus developed a relation between size and band gap for semiconductor nanoparticles and the word quantum dots was coined by physicists Daniel Chemla and David Miller [16,22]. This was an important advancement in the science and technology of quantum dots since semiconductor microelectronics was pushed down into the nanometer-size regime [22].

An important parameter that defines the limits of the size of a quantum dot is the exciton Bohr radius, an electrostatically bound electron–hole pair lowest energy state when a photon is absorbed by a semiconductor [4,15]. The exciton Bohr radius is a characteristic of the semiconducting material [15]. The exciton Bohr radius is of the order of a few nanometers and so are the diameters of most semiconductor quantum dots. The quantum size effect occurs when these nanostructures attain a size of the order of exciton Bohr radius. The charge carriers become spatially confined by the boundary (edges) and the energy difference between the electronic energy states of a semiconductor increases. A nanometer-sized nanocrystal has discrete electronic states and in all essential aspects is a molecule. The quantum confinement effect restricts the free movement of excitons in all directions in the quantum dot, and the nanocrystals demonstrate size-dependent absorption and fluorescence spectra with discrete electronic transitions [15]. The emission color shifts to higher energy as the size of the quantum dot decreases.

The size-tunable photoluminescence exhibited by quantum dots is exploited in a variety of optical, electrochemical, and biomedical applications such as in smart, luminescent materials, as fluorescent markers for bioimaging, and LEDs [3,13–15,23]. For example, the band gap in CdSe can be tuned from deep red (1.7 eV) to green (2.4 eV) by reducing the cluster diameter from 200 to 20 Å. The nanocrystal emission shifts with size and this narrow-band size tunable luminescence provides LEDs made of quantum dots, which offer a variety of means for tuning the output color without changing the fabrication method [14].

8.2.2 Graphene Quantum Dots

Though semiconductors have a characteristic finite exciton Bohr radius, graphene, due to the linear dispersion relation of the charge carriers has an infinite exciton Bohr radius [5,24]. This results in quantum confinement effect for graphene of any finite size resulting in many interesting phenomena not exhibited by graphene or other semiconductors [5,24]. In principle, the band gap of graphene can be tuned from 0 eV to that of benzene by varying their sizes [5].

Graphene quantum dots are zero-dimensional nanocrystals with single- up to five-layer (about 2.5 nm) formations of limited size (mostly 3 to 20 nm) fragments or pieces of graphene [1]. The shape of graphene quantum dots can vary from circular or elliptical to triangular, quadrate, and hexagonal. The edges can be zigzag or armchair type and can affect the luminescence and absorption bands of GQDs [2,3,25]. Generally, the surface groups such as epoxy, hydroxyl, carbonyl, and carboxyl groups of GQDs are similar to graphene. They have a similar crystalline nature as graphene [1]. A band gap exists in GQDs due to quantum confinement and edge effects, unlike graphene, which is a zero-band gap semiconductor [24,26].

The GQDs have the advantage over semiconductor quantum dots due to their superior properties like high chemical and photostability, low toxicity, and biocompatibility, which are especially important for biomedical applications [3].

8.3 Synthesis of Graphene Quantum Dots

The initial studies to engineer band gap in graphene nanostructures to study quantum confinement of charge carriers were based on fabrication of lithographically patterned graphene nanoribbon (GNR) and graphene quantum dot structures where the confinement of charge carriers creates an energy gap near the charge neutrality point [27,28]. These methods demonstrated the ability to engineer the band gap of graphene nanostructures by lithographic processes. However, though GNR- and GQD-based electronic devices can be created by electron beam lithography, this physical approach is limited by the need for expensive equipment and difficulties in obtaining smooth edges [1,2]. These drawbacks can be overcome by chemical routes that also allow surface functionalization of GQDs [3]. Several methods have been developed to prepare GQDs. These methods are classified into two categories, namely, top-down and bottom-up methods [3].

8.3.1 Top-Down Methods

The top-down approach involves breaking down carbonaceous material into GQDs by cleavage or exfoliation using chemical, electrochemical, or physical methods.

8.3.1.1 Exfoliation Using Acids

Exfoliation using acids involves exfoliation of sp^2-hybridized carbon materials such as carbon fibers [29], carbon nanotubes, graphene oxide [17], soot, coal [30], carbon black, and activated carbon by oxidation with strong acids [3]. The advantage of this method is the production of GQDs from readily available carbon sources by solution processing. In one of the studies (Figure 8.1) [30], coals ranging from bituminous coal to anthracite to coke were sonicated in nitric and sulfuric acids and heated for 24 hours. Different types of coals produced different types of fluorescent dots ranging from 2 to 40 nm. The synthesized graphene quantum dots, produced in up to 20% isolated yield from coal were soluble and fluorescent in aqueous solution, providing promise for applications in areas such as bioimaging, biomedicine, photovoltaics, and optoelectronics, in addition to being inexpensive additives for structural composites [30].

Acidic exfoliation of traditional micrometer-sized pitch-based carbon fibers leads to the creation of GQDs with different size distribution and different luminescent characteristics (Figure 8.2) [29]. The as-produced GQDs, in the size range of 1 to 4 nm, mostly possess semiconducting zigzag edge structure and one to three graphene layers. Luminescing GQDs of different sizes of 1–4 nm, 4–8 nm, and 7–11 nm with corresponding emission colors of blue, green, and yellow are obtained by varying processing temperature during synthesis of GQDs by exfoliation of carbon fibers [29]. Due to the luminescence stability, biocompatibility, low toxicity, and high water solubility, these GQDs are demonstrated to be excellent probes for high-contrast bioimaging and biosensing applications.

Using the top-down approach, cage opening of fullerene C_{60} was attempted by subjecting it to a mixture of concentrated sulfuric acid (H_2SO_4), sodium nitrate ($NaNO_3$), and potassium permanganate ($KMnO_4$) according to the modified Hummers method (Figure 8.3) [31]. This treatment of fullerene C_{60} with a mixture of strong acid and chemical oxidant induced the oxidation, cage-opening, and fragmentation processes of fullerene yielding very small GQDs about 2 to 3 nm in diameter and with strong luminescent properties and with the highest intensity at 460 nm under a 340 nm excitation wavelength. Further

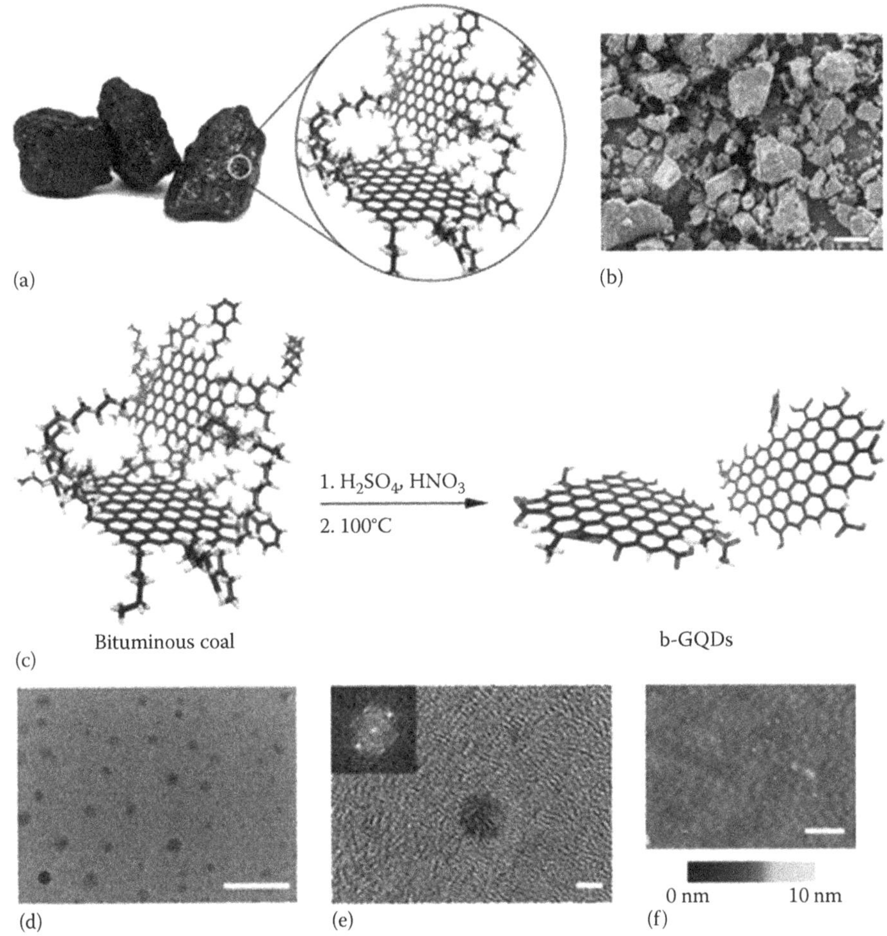

FIGURE 8.1
(a) Macroscale image and simplified illustrative nanostructure of coal. (b) SEM image of ground bituminous coal with sizes ranging from one to hundreds of microns in diameter. Scale bar, 50 μm. (c) Schematic illustration of the synthesis of bituminous coal GQDs (b-GQDs). (d) TEM image of b-GQDs showing a regular size and shape distribution. Scale bar, 20 nm. (e) HRTEM image of representative b-GQDs from (d); the inset is the 2D FFT image that shows the crystalline hexagonal structure of these quantum dots. Scale bar, 2 nm. (f) AFM image of b-GQDs showing height of 1.5–3 nm. Scale bar, 100 nm. (Reprinted by permission from Macmillan Publishers Ltd., *Nature Communications*, R. Ye, C. Xiang, J. Lin, Z. Peng, K. Huang, Z. Yan et al., *Nature Communications* 4, Article No. 2943, copyright 2013.)

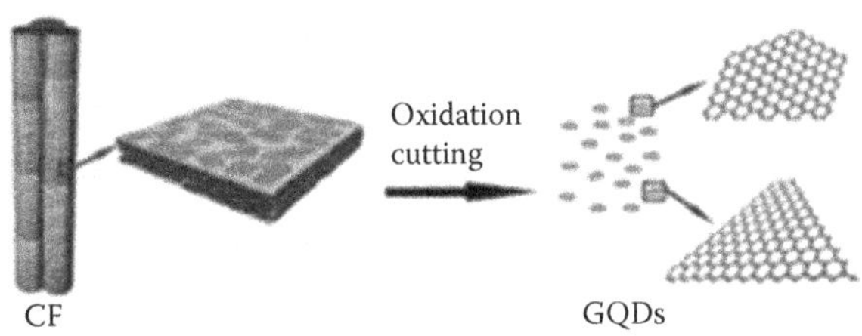

FIGURE 8.2
Representation scheme of oxidative cutting of carbon fibers into GQDs. (Reprinted with permission from J. Peng, W. Gao, B.K. Gupta, Z. Liu, R. Romero-Aburto, L. Ge et al., *Nano Letters* 12, no. 2: 844–849. Copyright 2012 American Chemical Society.)

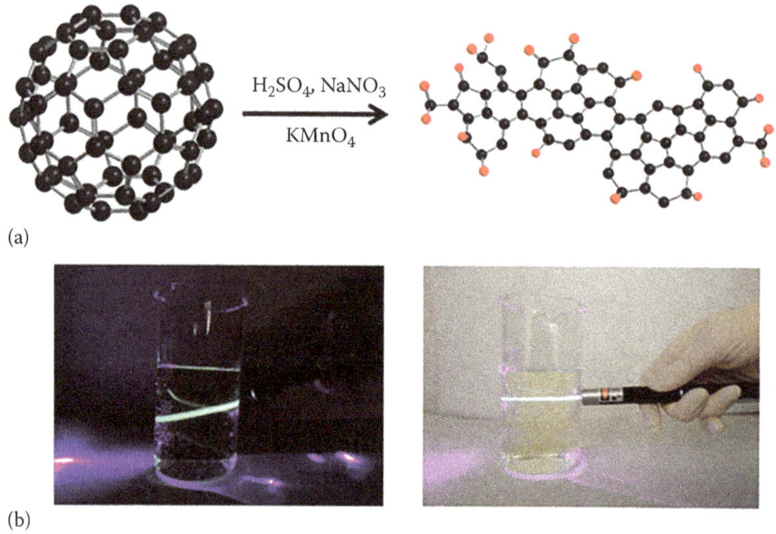

FIGURE 8.3
(a) Representation of oxidation and cage opening of fullerene C_{60} to obtain GQDs. (b) Luminescence of GQDs excited with a blue laser pointer (405 nm). (From C.K. Chua, Z. Sofer, P. Simek, O. Jankovský, K. Klímová, S. Bakardjieva, S. Hrdličková Kučková, and M. Pumera, *ACS Nano* 9, no. 3 (2015): 2548–2555.)

chemical treatments with hydrazine hydrate and hydroxylamine resulted in redshift and blueshift of the luminescence, respectively. The simplicity of this method in producing GQDs demonstrated potential for further development for integration into practical devices or applications including optoelectronics and biological labeling.

In another study, graphene oxide prepared by the modified Hummers method was used as starting material to obtain GQDs [17]. Two different types of fluorescent GQDs of sizes less than 10 nm were obtained by oxidative shearing of graphene oxide using $KMnO_4/H_2SO_4$ [17]. Depending on the duration of oxidation either bright blue luminescent or green luminescent GQDs were obtained, which possessed a size-dependent shift of emission.

8.3.1.2 Electrochemical Exfoliation

In this method graphene quantum dots are obtained by electrochemical cleavage of carbon precursors such as graphite rods, chemical vapor deposition (CVD)-grown graphene, or carbon nanotubes [1]. The process involves ionic liquid-assisted electrochemical exfoliation of the graphite or graphene or carbon nanotube electrode (anode) as a result of anodic oxidation of water and anionic intercalation from the ionic liquid, which releases GQDs. A water-miscible ionic liquid mixed with water is generally used as the electrolyte for the electrochemical exfoliation of graphite. Fluorescent GQDs exhibiting different emission colors like red, yellow, green, or blue have been synthesized by electrochemical exfoliation.

One of the approaches involved electrochemical exfoliation of graphite followed by reduction of the as-synthesized nanoscale GQDs with hydrazine at room temperature resulting in water-soluble GQDs of mostly single-layer graphene of uniform size 5–10 nm and a strong yellow fluorescence in high quantum yield of 14% [32]. The mechanism for the generation of the strong yellow luminescent GQDs was also proposed. The first step is electrochemical oxidative cleavage of the graphite anode. In this process, the O and OH radicals produced by anodic oxidation of water can serve as electrochemical "scissors" to cut carbon nanocrystals and form oxygenated groups. The resultant black solution is a mixture of graphene oxides and graphite oxide nanocrystals with the presence of carboxyl, OH, and carbonyl groups at the edge of the graphene sheet, while the basal plane is covered with mostly epoxide and OH groups. Electrostatic repulsion between these negatively charged groups in the electrochemically oxidized solution results in a stable solution. The second step involves reduction and modification of graphene with hydrazine at room temperature to yield a homogeneous colloidal solution with hydrazine reducing only epoxide and OH groups, and reacting with the carboxylic groups at the edge of graphene leading to the formation of hydrazide groups and yellow photoluminescence emission. The solution showed high photoluminescence stability and potential in cellular imaging [32].

In another approach, free-standing CVD-grown 3D graphene was electrochemically exfoliated using room temperature ionic liquid (1-butyl-3-methylimidazolium hexafluorophosphate [$BMIMPF_6$] in acetonitrile) as the electrolyte, to produce GQDs with a blue fluorescence [20]. The proposed mechanism of electrochemical cutting was attributed to three steps: (1) high electrical stress by the applied voltage; (2) the ability of PF_6^- to intercalate between the graphene layers; (3) intimate interaction (π–π or cation–π interaction) between $BMIM^+$ group and GQD [20]. High-resolution transmission electron microscopy (TEM) and atomic force microscopy (AFM) studies revealed uniform and highly crystalline mostly single-layered GQDs with average lateral diameter of 3 nm and a narrow size distribution. Among the different ionic liquids including 1-butyl-3-methyl imidazolium chloride (BMIMCl) and tetraethylammonium chloride (TEAC), $BMIMPF_6$ gave the highest yield. The tetrabutylammonium hexafluorophosphate can also effectively exfoliate GQDs from 3D graphene, suggesting the facilitating role of PF_6^- ions in cleaving graphene substrate. $BMIM^+$, on the other hand, assists to disperse GQDs in both organic solvents and water due to its amphiphilic property. The GQD aqueous dispersions were highly stable.

In one of the methods water-soluble, 3 nm uniform-sized GQDs with red emission were prepared by electrochemical exfoliation of graphite in $K_2S_2O_8$ solution. These GQDs with excellent photoluminescent properties and less cytotoxicity show great potential as biological labels for *in vivo* cellular imaging [33]. Electrochemical transformation of multiwalled carbon nanotubes (MWCNTs) to GQDs via unzipping of the carbon nanotubes has also been successfully used to synthesize GQDs [34].

8.3.1.3 Hydrothermal/Solvothermal Synthesis

The hydrothermal/solvothermal (H/S) synthesis is one of the most frequently used methods to synthesize GQDs. Xu et al. [35] have reviewed the work on hydrothermal/solvothermal synthesis of GQDs.

The hydrothermal route was the method used by Pan et al. [2] for the first chemical synthesis of GQDs (Figure 8.4). Water-soluble blue luminescent quantum dots were obtained by this method. Pan et al. used thermally reduced graphene oxide sheets as the starting material for GQDs [2]. Before the hydrothermal treatment, thermally reduced graphene oxide is subjected to acidic oxidation using nitric acid or ozone under mild sonication. As a result, oxygen-containing functional groups such as epoxy (C-O-C), carbonyl (C = O), hydroxyl (OH), or carboxyl (COOH) are introduced on the basal plane and at the edges of graphene lattice [2,3]. The presence of these groups makes graphene sheets soluble in water. This is followed by hydrothermal treatment of oxidized graphene sheets (200°C) during which deoxidization occurs. The changes in the graphene structure before and after the hydrothermal treatment can be monitored by TEM in terms of changes in the interlayer spacing. The

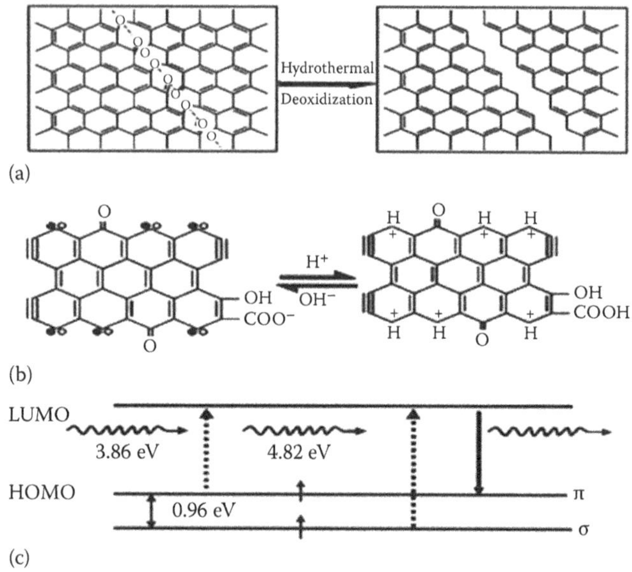

FIGURE 8.4
(a) Mechanism for the hydrothermal cutting of oxidized graphene sheets (GSs) into GQDs: a mixed epoxy chain composed of epoxy and carbonyl pair groups (left) is converted into a complete cut (right) under the hydrothermal treatment. (b) Models of the GQDs in acidic (right) and alkali (left) media. The two models can be converted reversibly depending on pH. The pairing of σ (●, filled circles) and π (○, open circles) localized electrons at carbene-like zigzag sites and the presence of triple bonds at the carbene-like armchair sites are represented. (c) Typical electronic transitions of triple carbenes at zigzag sites observed in the optical spectra of GQDs. (D. Pan, J. Zhang, Z. Li, and M. Wu, *Advanced Materials*. 2010. 22. No. 6. 734–738. Copyright Wiley-VCH Verlag GmbH & Co. KGaA. Reproduced with permission.)

Fourier transform infrared spectroscopy (FTIR) spectrum is a useful guide to confirm presence and weakening or disappearance of the strongest vibrational absorption band of C = O/COOH at 1720 cm^{-1} and the vibration band of epoxy groups at 1052 cm^{-1}. The deoxidization process takes place in alkaline medium (NaOH, ammonia) when fragile epoxy and carbonyl groups are attacked leading to cutting of the graphene sheet. The diameter or size distribution of GQDs obtained from TEM images showed a diameter distributed in the range of 5 to 13 nm. The corresponding height profiles obtained from AFM images of GQDs prepared under mild alkaline conditions (pH = 8) showed topographic heights mostly between 1 and 2 nm with more than 85% of the GQDs consisting of one to three layers. By using strong alkaline hydrothermal conditions (pH = 12), well-crystallized green fluorescent GQDs with an average diameter of 3 nm and two to three graphene layers have been obtained [5].

Strongly green fluorescent GQDs from graphite oxide were obtained by solvothermal synthesis using dimethylformamide (DMF) as the solvent to dissolve graphite oxide followed by sonication and heating at 200°C for a few hours to get GQDs. The average diameter of GQDs was 5.3 nm (from TEM) and were mostly single layer or bilayer (average height 1.2 nm from AFM). The GQDs possessed strong fluorescence with a high photoluminescence quantum yield of 11.4%. The GQDs were soluble in water and polar organic solvents [36].

By using oxidized graphene sheets and ammonia, optically tunable amine-functionalized graphene quantum dots (NH_2-GQDs) with diameters 2.5 nm and thickness 1.1 nm have been synthesized by this method [37]. The luminescence of GQDs from violet to yellow could be controlled by varying the ammonia concentration [37].

8.3.1.4 Microwave and Ultrasonic Routes

Microwave-assisted synthesis combines the advantages of both the hydrothermal and microwave processes [4]. Microwave-assisted exfoliation of graphene nanosheets in acidic conditions is a facile route for synthesis of GQDs [38]. Microwave irradiation shortens the reaction time by rapid and uniform heating of the reaction medium, thus offering a simple and efficient synthetic route to produce GQDs. Figure 8.5 [38] shows the schematic of basic steps involved in the synthesis procedure. Greenish-yellow luminescent GQDs (g GQD) of average size 4–5 nm were obtained by acidic exfoliation. The as-synthesized GQDs were further moderately reduced with $NaBH_4$, and blue GQDs were produced with the same dimensions. The quantum yield of blue and green GQDs was 23% and 12%, respectively.

In a study on doped GQDs, nitrogen-doped GQDs (N-GQDs) were prepared by mixing 3 wt% of glucose dissolved in aqueous ammonia (25%) at room temperature. The homogeneous solution was heated in a microwave reactor (300 W power) for 5 min at 180°C. The GQDs obtained were 6 nm in size and exhibited excitation-dependent luminescence [39].

Highly ordered, uniformly sized GQDs were prepared from amorphous activated carbon and hydrogen peroxide raw materials using microwave and hydrothermal treatment [40]. The route enabled the difficult transformation from amorphous carbon to highly ordered GQDs for the first time while completely avoiding the use of concentrated sulfuric acid. Figure 8.6 [40] shows a schematic of the hydrothermal-microwave route.

The synthesized GQDs with average diameter of 7 nm and with a thickness up to five layers showed strong blue fluorescence. The evolution of the XRD (x-ray powder diffraction) spectra in the sequence raw activated carbon, activated carbon after microwave treatment for 3 minutes, activated carbon after microwave treatment for 6 minutes, and finally

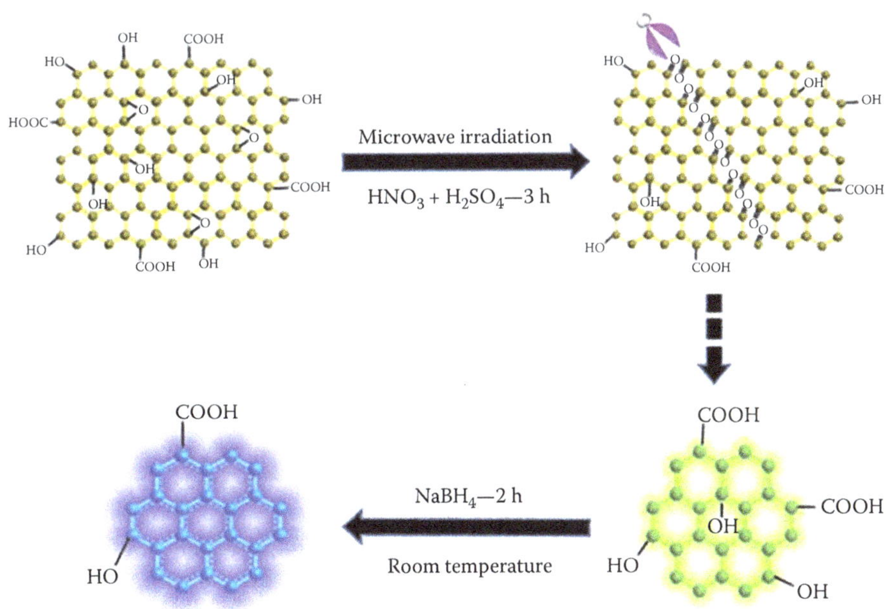

FIGURE 8.5
Schematic representation of mechanism of microwave-assisted synthesis to obtain green and blue GQDs. (L.L. Li, J. Ji, R. Fei, C.Z. Wang, Q. Lu, J.R. Zhang, L.P. Jiang, and J.J. Zhu: *Advanced Functional Materials*. 2012. 22. No. 14. 2971–2979. Copyright Wiley-VCH Verlag GmbH & Co. KGaA. Reproduced with permission.)

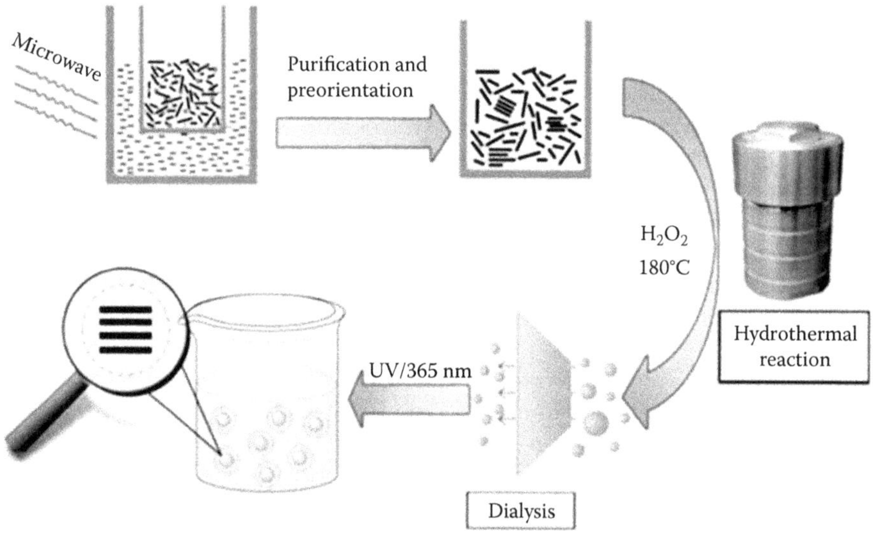

FIGURE 8.6
Schematic of the hydrothermal-microwave route. (From Y. Huang, C. Bai, K. Cao, Y. Tian, Y. Luo, C. Xia, S. Ding, Y. Jin, L. Ma, and S. Li, *RSC Advances* 4, no. 81 (2014): 43160–43165. Reproduced by permission of The Royal Society of Chemistry.)

GQDs revealed that microwave irradiation leads to a preordering process of activated carbon and microwave-assisted removal of organic impurities during rapid, uniform, volumetric, and energy-efficient heating by microwave irradiation [40].

Exfoliation of layered carbon structures by high energy ultrasonic waves is a useful technique to produce GQDs. Luminescent graphene quantum dots of 3 nm size in high quantum yield were prepared using ultrasonication with only graphene oxide and $KMnO_4$ [41].

8.3.2 Bottom-Up Methods

Bottom-up methods involve synthesis of GQDs using organic synthesis techniques such as pyrolysis or carbonization of hydrocarbons or stepwise condensation or fusion of small aromatic molecules.

8.3.2.1 Catalytic Decomposition of Fullerene C_{60}

Fullerenes can be opened up at high temperature using ruthenium as a catalyst to produce GQDs [12]. In this method quantum dots are generated by decomposing C_{60} molecules at high temperatures on a ruthenium metal surface. The metal acts as a catalyst and causes the C_{60} to break down into carbon clusters. Since each fullerene molecule has the same shape and size, by carefully controlling the density of the clusters on the surface, cluster aggregation can be controlled and different-shaped clusters (flower-shaped and hexagonal) merge and crystallize into geometrically well-defined, hexagonal-shaped GQDs at temperatures of around 825 K. Scanning tunneling microscopy is a useful tool to observe how the carbon clusters diffuse onto the metal surface and how they aggregate to form quantum dots [12].

8.3.2.2 Pyrolysis or Carbonization

Carbonization of organic precursors is an effective bottom-up strategy to synthesize GQDs [3]. These methods generally involve heating small organic precursors and subsequent condensation and nucleation to form larger GQDs. Several organic precursors that have been used include citric acid [42], pyrene [43], and L-glutamic acid [44].

Figure 8.7 [45] shows multicolored photoluminiscent GQDs prepared from hexa-peri-hexabenzocoronene (HBC). HBC is a polycyclic aromatic hydrocarbon that resembles nanoscaled fragments of graphene that stack via π–π interactions. This method produced monodisperse disk-like GQDs about 60 nm in size and 2–3 nm in thickness [45].

In another study, highly fluorescent GQDs were synthesized using a one-step pyrolysis of a natural amino acid, L-glutamic acid, using only a heating mantle [44]. The amino acid served as a precursor to synthesize N-doped GQDs since the amino group on the L-glutamic acid surface provided active nitrogen groups for surface modification of GQDs. The GQDs showing strong blue, green, and red luminescence under the irradiation of ultraviolet, blue, and green light were obtained, respectively, and extending up to near infrared (NIR) with excellent quantum yield of 55%. The residue L-glutamic acid from pyrolysis could act as a surface passivation agent, resulting in the extremely stable and highly fluorescent GQDs in solution.

In another approach, instead of conventional high-temperature carbonization methods, concentrated H_2SO_4 was used to carbonize hydrophobic polycyclic aromatic hydrocarbons (PAHs), such as pyrene, benzo[a]pyrene, or naphtho[2,3–1]pyrene, and introduce hydrophilic functional groups (Figure 8.8) [43]. During the carbonization process, small graphite fractions could assemble into larger graphitic structures. After the acid treatment, the resulting artificial graphite was exfoliated by sonication and graphene oxide solution

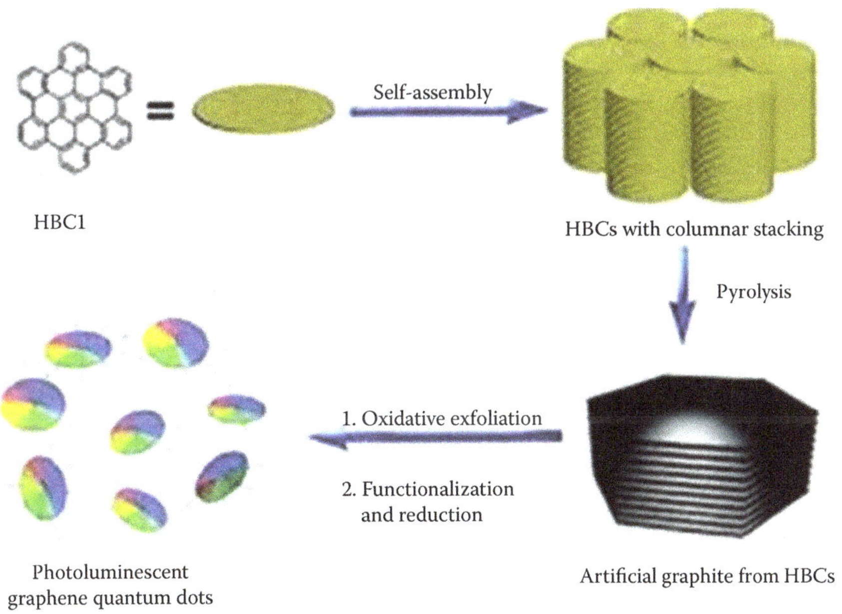

FIGURE 8.7
Schematic representation of multicolored photoluminiscent GQDs prepared from hexa-peri-hexabenzocoronene (HBC). (Reprinted with permission from R. Liu, D. Wu, X. Feng, and K. Müllen, *Journal of the American Chemical Society*, 133, no. 39: 15221–15223. Copyright 2011 American Chemical Society.)

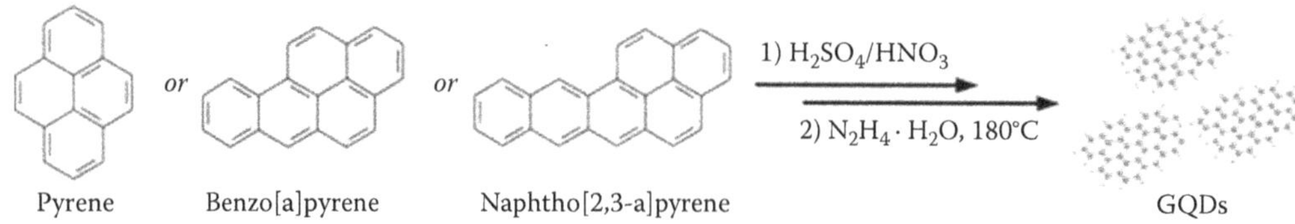

FIGURE 8.8
Schematic illustration of the GQDs synthesis from commercially available PAHs. (L. Zhou, J. Geng, and B. Liu: *Particle & Particle Systems Characterization*. 2013. 30. No. 12. 1086–1092. Copyright Wiley-VCH Verlag GmbH & Co. KGaA. Reproduced with permission.)

was obtained. The reduction of graphene oxide by hydrazine hydrate treatment was carried out under the hydrothermal condition to yield water-soluble GQDs in solution. Acid treatment of PAHs followed by hydrothermal reduction with hydrazine hydrate afforded graphitic GQDs with numerous oxygen-containing functional groups resulting in excellent water solubility of GQDs. The obtained photoluminescent GQDs having sizes 5–10 nm and thicknesses 0.5–2 nm (one to three graphene layers) exhibited stable and tunable fluorescence, showed low toxicity, and were effective for ferric ions (Fe^{3+}) and hydrogen peroxide sensing [43]. The fluorescence quantum yields of P-GQDs, B-GQDs, and N-GQDs obtained from pyrene, benzo[a]pyrene, or naphtho[2,3–l]pyrene, respectively, were measured to be 11.7%, 4.8%, and 2.2%. The wavelength of the strongest emission peak and the range of emission wavelength for these GQDs can be easily tuned by choosing different PAH precursors.

8.3.2.3 Organic Synthesis Route

Graphene quantum dots can also be synthesized via stepwise solution chemistry [3,5]. Large, stable colloidal graphene quantum dots with uniform and tunable size and shape were obtained by stepwise-solution-based organic synthetic route [46]. Graphene quantum dots containing 168, 132, and 170 conjugated carbon atoms were obtained through fusion of graphene moieties obtained by oxidative condensation of aryl groups of polyphenylene dendritic precursors through stepwise solution chemistry (Figure 8.9) [46]. Trialkyl phenyl groups covalently attached to the edge of graphene rendered them with excellent solubility in common organic solvents, such as toluene, tetrahydrofuran (THF), and chloroform.

In another bottom-up approach, highly fluorescent hydroxyl- and amine-functionalized water-soluble GQDs in high yield up to 63% and long-term photostability were synthesized under milder hydrothermal conditions [19]. The method was based on water-phase molecular fusion of pyrene. Pyrene and substituted derivatives such as 1,3,6-trinitropyrene

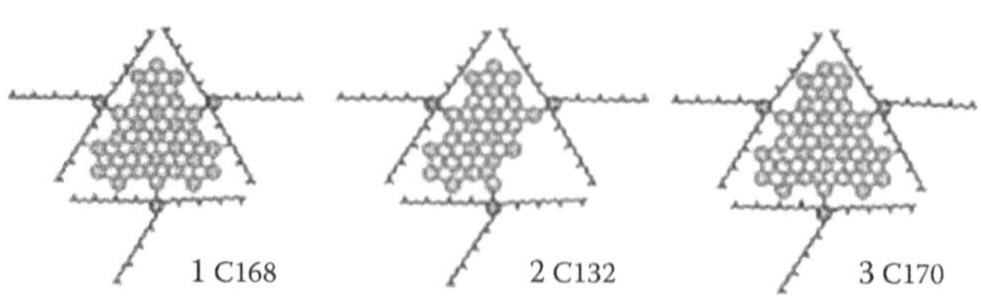

FIGURE 8.9
Schematic representation of different-sized quantum dots obtained by the bottom-up approach. (Reprinted with permission from X. Yan, X. Cui, and L.S. Li, *Journal of the American Chemical Society* 132, no. 17: 5944–5945. Copyright 2010 American Chemical Society.)

consist of four perifused benzene rings with a unique carbon skeleton similar to the primitive cell of graphene. This synthetic design was based on the controlled fusion or condensation of such graphene-cell molecules into single-crystalline GQDs [19]. The synthesis involved nitration of pyrene to 1,3,6-trinitropyrene in nitric acid followed by hydrothermal treatment in alkaline aqueous solutions, where alkaline species play a crucial role in control over the size, edge functionalization (-OH, $-NH_2$, $-NHNH_2$), and optical properties of GQDs. The three nitro groups in substituted pyrene were sites for nucleophilic substitution reactions with alkaline species such as -OH, NH_3, and NH_2NH_2 added to the hydrothermal media. Compared with the multistep organic-phase fusion, the water-phase fusion is simpler, greener, milder, lower in cost, and higher in overall product yield [19]. Moreover, the resulting water-soluble GQDs were almost perfect single crystals whose edge sites were well passivated by short electron-donating moieties rather than by long-chain, electron-accepting phenyl moieties. Green-, yellow-, cyan-, and blue-colored GQDs were obtained depending on the type of functional group at the edge site.

8.4 Characterization of Graphene Quantum Dots

Graphene quantum dots can be characterized by a variety of techniques such as x-ray powder diffraction (XRD), Raman spectroscopy, Fourier transform infrared spectroscopy (FTIR), x-ray photoemission spectroscopy (XPS), atomic force microscopy (AFM), and transmission electron microscopy (TEM).

XRD is a useful tool to characterize graphitization in the sample. Identification of the characteristic (002) peak in the XRD pattern of GQDs reveals the extent of graphitization. The typical value for (002) interlayer spacing is 3.34 Å in bulk graphite [2] and a value close to this reveals the presence of a graphitic lattice in the sample and thus reflects crystallinity of the synthesized GQDs.

Raman spectroscopy also gives a measure of the crystalline quality of synthesized GQDs by analyzing the characteristic ordered G band (1580 cm^{-1}), the disordered D band (1350 cm^{-1}), and their relative intensity [2,32].

FTIR spectra of GQDs give useful structural information about the graphene structure and oxidation or functionalization of the same by detection of the characteristic vibrational bands of sp^2 hybridized carbon atoms (1600 cm^{-1}) and presence or absence of functional groups such as carboxyl, hydroxyl, carbonyl, epoxy, and amines in the sample. Additional information about chemical composition of GQDs, types of functional groups, and presence of graphitic-like sp^2 carbons is obtained from XPS.

Thickness of GQDs can be measured by AFM. Height profile and height distribution in the AFM images reveal thickness of the sample, hence the number of layers in the GQDs. The number of layers can vary from one to five or even more depending upon the synthesis methods. Size distribution or the lateral diameter of GQDs is revealed by TEM and the graphitic lattice of GQDs can be clearly resolved under high-resolution TEM (HRTEM) [20]. The hexagonal lattice plane spacing and hence the honeycomb lattice structure visible in HRTEM images serves as evidence for the formation of GQDs and their uniformity or crystallinity.

Wang et al. [19] synthesized hydroxyl- and amine-functionalized GQDs via alkali-mediated hydrothermal molecular fusion using an active PAH molecule as the precursor. A detailed structural analysis (Figures 8.10 and 8.11) [19] and interpretation of the GQDs

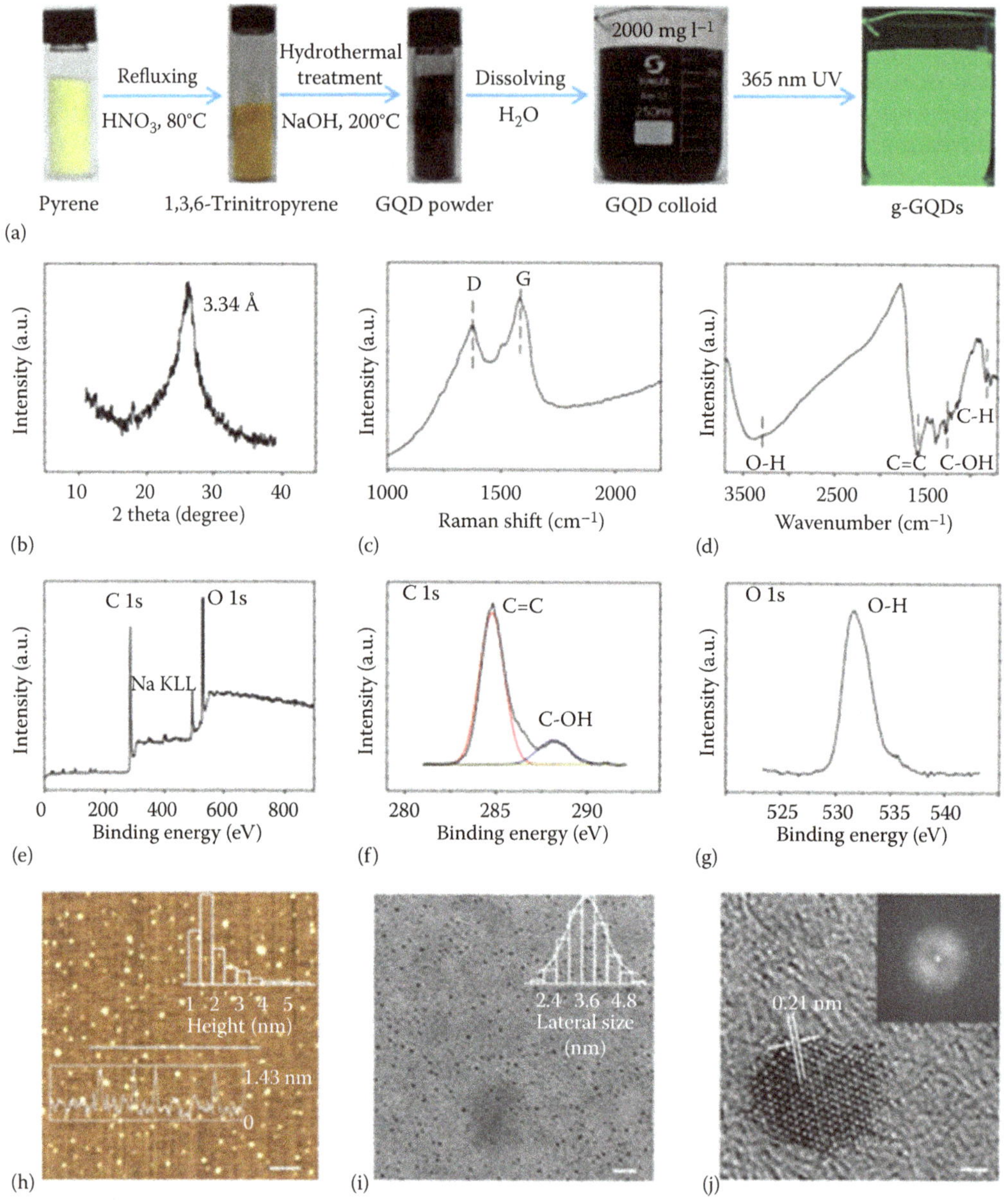

FIGURE 8.10
Preparation and structural characterization of OH-GQDs (green fluorescent GQDs [g-GQDs]). (a) Synthetic procedure. (b) XRD pattern. (c) Raman spectrum. (d) FT-IR spectrum. (e) Survey XPS spectrum. (f) High-resolution C 1 s spectrum. (g) High-resolution O 1 s spectrum. (h) AFM image. Insets: height profile along the white line and height distribution. Scale bar, 500 nm. (i) TEM image. Inset: lateral size distribution. Scale bar, 20 nm. (j) HRTEM image. Inset: FFT pattern. Scale bar, 1 nm. (Reprinted by permission from Macmillan Publishers Ltd., *Nature Communications*, L. Wang, Y. Wang, T. Xu, H. Liao, C. Yao, Y. Liu et al., *Nature Communications* 5, Article No. 5357, copyright 2014.)

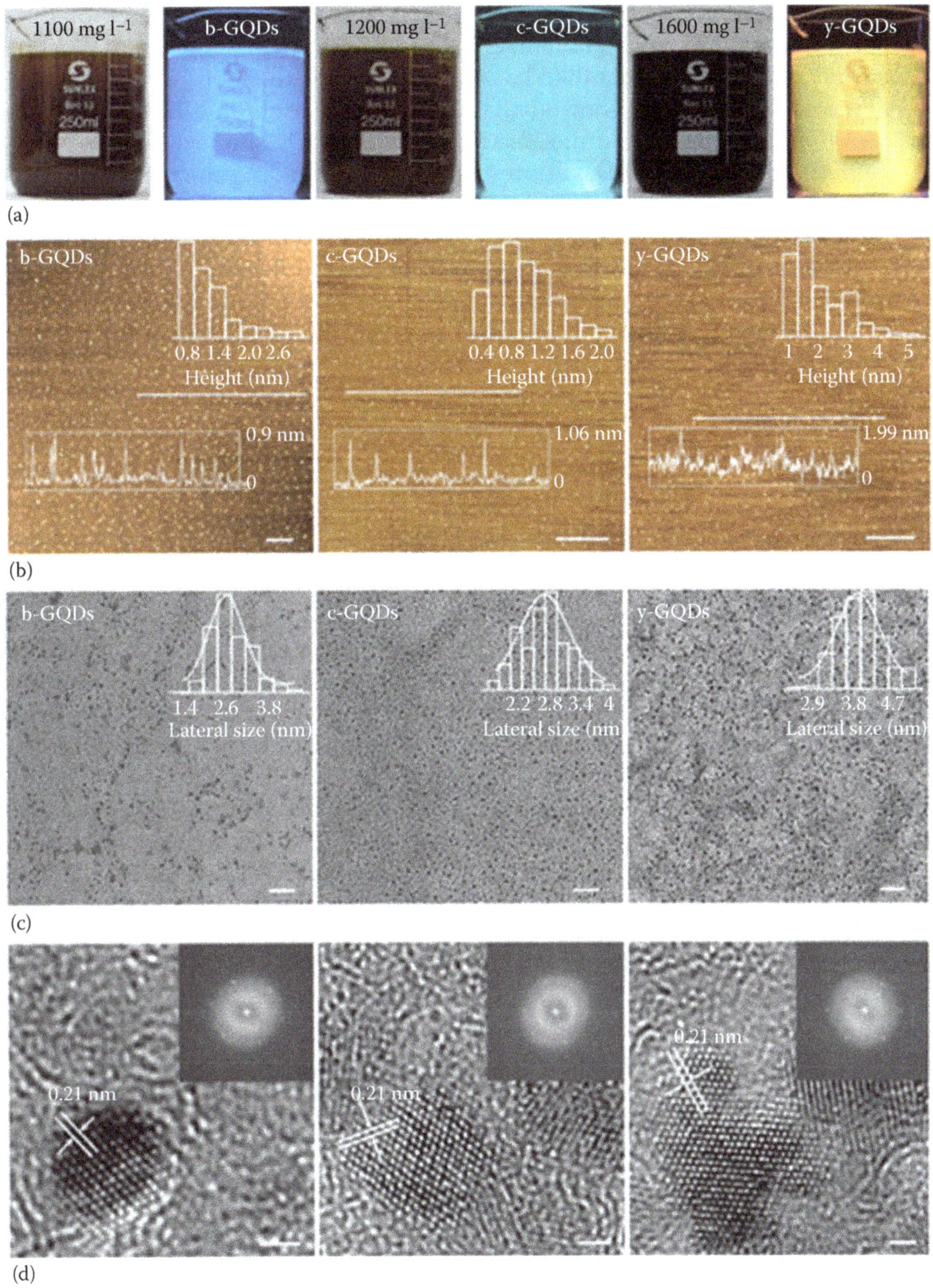

FIGURE 8.11

Photographs of amine-functionalized GQDs showing yellow (y-GQDs) fluorescent NH_2-GQDs prepared using ammonia solution, blue (b-GQDs) and cyan (c-GQDs) fluorescent $NHNH_2$-GQDs prepared using ammonia and hydrazine hydrate solution. (a) Photographs of aqueous colloids of b-, c- and y-GQDs taken under (right) UV and (left) visible lights. (b) AFM images of b-, c- and y-GQDs. Insets: height profiles along the white lines and height distributions. Scale bars, 2 µm, 500 nm, and 2 µm, respectively. (c) Their TEM images. Insets: lateral size distributions. Scale bar, 20 nm. (d) Their HRTEM images. Insets: FFT patterns. Scale bar, 1 nm. (Reprinted by permission from Macmillan Publishers Ltd., *Nature Communications*, L. Wang, Y. Wang, T. Xu, H. Liao, C. Yao, Y. Liu et al., *Nature Communications* 5, Article No. 5357, copyright 2014.)

using XRD, XPS, Raman spectroscopy, FTIR, AFM, and TEM revealed high-quality and almost defect-free features of functionalized GQDs, such as their single-crystalline nature and edge functionalization rather than basal plane functionalization.

The AFM and TEM images showed that the quantum dots in all the samples were uniform and well dispersed. Amine-functionalized GQDs showed yellow (y-GQDs), cyan (c-GQDs), and blue (b-GQDs) fluorescent colors depending upon the reaction conditions used [19]. A survey of their size showed that the yellow, cyan, and blue fluorescent amine-functionalized GQDs had gradually decreasing average lateral sizes (3.8 ± 0.5, 2.9 ± 0.5, 2.6 ± 0.6 nm) [19]. According to the theory of quantum dots the different fluorescent colors depend on the size of the quantum dot, which is a result of the quantum confinement effect. These studies prove the quantum confinement effect experimentally.

8.5 Properties of Graphene Quantum Dots

Opening of a band gap and quantum confinement and edge effects lead to novel optical properties and photoluminescence (PL) in graphene quantum dots [3]. Dispersions of GQDs generally appear light brown in visible light but show fluorescent colors under ultraviolet (UV) irradiation [2,20]. The GQDs efficiently absorb UV light with strong UV absorption in the 230 nm wavelength corresponding to the π-π^* transition of the aromatic sp^2 domains but also show new absorption bands that are responsible for the PL behavior that results in emission of bright colors (Figure 8.12a–c) [2].

Most GQDs show excitation-dependent emission or PL behavior [2,3,20], that is, the PL maximum redshifts and becomes lower in intensity as the excitation wavelength increases [2,3], with the strongest peak excited at the absorption band (Figure 8.12b). The PL excitation (PLE) spectrum recorded with the strongest luminescence exhibits sharp peaks (Figure 8.12c), which correspond to the new absorption bands that appear in the UV-visible spectra of GQD (Figure 8.12a), hence new transitions other than the commonly observed π-π^* transitions are involved. These new electronic transitions responsible for observed luminescence are σ-π^* and π-π^* transitions shown in Figure 8.4 [2] that originate from free zigzag edges with a carbene-like triplet ground state [2,20].

The GQDs show pH-dependent PL (Figure 8.12d) [2]. Under alkaline conditions, the GQDs emit strong PL, whereas, under acidic conditions, the PL is nearly completely quenched. If pH is switched repeatedly between 13 and 1, the PL intensity varies reversibly. Under acidic conditions, the free zigzag sites of the GQDs are protonated, forming a reversible complex between the zigzag sites and H^+. Thus the emissive triple carbene state is broken and becomes inactive in PL. However, under alkaline conditions, the free zigzag sites are restored, thereby leading to the restoration of PL [2].

Different colors of graphene quantum dots have been synthesized, ranging from UV to red, and most commonly blue and green with different quantum yields [3]. The PL quantum yield (QY) of GQDs is high due to their crystalline structure [3]. Reaction process parameters influence the photoluminescence properties of GQDs. The OH-GQDs synthesized by mild hydrothermal treatment in aqueous solution emit bright green fluorescence when irradiated by a UV light [19]. By varying the hydrothermal reaction conditions to produce amine-functionalized GQDs, yellow, cyan and blue fluorescent colloidal solutions have been obtained [19]. For GQDs synthesized from different polycyclic hydrocarbons,

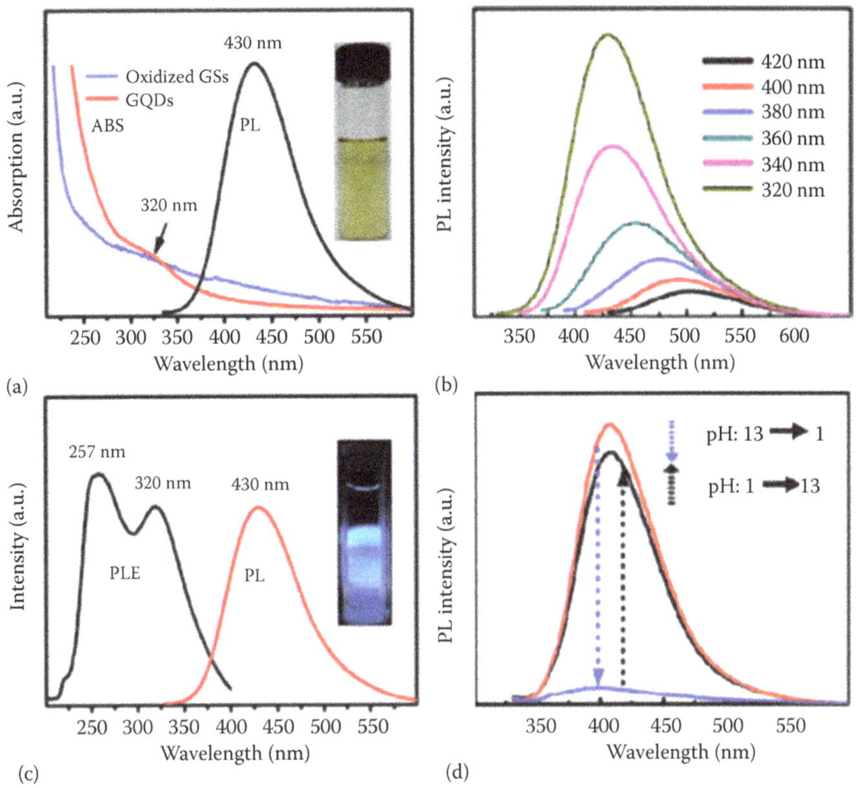

FIGURE 8.12
(a) UV–vis absorption and PL (at 320 nm excitation) spectra of the GQDs dispersed in water; UV–vis absorption spectrum of oxidized GSs. Inset: Photograph of the GQD aqueous solution taken under visible light. (b) PL spectra of the GQDs at different excitation wavelengths. (c) PLE spectrum with the detection wavelength of 430 nm and PL spectrum excited at 257 nm. Inset: Photograph of the GQD aqueous solution taken under UV light (shows bright blue luminescence) in a fluorescence spectrophotometer. (d) pH-dependent PL spectra when pH is switched between 13 and 1. (D. Pan, J. Zhang, Z. Li, and M. Wu: *Advanced Materials*. 2010. 22. No. 6. 734–738. Copyright Wiley-VCH Verlag GmbH & Co. KGaA. Reproduced with permission.)

the maximum emission peak showed a redshift with increased size of PAH precursors due to the different shapes and sizes of the prepared GQDs from different-sized PAHs [43]. Studies show that though oxygen-containing functional groups on GQDs render them hydrophilic and provide sites for further chemical functionalization, they also serve as emissive traps leading to low quantum yield. Li et al. [38] showed that quantum yield of GQDs almost doubled and the emission color changed to blue by reducing greenish-yellow luminescent oxidized GQDs. The oxidation of GQDs redshifts the emission. Excitation-dependent luminescence and up-conversion photoluminescence (i.e., two-photon luminescence leading to emission of shorter wavelength) was observed in N-doped GQDs [39]. Up-conversion is desirable for *in vivo* imaging because of deep-tissue penetration ability of long excitation wavelength (e.g., near infrared) [3]. Compared to conventional organic and inorganic fluorophores, the GQDs show nonblinking photoluminescence and excellent photostability, which are important for single-molecule tracking and long-term real-time imaging, respectively [3]. Photostability of amine-functionalized GQDs was compared with that of semiconductor quantum dots [19]. The conventional semiconductor QDs such as CdSe/ZnS QDs show a typical photobleaching effect that leads to the attenuation of their PL intensity and brightness on UV exposure.

Photoluminescence of GQDs can be tailored to obtain luminescent dots with different emission colors of blue, green, and yellow through varying the size of the GQDs by changing reaction temperature. As processing temperature during synthesis of GQDs by exfoliation of carbon fibers increased from 80°C to 100°C to 120°C, the UV-visible spectrum showed a blue shift indicating reaction temperature could tune the size of as-prepared GQDs and affect their absorption properties [30]. Highly oxygenated GQDs synthesized by cage opening of fullerene C_{60} showcased their broad prospects for chemical modifications through successful functionalization reactions [31]. The luminescence properties varied according to the types of chemical treatments. While as-synthesized oxidized GQDs exhibited strong luminescence properties when excited at 340 nm, hydroxylamine-functionalized GQDs showed a blueshift of the emission maximum, whereas hydrazine-reduced GQDs showed a redshift of the emission maximum.

Besides striking photoluminescence behavior, graphene quantum dots also possess other properties such as electrochemiluminescence behavior, electrochemical properties, low toxicity, water solubility, and biocompatibility [3].

8.6 Applications of Graphene Quantum Dots

The optical properties of GQDs, especially their striking PL property, are the basis of many of the applications of GQDs in optoelectronic and biomedical applications [1,3,5,19,20,32,33,43]. These luminescent carbon nanomaterials are advantageous over fluorescent dye molecules, fluorescent proteins, and semiconductor QDs owing to their stable photoluminescence, excellent biocompatibility, water solubility, and low cytotoxicity, which is important for biomedical applications [3,19,43]. Some of these potential applications include *in vitro* and *in vivo* fluorescence bioimaging, real-time biological tracking, drug delivery, photodetectors, LEDs, solar cells, and various types of smart sensors [1]. Many studies have been carried out on probing the use of GQDs for these applications [1,3,5,19,20,32,33,43].

The highly fluorescent $BMIM^+$-functionalized GQDs synthesized by electrochemical method were studied for optical detection of Fe^{3+} (Figure 8.13) [20]. The detection of Fe^{3+}

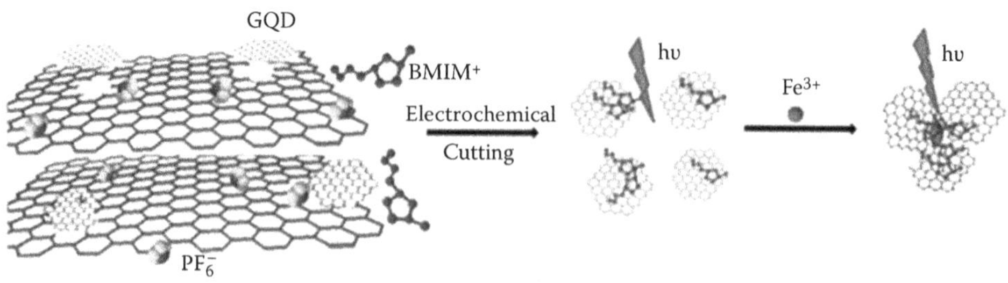

FIGURE 8.13
Schematic illustration of GQD synthesis from 3D graphene and mechanism of Fe^{3+} detection. (A. Ananthanarayanan, X. Wang, P. Routh, B. Sana, S. Lim, D.H. Kim, K.H. Lim, J. Li, and P. Chen: *Advanced Functional Materials*. 2014. 24. No. 20. 3021–3026. Copyright Wiley-VCH Verlag GmbH & Co. KGaA. Reproduced with permission.)

in biological systems as well as in the environment is extremely important. The $BMIM^+$-functionalized GQDs have been studied for their ability to be used as fluorescence sensors for Fe^{3+} based on the fluorescence quenching effect of Fe^{3+} [20]. The pyridinic nitrogen of $BMIM^+$ enhances fluorescence of GQDs by imposing n-doping-like characteristics to GQD. The presence of Fe^{3+} causes significant quenching of GQD fluorescence by bridging

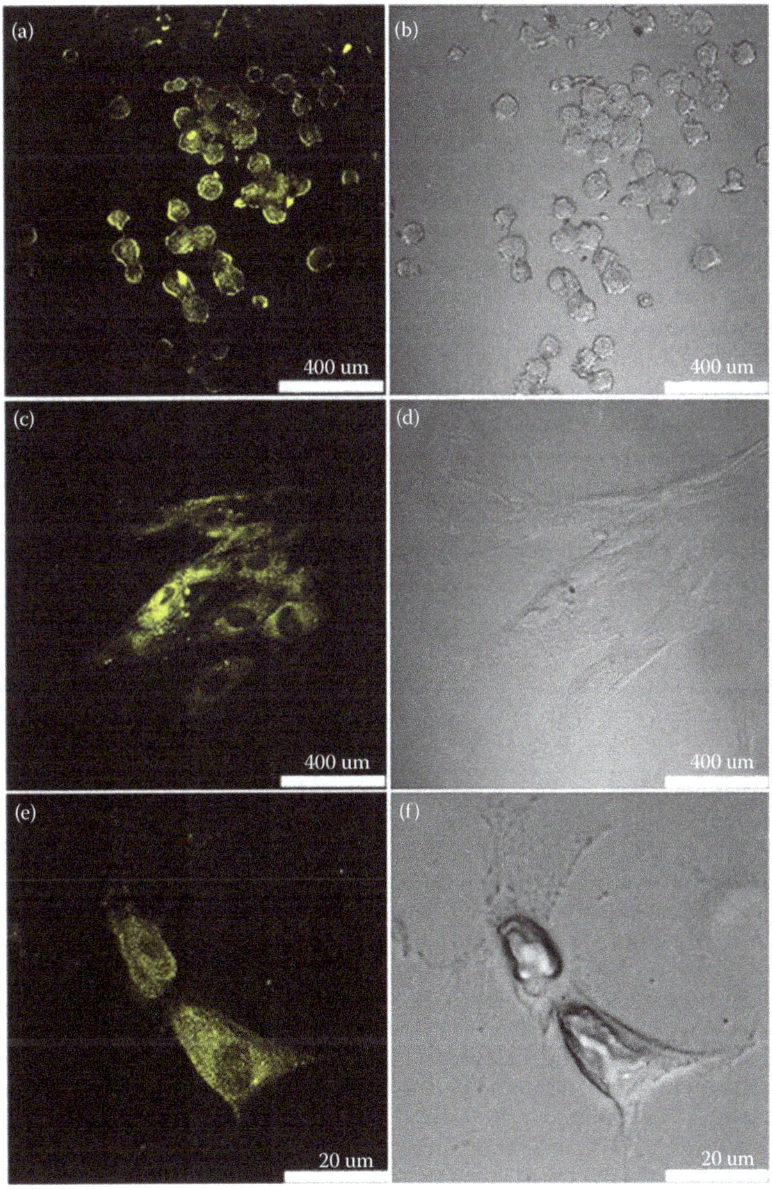

FIGURE 8.14
Confocal fluorescence microscopy images of three different kinds of stem cells, neurospheres cells (a), pancreas progenitor cells (c), and cardiac progenitor cells (e) with the fluorescent GQDs incorporated at the excitation wavelength of 405 nm and corresponding images under bright field (b, d, f). (From M. Zhang, L. Bai, W. Shang, W. Xie, H. Ma, Y. Fu et al., *Journal of Materials Chemistry* 22, no. 15 (2012): 7461–7467. Reproduced by permission of The Royal Society of Chemistry.)

several BMIM$^+$-functionalized GQDs together by coordination due to the imidazole ring of BMIM$^+$, which has a high binding affinity for Fe^{3+} ions (Figure 8.13). This binding induces aggregation of GQDs and resulting quenching of fluorescence. The studies showed ability of GQDs for sensitive and specific optical detection of Fe^{3+} ions [20]. Photoluminescent GQDs synthesized from PAH showed low toxicity and good photostability suggesting promising fluorescent probes for cellular imaging [43]. Additionally, sensitivity of the fluorescence of these GQDs in the detection of Fe^{3+} and H_2O_2 (hydrogen peroxide) showed their capability as sensors for effective Fe^{3+} and hydrogen peroxide sensing [43]. The use of water-soluble fluorescent GQDs synthesized by electrochemical method as a biological label for cells showed that GQDs easily penetrated into the stem cells and no intensity change of the fluorescence in the fluorescence microscopy images could be detected, indicating high photostability of the GQDs in the cells (Figure 8.14) [32]. While semiconductors QDs like CdS are cytotoxic, the GQDs showed biocompatibility.

Feasibility of using GQDs obtained from pyrolysis of L-glutamic acid as a fluorescent biomarker was investigated through *in vitro* and *in vivo* fluorescence imaging [44]. The results showed that the GQDs could be promising candidates for bioimaging. Use of graphene quantum dots as a fluorescent labeling agent for cell imaging in biomedical applications was also probed by Wang et al. [19]. The GQDs in cells showed strong photoluminescence, good photostability, and low cytotoxicity suggesting potential as higher-quality fluorescent probes for bioimaging for future clinical diagnostic applications. Graphene quantum dots synthesized by microwave and hydrothermal means were studied for their cell biocompatibility and biolabeling applications [40]. Sun et al. [1] have extensively reviewed the studies on graphene quantum dots for sensing such as electronic sensors, PL sensors, electrochemical sensors, electrochemiluminescence sensors for physical analysis, chemical analysis, and bioanalysis.

References

1. H. Sun, L. Wu, W. Wei, and X. Qu. Recent advances in graphene quantum dots for sensing. *Materials Today* 16, no. 11 (2013): 433–442.
2. D. Pan, J. Zhang, Z. Li, and M. Wu. Hydrothermal route for cutting graphene sheets into blue-luminescent graphene quantum dots. *Advanced Materials* 22, no. 6 (2010): 734–738.
3. X.T. Zheng, A. Ananthanarayanan, K.Q. Luo, and P. Chen. Glowing graphene quantum dots and carbon dots: Properties, syntheses, and biological applications. *Small* 11, no. 14 (2015): 1620–1636.
4. P.N. Joshi, S. Kundu, S.K. Sanghi, and D. Sarkar. 2016. Graphene quantum dots—From emergence to nanotheranostic applications. In *Smart Drug Delivery System*, edited by A.D. Sezer, Chap. 7. InTech, 2016.
5. L. Li, G. Wu, G. Yang, J. Peng, J. Zhao, and J.J. Zhu. Focusing on luminescent graphene quantum dots: Current status and future perspectives. *Nanoscale* 5, no. 10 (2013): 4015–4039.
6. K.S. Novoselov, A.K. Geim, S.V. Morozov, D. Jiang, Y. Zhang, S.A. Dubonos, I.V. Grigorieva, and A.A. Firsov. Electric field effect in atomically thin carbon films. *Science* 306, no. 5696 (2004): 666–669.
7. K.S.A. Novoselov, A.K. Geim, S. Morozov, D. Jiang, M. Katsnelson, I. Grigorieva, S. Dubonos, and A. Firsov. Two-dimensional gas of massless Dirac fermions in graphene. *Nature* 438, no. 7065 (2005): 197–200.
8. M.J. Allen, V.C. Tung, and R.B. Kaner. Honeycomb carbon: A review of graphene. *Chemical Reviews* 110, no. 1 (2009): 132–145.

9. J. Du and H.M. Cheng. The fabrication, properties, and uses of graphene/polymer composites. *Macromolecular Chemistry and Physics* 213, no. 10–11 (2012): 1060–1077.
10. A.K. Geim and K.S. Novoselov. The rise of graphene. *Nature Materials* 6, no. 3 (2007): 183–191.
11. B. Guo, L. Fang, B. Zhang, B and J.R. Gong. Graphene doping: A review. *Insciences Journal* 1, no. 2 (2011): 80–89.
12. J. Lu, P.S.E. Yeo, C.K. Gan, P. Wu, and K.P. Loh. Transforming C60 molecules into graphene quantum dots. *Nature Nanotechnology* 6, no. 4 (2011): 247–252.
13. J.J. Zhu, J.J. Li, H.P. Huang, and F.F. Cheng. *Quantum Dots for DNA Biosensing*, vol. 165, pp. 341–346. Springer, 2013.
14. A.P. Alivisatos. Semiconductor clusters, nanocrystals, and quantum dots. *Science* 271, no. 5251 (1996): 933.
15. A.M. Smith and S. Nie. Semiconductor nanocrystals: Structure, properties, and band gap engineering. *Accounts of Chemical Research* 43, no. 2 (2009): 190–200.
16. J. Cox. A quantum paintbox. *Chemistry in Britain* 39, no. 9 (2003): 21–25.
17. T. Fan, W. Zeng, W. Tang, C. Yuan, S. Tong, K. Cai, Y. Liu, W. Huang, Y. Min, and A.J. Epstein. Controllable size-selective method to prepare graphene quantum dots from graphene oxide. *Nanoscale Research Letters* 10, no. 1 (2015): 1–8.
18. D. Wang, J.F. Chen, and L. Dai. Recent advances in graphene quantum dots for fluorescence bioimaging from cells through tissues to animals. *Particle & Particle Systems Characterization* 32, no. 5 (2015): 515–523.
19. L. Wang, Y. Wang, T. Xu, H. Liao, C. Yao, Y. Liu et al. Gram-scale synthesis of single-crystalline graphene quantum dots with superior optical properties. *Nature Communications* 5, Article No. 5357 (2014).
20. A. Ananthanarayanan, X. Wang, P. Routh, B. Sana, S. Lim, D.H. Kim, K.H. Lim, J. Li, and P. Chen. Facile synthesis of graphene quantum dots from 3D graphene and their application for Fe3+ sensing. *Advanced Functional Materials* 24, no. 20 (2014): 3021–3026.
21. A.I. Ekimov and A.A. Onushchenko. Quantum size effect in three-dimensional microscopic semiconductor crystals. *SPIE Milestone Series*, MS 180 (2005): 3.
22. T. Davis. Biography of Louis E. Brus. *Proceedings of the National Academy of Sciences of the United States of America* 102, no. 5 (2005): 1277–1279.
23. B.A. Kairdolf, A.M. Smith, T.H. Stokes, M.D. Wang, A.N. Young, and S. Nie. Semiconductor quantum dots for bioimaging and biodiagnostic applications. *Annual Review of Analytical Chemistry (Palo Alto, Calif.)* 6, no. 1 (2013): 143.
24. C.O. Kim, S.W. Hwang, S. Kim, D.H. Shin, S.S. Kang, J.M. Kim, C.W. Jang, J.H. Kim, K.W. Lee, S.H. Choi, and E. Hwang. High-performance graphene-quantum-dot photodetectors. *Scientific Reports* 4, Article No. 5603 (2014).
25. A.D. Güçlü, P. Potasz, and P. Hawrylak. Excitonic absorption in gate-controlled graphene quantum dots. *Physical Review B, Condensed Matter and Materials Physics* 82, no. 15 (2010): 1–5.
26. Z. Jin, P. Owour, S. Lei, and L. Ge. Graphene, graphene quantum dots and their applications in optoelectronics. *Current Opinion in Colloid & Interface Science* 20, no. 5 (2015): 439–453.
27. M.Y. Han, B. Özyilmaz, Y. Zhang, and P. Kim. Energy band-gap engineering of graphene nanoribbons. *Physical Review Letters* 98, no. 20 (2007): 206805.
28. C. Stampfer, J. Güttinger, F. Molitor, D. Graf, T. Ihn, and K. Ensslin. Tunable Coulomb blockade in nanostructured graphene. *Applied Physics Letters* 92, no. 1 (2008): 012102.
29. J. Peng, W. Gao, B.K. Gupta, Z. Liu, R. Romero-Aburto, L. Ge et al. Graphene quantum dots derived from carbon fibers. *Nano Letters* 12, no. 2 (2012): 844–849.
30. R. Ye, C. Xiang, J. Lin, Z. Peng, K. Huang, Z. Yan et al. Coal as an abundant source of graphene quantum dots. *Nature Communications* 4, Article No. 2943 (2013).
31. C.K. Chua, Z. Sofer, P. Simek, O. Jankovský, K. Klímová, S. Bakardjieva, S. Hrdličková Kučková, and M. Pumera. Synthesis of strongly fluorescent graphene quantum dots by cage-opening buckminsterfullerene. *ACS Nano* 9, no. 3 (2015): 2548–2555.

32. M. Zhang, L. Bai, W. Shang, W. Xie, H. Ma, Y. Fu et al. Facile synthesis of water-soluble, highly fluorescent graphene quantum dots as a robust biological label for stem cells. *Journal of Materials Chemistry* 22, no. 15 (2012): 7461–7467.
33. X. Tan, Y. Li, X. Li, S. Zhou, L. Fan, and S. Yang. Electrochemical synthesis of small-sized red fluorescent graphene quantum dots as a bioimaging platform. *Chemical Communications* 51, no. 13 (2015): 2544–2546.
34. D.B. Shinde and V.K. Pillai. Electrochemical preparation of luminescent graphene quantum dots from multiwalled carbon nanotubes. *Chemistry—A European Journal* 18, no. 39 (2012): 12522–12528.
35. M. Xu, Z. Li, X. Zhu, N.T. Hu, H. Wei, Z. Yang, and Y.F. Zhang. Hydrothermal/solvothermal synthesis of graphene quantum dots and their biological applications. *Nano Biomedicine and Engineering* 5 (2013): 65–71.
36. S. Zhu, J. Zhang, C. Qiao, S. Tang, Y. Li, W. Yuan et al. Strongly green-photoluminescent graphene quantum dots for bioimaging applications. *Chemical Commununications* 47, no. 24 (2011): 6858–6860.
37. H. Tetsuka, R. Asahi, A. Nagoya, K. Okamoto, I. Tajima, R. Ohta, and A. Okamoto. Optically tunable amino-functionalized graphene quantum dots. *Advanced Materials* 24, no. 39 (2012): 5333–5338.
38. L.L. Li, J. Ji, R. Fei, C.Z. Wang, Q. Lu, J.R. Zhang, L.P. Jiang, and J.J. Zhu. A facile microwave avenue to electrochemiluminescent two-color graphene quantum dots. *Advanced Functional Materials* 22, no. 14 (2012): 2971–2979.
39. C.M. Luk, M.K. Tsang, C.F. Chan, and S.P. Lau. Two-photon fluorescence in n-doped graphene quantum dots. *World Academy of Science, Engineering and Technology, International Journal of Chemical, Molecular, Nuclear, Materials and Metallurgical Engineering* 8, no. 12: 1354–1357.
40. Y. Huang, C. Bai, K. Cao, Y. Tian, Y. Luo, C. Xia, S. Ding, Y. Jin, L. Ma, and S. Li. Chaos to order: An eco-friendly way to synthesize graphene quantum dots. *RSC Advances* 4, no. 81 (2014): 43160–43165.
41. Y. Zhu, G. Wang, H. Jiang, L. Chen, and X. Zhang. One-step ultrasonic synthesis of graphene quantum dots with high quantum yield and their application in sensing alkaline phosphatase. *Chemical Communications* 51, no. 5 (2015): 948–951.
42. Y. Dong, J. Shao, C. Chen, H. Li, R. Wang, Y. Chi, X. Lin, and G. Chen. Blue luminescent graphene quantum dots and graphene oxide prepared by tuning the carbonization degree of citric acid. *Carbon* 50, no. 12 (2012): 4738–4743.
43. L. Zhou, J. Geng, and B. Liu. Graphene quantum dots from polycyclic aromatic hydrocarbon for bioimaging and sensing of Fe3+ and hydrogen peroxide. *Particle & Particle Systems Characterization* 30, no. 12 (2013): 1086–1092.
44. X. Wu, F. Tian, W. Wang, J. Chen, M. Wu, and J.X. Zhao. Fabrication of highly fluorescent graphene quantum dots using L-glutamic acid for in vitro/in vivo imaging and sensing. *Journal of Materials Chemistry C* 1, no. 31 (2013): 4676–4684.
45. R. Liu, D. Wu, X. Feng, and K. Müllen. Bottom-up fabrication of photoluminescent graphene quantum dots with uniform morphology. *Journal of the American Chemical Society* 133, no. 39 (2011): 15221–15223.
46. X. Yan, X. Cui, and L.S. Li. Synthesis of large, stable colloidal graphene quantum dots with tunable size. *Journal of the American Chemical Society* 132, no. 17 (2010): 5944–5945.

9

Amorphous Nanocarbons

9.1 Introduction

Amorphous carbon, or free, reactive carbon, is an allotrope of carbon that does not have any crystalline structure [1]. Unlike other group IV amorphous materials, amorphous carbon has sp^2 (graphite-like) as well as sp^3 (diamond-like) hybrid bonds, making it a much more complex system. There are many forms of sp^2-bonded carbons with various degrees of graphitic ordering, ranging from microcrystalline graphite to glassy carbon. In general, an amorphous carbon can have any mixture of sp^3, sp^2, and even sp^1 sites, with the possible presence of up to 60 wt.% hydrogen. True amorphous carbon has localized π electrons (as opposed to the aromatic π bonds in graphite), and its bond lengths are inconsistent with any other allotrope of carbon. These dangling carbon bonds are quite reactive and may be stabilized by terminating dangling-π bonds with hydrogen. These materials are then called hydrogenated amorphous carbons. The hydrogenated amorphous carbons (*a*-C:H) have a rather small C–C sp^3 content. Diamond-like carbons (DLCs) with higher sp^3 content are termed tetrahedral amorphous carbon (*ta*-C) and its hydrogenated analogue (*ta*-C:H).

The primary method for characterizing amorphous carbon is through the ratio of sp^2 to sp^3 hybridized bonds present in the material. Graphite consists purely of sp^2-hybridized bonds, whereas diamond consists purely of sp^3-hybridized bonds. Experimentally, sp^2 to sp^3 ratios can be determined by comparing the relative intensities of various spectroscopic peaks (including electron energy loss spectroscopy [EELS], x-ray photoelectron spectroscopy [XPS], and Raman spectroscopy) to those expected for graphite or diamond.

Another crucial parameter is the degree of clustering of the sp^2 phase. Because the sp^2 phase has smaller band gaps than the sp^3 phase, the optical and electronic properties of amorphous carbon are dominated by the content and structure of the sp^2 phase. Amorphous carbons therefore with the same sp^3 and H content show different optical, electronic, and mechanical properties according to the clustering of the sp^2 phase.

Coal and soot or carbon black are informally called amorphous carbon. While entirely amorphous carbon can be produced, most amorphous carbon actually contains microscopic crystals of graphite-like, or even diamond-like carbon. The coal industry divides coal into various grades depending on the amount of carbon present in the sample compared to the amount of impurities. The highest grade, anthracite, is about 90% carbon and 10% other elements. Bituminous coal is about 75%–90% carbon, and lignite is the name for coal that is around 55% carbon.

With the development of modern thin-film deposition and growth techniques in the latter half of the 20th century, such as chemical vapor deposition, sputter deposition, and cathodic arc deposition, it has become possible to fabricate truly amorphous carbon materials.

9.2 Carbon Black

Carbon-derived powders and particles comprise a family of synthetic materials of nanometer dimension, known under the generic term of carbon black, made by incomplete combustion (insufficient supply of air) of many organic substances in solid, liquid, or gaseous form. The carbon particles are aggregates of graphite nanocrystals, each only a few unit cells in size and so small that they are generally not measurable by diffraction techniques due to the broad diffraction pattern. The physical properties of these materials are essentially determined by the size, shape, porosity, and surface area. They are materials essentially composed of elemental carbon in the form of near spherical particles of colloidal sizes (i.e., ~1 to 100 nm). Figure 9.1a shows a sample of as-produced carbon black material.

The carbon blacks are rarely found as individual, separate spherical particles, but generally exist as aggregates of coalesced, fused elementary particles shown in Figure 9.1b. The generic term carbon black refers to a group of industrial products consisting of furnace black, channel black, thermal black, and lampblack. An exception is found in the thermal black, which consists mostly of single spherical particles. The aggregates, however, have a tendency to agglomerate in dispersion into larger units, which may be redispersed by shear. However, the particles reagglomerate on standing. The particles of a given carbon black do not have the same diameter but show a size distribution.

9.2.1 Mechanism of Carbon Black Formation

Carbon is formed from hydrocarbon in two chemical processes: pyrolysis and incomplete combustion. In the first, hydrocarbon and air are mixed before leaving the burner (premixed flames), while in the second the oxygen diffuses in the gaseous hydrocarbon stream after leaving the burner (diffusion flames). Two types of carbons are thus obtained from hydrocarbons. The first type is formed by deposition on the walls of the reactor and is known as pyrocarbon or pyrolytic carbon, while the second type is formed in the bulk of the gas phase. The latter type is characterized by spherical or near-spherical particles, each containing 10^5 to 10^{10} carbon atoms, and is known as soot or carbon black.

(a) (b)

FIGURE 9.1
Carbon black powder: (a) bulk as-produced and (b) agglomerated particles.

As early as 1866, Berthelot and Anderson advanced the theory according to which acetylene is the main intermediate in carbon formation assuming that carbon is obtained by polymerization and dehydrogenation of acetylene. In 1955 Gaydon and Wolfhard [2] suggested that C_2 plays a part in carbon particle nucleation in the flame by means of reactions such as the following:

$$C_2 + C_2H_2 \rightarrow C_4 + H_2 \tag{9.1}$$

$$C_4 + C_2H_2 \rightarrow C_6 + H_2 \tag{9.2}$$

It is practically certain that the carbon formation requires the presence of intermediate nonsaturated aliphatic compounds, but not necessarily acetylene itself.

9.2.1.1 *Theory of Polymerization of Hydrocarbons*

The period of growth of a particle containing about 50,000 carbon atoms is about 1–10 msec. Such rapid growth can only occur by means of very fast reactions. The temperature of these reactions must be between 800°C to 1300°C. The molecules undergoing the reaction that transforms hydrocarbons into carbon must therefore possess, in addition to thermal stability, high reactivity and should easily react with one another. Polyethylenic radicals, in view of their ease of Diels-Alder additions, appear to conform readily to these prerequisites. The temperatures required for growth of carbon deposit are generally much lower than those needed for nucleation of the carbon particles.

The characteristics of carbon black vary depending on the manufacturing process, and therefore carbon black is classified by manufacturing process.

9.2.2 Manufacture of Carbon Black

Carbon blacks have been manufactured by means of three major processes. For more detailed understanding of the manufacturing processes, the work of Kuhner and Voll is of particular merit [3].

9.2.2.1 *Furnace Black Process*

Carbon black produced with the furnace process, which is the most commonly used method now, is called "furnace black." This method forms carbon black by blowing petroleum oil or coal oil as raw material (feedstock oil) into high-temperature gases to combust them partially. This method is suitable for mass production due to its high yield, and allows wide control over its properties such as particle size or structure. This is currently the most common method used for manufacturing carbon black for various applications from rubber reinforcement to coloring. The process can be further subdivided into the gas furnace process and oil furnace process.

9.2.2.1.1 Gas Furnace Process

In the gas furnace process a diffusion flame is created by burning part of the gas with added air. The rest of the gas is thermally decomposed in the flame, forming the desired carbon black. The gas used is natural gas, consisting predominantly of methane. The quality and the yield of the product are determined by the temperature of the flame, ~1300°C

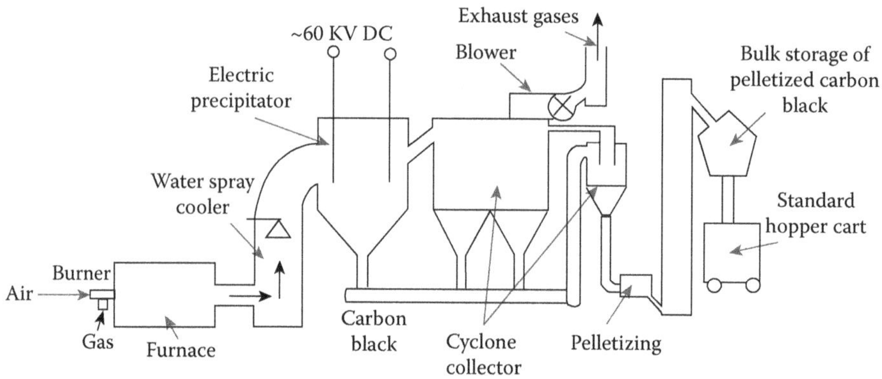

FIGURE 9.2
Flow diagram of gas furnace process.

to 1400°C, regulated by the ratio of the gas and the air, usually about 1:5 under the turbulent conditions of the gas mixture. Generally, the yields are of the order of 25% to 40% for the larger particle size blacks. Figure 9.2 shows a flow diagram for a typical gas furnace process.

9.2.2.1.2 Oil Furnace Process

Most of the carbon blacks available today are produced by this process [4]. In the oil furnace process liquid fuel is injected into the flame to produce carbon black. This is a very rapid continuous process in which the oil is converted to carbon black aggregates in the fraction of a second at high temperatures.

There are four basic stages of carbon black formation:

1. Atomization/vaporization/pyrolysis
2. Nucleation: Molecule, particle, surface growth
3. Agglomeration of particles: Aggregate growth due to van der Waals forces
4. Carbonization/dehydrogenation: Secondary reactions

As illustrated in Figure 9.3, during the first stage the feed stock oil must be completely vaporized and pyrolyzed, down to C1 and C2 groups. If the oil converts directly to carbon, the resulting particles comprise "coke," which are much larger in size and have very different microstructure compared to carbon black.

The nucleation stage is the least understood step. The nuclei are formed from polyaromatic hydrocarbons. Surface growth accounts for about 95% of the mass of the carbon black and is thought to be based on acetylene species [5]. Aggregate growth is largely the result of particles collision and partial coalescence.

Surface growth continues during the aggregate growth stage and the carbon black particles become bonded together by the layers of carbon deposited. In the final step dehydrogenation continues after the aggregates are fully formed.

The flame itself is generally produced from natural gas, but occasionally from liquid fuels. The temperatures reached in the flame are 1200°C to 1600°C dependent on the type

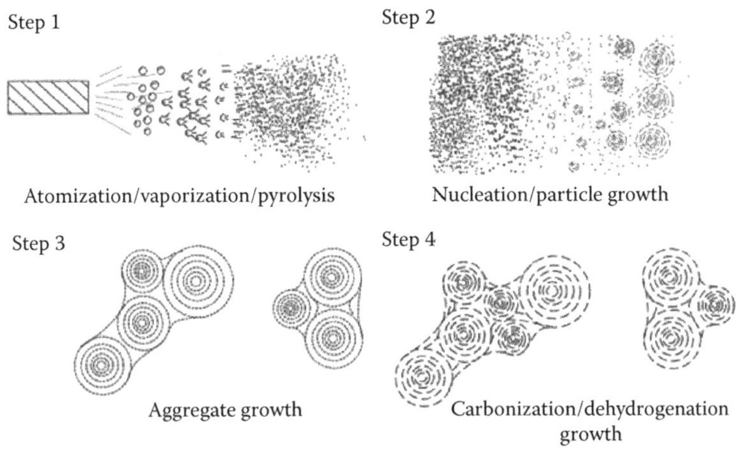

FIGURE 9.3
Different steps in carbon black formation from oil process.

of carbon black manufactured. The finer particle-size blacks require a higher flame temperature, achieved by means of higher combustion ratio. The feedstock used in the oil must fulfill a number of requirements, the aromatic contents being the most important criteria, since it largely controls the ultimate yield of the carbon black. It also indicates the importance of the boiling point, since a good feedstock should be completely vaporized in a fraction of a second at prevailing conditions in the furnace.

Chemically speaking, highly aromatic hydrocarbons of molecular weight up to 500 fulfill the requirements. Particularly, polynuclear aromatics are very useful, while short side chains with many double bonds enhance the quality of feedstock. Napthelene and anthracene oils are economically feasible. Coal tar can also be used. Yields in the oil furnace process are 30% to 50%, with actual quantity of carbon feed stocks nearly 85% to 90%. The average diameter of carbon blacks made with the furnace process vary from 20 to 80 nm.

Blacks of larger diameter are not made with the furnace process but from the thermal process. The collected black has a density of 0.03–0.08 g/cc and is not suitable for transportation in bulk. It is therefore changed into free-flowing pellets at a density of 0.3 to 0.5 g/cc by a pelletization process. Carbon black powders may be considered as dispersion in air, justified by the very large void volume, which may be as high as 98% in uncompressed blacks and still about 80% in densified pelletized blacks.

While there are distinct differences in morphology and surface chemistry among carbon blacks made by different manufacturing processes, the basic mechanism of formation is essentially the same [6–9].

9.2.2.2 Channel Process

With the discovery of natural gas in the United States, channel black became the most important carbon black in the period following its first manufacture in 1872 and reached its peak in 1920s and 1930s. The carbon source in the channel process is natural gas from which the lower boiling gasoline fractions have been removed by means of absorption in mineral seal oils. This method forms carbon black by bringing partially combusted fuel, which is generated with natural gas as raw material, into contact with channel steel

(H-shaped steel). Thousands of small flames of natural gas impinge on a cool metallic surface. The channels are moved in slow reciprocating motion over stationary scrappers, thereby removing the black deposits. The carbon black formed on the cool surface is then exposed to high temperature in air to oxidize the surface of each particle. These particles are in the form of small spheroids. The channel carbon black has the smallest particle size (~10 nm), the highest surface area, and the highest volatile content of all carbon blacks.

The yield of the channel process is extremely low, less than 5% of the available carbon in the feedstock is recovered as channel black. The result is that the channel black is not economically competing with furnace black of the same particle size.

There are yield and environment issues around this method, and therefore it has lost the leading role as the mass production process to the furnace process. This method, however, provides carbon black with many functional groups on the surface, and is being used in some painting applications. A number of higher prized grades of channel blacks are manufactured as color blacks. Such blacks are prepared under conditions of excessive oxidation in air or after-oxidation by chemical means. The black contains 2–5 wt.% of oxygen and is unique among the different carbon black types. Such color blacks are used in high quality printing inks and lacquers. They are characterized with jet-blackness and low oil properties.

9.2.2.3 Thermal Process

In the thermal process, the carbon black is formed by the thermal decomposition of natural gas, methane, or oil, in the absence of air in a preheated firebrick-lined chamber. The thermal process utilizes natural gas or oil as the source of carbon. It is a cyclic process in which the feedstock, predominantly methane, is decomposed into carbon and hydrogen at about 1300°C. The hydrogen gas produced during the decomposition of the feedstock becomes part of the purge gas and is recovered along with carbon black. The process produces a coarser grade than the channel process with particle size up to 500 nm and lower surface area.

Medium thermal blacks have particle sizes 200 to 500 nm, while fine thermal blacks have diameters of 120 nm to 150 nm. Fine thermal blacks are made by diluting the natural gas feedstock with process hydrogen. Yields are about 40% to 50%. They are used in cross-linked polyethylene and rubber applications requiring very high filler volume fractions.

9.3 Physical Properties of Carbon Black

The key to understanding the nature of carbon blacks is the recognition of their fundamental properties. The four fundamental properties of carbon black are particle size, structure (aggregate and shape), porosity, and surface functionality. Electron microscopy of carbon black particles shows that they have a complicated structure, with some spherical particles fused together. As shown schematically in Figure 9.4, the size of these particles is called "particle size" and the size of the particle chain is called its "structure." Various functional groups such as the hydroxyl(-OH) or carboxyl(-COOH) group are found on the surface of carbon black, and their amount or composition is sometimes referred to as "surface

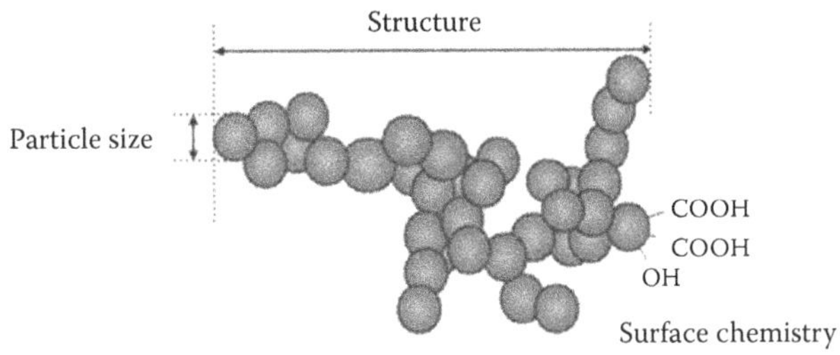

FIGURE 9.4
Micrograph of carbon black aggregate.

chemistry." The three main properties have a large effect on practical properties such as blackness and dispersibility when they are mixed with inks, paints, or resins.

9.3.1 Particle Size

The diameter of spherical particles is the fundamental property that largely affects blackness and its ability to disperse when carbon black is mixed with resins or as reinforcement in elastomers. In general, the smaller the particle size, the higher the blackness of carbon black becomes. Dispersion, however, becomes difficult due to an increase in coagulation/van der Waals force. The measurement of carbon black particle size is complex and very much dependent on the technique used.

9.3.2 Microstructure of Carbon Black

Like particle size, the size of the structure also affects the blackness and dispersability of carbon black. Generally, the increase of structure size improves dispersability but lowers blackness. Carbon black with a larger structure in particular shows an excellent conductive property.

Normally carbon black has been considered to be polycrystalline solids. The high-resolution electron microscopy has made it possible to extend our knowledge of the internal organization of carbon black particles. The first x-ray diffraction studies of carbon black and soot particles were carried out by Warren in the 1930s and 1940s [10,11] who demonstrated the presence of layer planes of graphite-like carbon. The first x-ray diffraction patterns of carbon black shows two or three diffused rings at about the same position as the most intense graphitic diffraction peak. The results show that carbon black consists of bundles of parallel-oriented graphitic planes at distances of 3.5 to 3.8 Å. However, no evidence of the three-dimensional graphite structure was detected. For carbon blacks, the crystallize size La usually falls in the range 1–3 nm, whereas Lc is usually of the order of 1.5 nm.

A typical powder diffraction pattern of a particular grade of carbon black (N774) as produced and graphitized is shown in Figure 9.5 [12]. It can be seen that after heat treatment at 2700°C the peak profile becomes sharper. The diffractogram shows strong overlap of the peaks for both samples [13].

The microstructural model derived from x-ray diffraction analysis of carbon black postulates that the crystallites in the bulk of the particles are connected to one another by single planes or poorly organized carbons, as shown in the schematic model of Figure 9.6 [13].

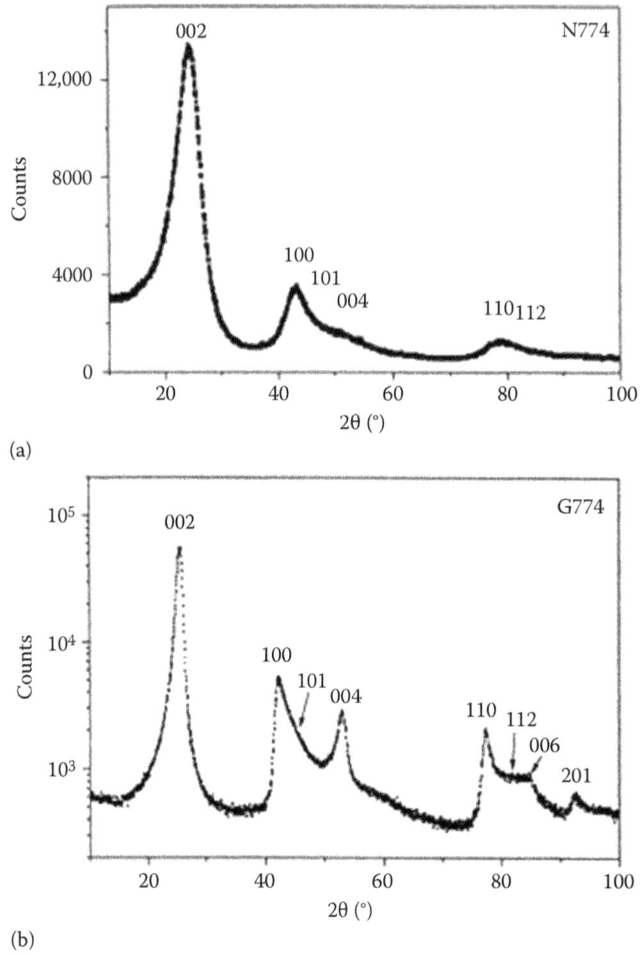

FIGURE 9.5
Powder diffraction pattern of N774 carbon black: (a) as-produced and (b) after HTT at 2700°C. (Reprinted from *Carbon* 40, no. 6, T. Ungar, J. Gubicza, G. Ribarik, C. Pantea, and T. Waldek Zerda. Microstructure of carbon blacks determined by x-ray diffraction profile analysis, 929–937. Copyright 2002, with permission from Elsevier.)

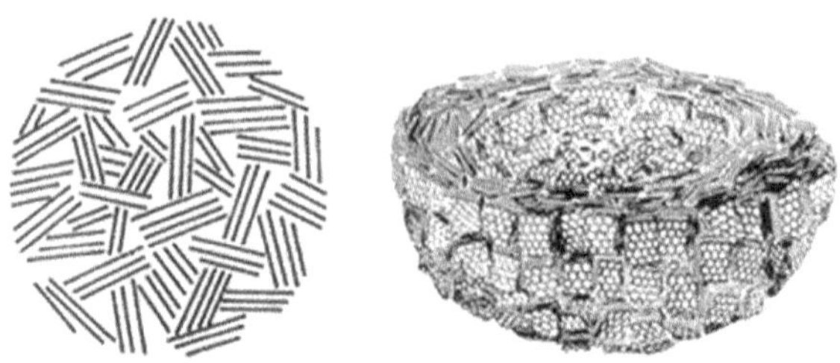

FIGURE 9.6
Schematic model of carbon black microstructure. (From R.D. Heidenreich, W. Hess, and L. Ban, *Journal of Applied Crystallography* 1, no. 1 (1968): 1–19.)

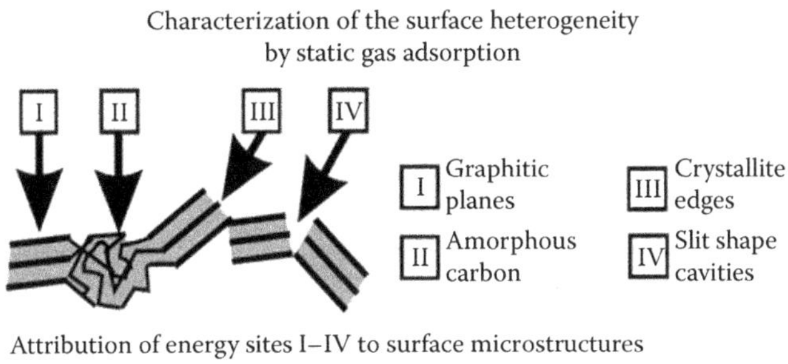

FIGURE 9.7
Schematic model of carbon black surface heterogeneity. (From A. Schröder, *Thesis*, 2000.)

Carbon black therefore is an intermediate structural form between graphite and two-dimensional structure with no three-dimensional ordering.

Figure 9.7 shows a schematic of the surface microstructure or heterogeneity of carbon black in terms of graphitic planes separated with amorphous carbon. Crystalline edges are exposed with slit-type porosity [14].

Upon heat treatment of carbon blacks to high temperatures, up to 3000°C in an inert atmosphere, a limited crystallite growth is observed and it is generally recognized that an interplanar distance of 3.44 Å is a threshold value above which carbons are not graphitizable. Such carbons are known as hard carbons. Carbon blacks are therefore classified as hard carbons.

9.3.2.1 Electron Microscopy

The resolving power of the optical microscope is not high enough to permit the resolution of the particles of carbon black. Early investigators generally observed aggregates, agglomerates, and flocculates instead of particles. Electron microscopy played an important role to show that the gas carbon blacks approximate particle dimensions of 50 to 60 nm. It became soon abundantly clear the shape of the elementary carbon black particle is nearly spherical.

An electron micrograph of a carbon black have aggregate is shown in Figure 9.8. The linear nature of its shape and uniformity of the particles within this aggregate are striking.

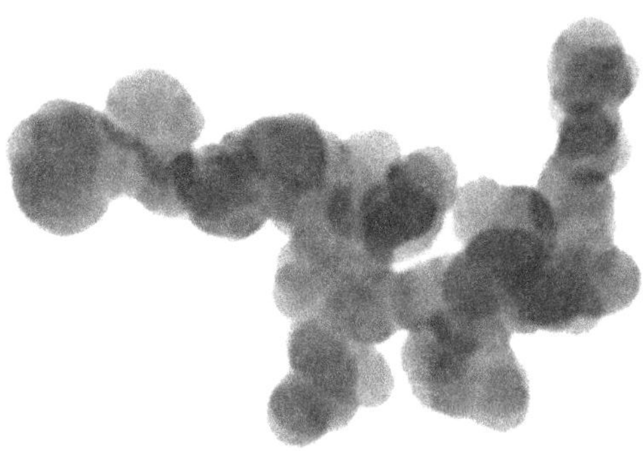

FIGURE 9.8
Micrograph of carbon black aggregate. (© Mitsubishi Chemical Corporation.)

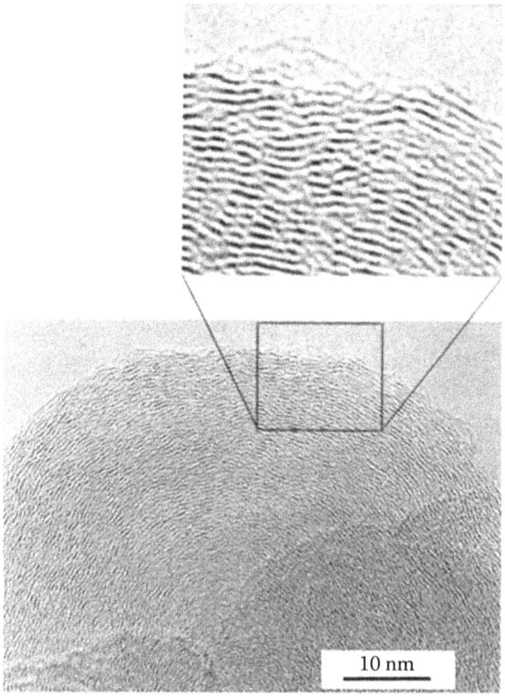

FIGURE 9.9
HRTEM image of commercial carbon black particle. Enlarged area shows individual graphene layers. (From P. J. F. Harris, *Critical Reviews in Solid State and Materials Sciences*, 30: 235–253, 2005.)

Even though the distribution of particle sizes is quite broad, the variability of particle size within an aggregate is relatively small.

In high-resolution electron microscopy (HRTEM) all interplanar areas are imaged including those from curved and bent layer planes, as shown in Figure 9.9. Such distorted layers are generally more widely spaced than parallel planes. By including all layer planes, an average d-spacing of 3.53 Å could be ascertained.

The effect of high temperature heat treatments on carbon materials may provide clues to the structure of the original carbons. It is well established that such treatments transform carbon black particles into faceted particles that sometimes appear to have closed shell-like structures. The presence of faceted planes and closed structures in the heat-treated carbon is indicative of the presence of pentagonal rings, and suggests that fullerene-like elements may have been present in the original particles (Figure 9.10).

9.3.3 Surface Area, Porosity, and Adsorption Properties of Carbon Black

9.3.3.1 Surface Area

The surface is one of the most important characteristics of carbon black, particularly since the particles are very small and the specific surface area, therefore, very large. Since it is impossible to give a complete picture of the texture at each point of the surface, it is necessary to accept a more limited description of surface morphology. In fact, we may determine the specific surface area, the value of the surface area per gram of material as well as the total porosity, the total pore volume fraction. The determination of all these quantities is

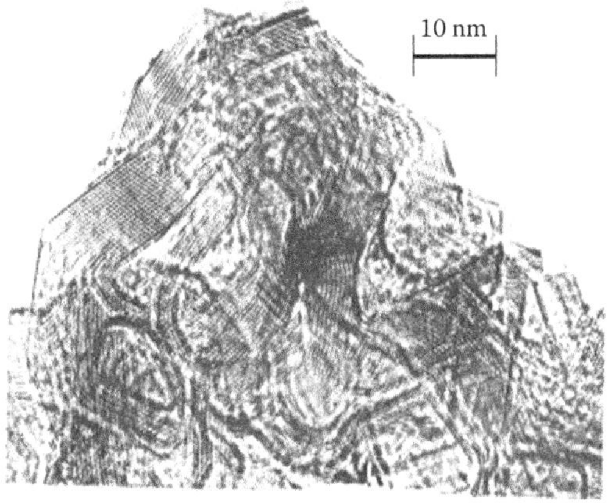

FIGURE 9.10
High-resolution phase contrast electron micrograph of graphitized carbon black. (From J.-B. Donnet, R.C. Bansal, and M.-J. Wang. *Carbon Black*, 2nd ed. Marcel Dekker, New York, 1993.)

primarily based on gas adsorption measurements, for example, BET surface area determination. Different gases may be used, such as nitrogen, argon, or helium, depending upon the pore sizes one is looking for. Adsorption from solution is widely used for the determination of the surface area of carbon blacks such as iodine adsorption method ASTMD1510.

9.3.3.2 Porosity

Since carbon black forms in concentric layers (similar to an onion), the outermost layers are under less strain and are more stable than inner layers. For this reason, once the reaction gases gain entry to the interior of the particles through pores they will preferentially hollow out the core.

9.3.3.3 Surface Activity

Surface activity is a general term describing the surface properties of carbon black to interact with its surroundings. It is the most difficult of the fundamental properties of carbon black to characterize. Some of the more important measurements include hydrogen contents, crystallite size, oxygen contents, and heats of adsorption. Hydrogen content is highly significant but may only be an indicator of relative surface activity, that is, when the hydrogen is removed from a carbon black (at elevated temperature) the graphite layers are free to join into a more continuous and inert surface microstructure. This results in increased crystallite dimensions La and Lc. Therefore, large Lc values and low hydrogen content are indicative of low surface activity.

In polymers, more specifically, elastomers where reinforcing effects are important, surface activity can be defined as the ability of the carbon black surface to crosslink or immobilize the polymer. This property limits slippage and dewetting under strain and is a requirement for developing high modulus and tensile strength.

In liquid vehicle systems, such as paints and inks, surface activity is often more closely associated with the stability of the final dispersion as it is used in the printing or coating process. Carbon black surface contains various oxygen functional groups (acidic), which interacts specifically with the functional groups present in the solvent. The affinity of carbon black with inks or paint varnishes changes depending on the type and amount of the functional groups. In liquid systems, high carbon black–matrix interaction enhances the microscopic dispersion of the black, hence reducing viscosity (increasing flow) and enhancing dispersion stability.

9.4 Electrical and Thermal Properties

9.4.1 Electrical Resistivity

Carbon black may be considered as a semiconductor. Their electric properties vary widely from those of other forms of carbon. Mrozowski studied the variation of the electrical resistivity of ground carbons with pressure up to 1000 kg/cm^2 and found generally a linear relationship between the logarithm of the resistivity and the logarithm of the pressure. The results are interpreted on the basis of Holm's contact theory, describing current conductance in terms of through-going paths in the powder. The application of pressure influences the contact area between the particles of the powders by elastic as well as yielding deformation. Special carbon blacks with high electrical conductivities, also in elastomers, combine a high structure with a high porosity and a small particle size, such as Cabot's XC 72 black. An extremely highly conductive black, Ketjen black EC (AK Z0, Holland) consisting predominantly of hollow particles but with moderate structure, emphasizes the importance of interparticle distance for electrical conductivity.

9.4.2 Thermal Conductivity

Whereas the electrical conductivity of carbon blacks has received a great deal of attention, very little has been published about heat conductivity. The heat conduction data for carbon black differ from those of graphite's. Various branches of industry make wide use of technical-grade carbon black as a technological ingredient as thermal insulator, a thermostable material in the tier industry. The effective thermal conductivity of carbon black was determined in 1 mm thick layers [15] and the thermal conductivities in the range 0.3 to 0.4 W/m.K were reported.

9.5 Applications of Carbon Black

Carbon black is truly a remarkable material. It is found to be a raw material of prime importance in the manufacture of rubber tires, without which automobile and air transportation would not be possible. It has fascinated not only those engaged in the technology and application of carbon black, but it has also inspired a great many purely fundamental investigations in research laboratories and institutes.

A primary use of carbon black is as a filler in rubber to improve the strength, stiffness, hardness, and wear and heat resistance. Carbon-black fillers are found in practically every rubber product.

Carbon blacks are currently sold in the form of hundreds of commercial grades, which vary in their particle size and shape, porosity, and surface chemistry. The current worldwide market of carbon black is in excess of 10 million metric tons per year. Carbon black is also in the top 50 industrial chemicals manufactured worldwide, based on annual tonnage. Approximately 90% of carbon black is used in rubber applications, 9% as a pigment, and the remaining 1% as an essential ingredient in hundreds of diverse applications.

The reinforcing action of carbon black in rubber was first discovered in the early 1900s. Improvements of strength and abrasion resistance are two of the most obvious ways in which carbon black reinforces elastomers.

The most important carbon black properties for inks, coatings, and so forth are related to final product appearance (blackness, tone and tint), UV protection, and electrical conductivity.

Carbon black is also the base of most black typographical inks. This is a large market since an average of 16 kg of ink, containing 11%–13% by weight of carbon black, is required for every ton of newsprint. Other applications are found in paints, enamels, lacquers, molded carbon, and electric conductive agents of high-technology materials.

Modern carbon black products are direct descendants of early "lamp blacks" first produced by the Chinese over 3500 years ago. These early lampblacks were not very pure and differed greatly in their chemical composition from current carbon blacks.

9.5.1 Carbon Black Reinforcement of Elastomers

The reinforcing action of carbon black in rubber was first discovered in the early 1900s. Improvements of mechanical strength and of abrasion resistance are two of the most obvious ways in which carbon black affects elastomer reinforcement to impart the necessary durability and strength to these products for longer lifetime and greatly improved performance by distributing and absorbing stress applied to a rubber component. For example, an optimum loading of carbon black can enhance the tensile strength of synthetic, noncrystalline elastomers more than tenfold. A similar performance improvement is possible for the wear resistance of tire treads. Carbon black also reinforces by making rubber much stiffer and less readily deformable under stress. Almost 80% of carbon black production is consumed in the automotive or tire industry.

For a given loading of carbon black, diminishing particle size provides a greater total interface between the carbon black and polymer and also causes the stress distribution to be more uniform when the rubber is deformed. Similar effects are seen with carbon blacks exhibiting a narrow particle size distribution.

Carbon black can also impart electrical conductivity/resistivity to a rubber compound for dissipating static charge in dynamic applications such as automotive belts and green tires.

9.5.2 Strength

Up to a maximum of each grade of carbon black, the increase in carbon black loading, or the number of carbon black aggregates contributes to an increase in the length of the rupture path, which enhances tensile strength. Above this point there are restrictions concerning the ability to wet effectively the carbon black aggregates with polymer. Therefore, further increase in the carbon black loading simply creates flaws that initiate the onset of tensile failure.

9.5.3 Abrasion

Abrasion is one of the most complex rubber failure properties and may simultaneously involve tensile, tear, and fatigue failure in conjunction with chemical and thermal degradation process. Carbon black particle size, structure, and surface activity are all important to abrasion resistance.

9.5.4 Coloring Agent for Ink and Paints

Carbon black is a basic material with a long history, and it has been used as a coloring agent (pigment) since the days before Christ. It has higher tinting strength (blackness) compared to iron black or organic pigments, and is widely used for newspaper inks, printing inks, India inks, and paints. Nowadays there is large consumption of carbon black as black pigment for inkjet ink or toners.

9.5.5 Blackness, Tint, and Tone

The key to understanding any color problem lies in a knowledge of what happens to the energy distribution of incident light. Carbon black is a semiconductor and has a gap of forbidden energies separating the filled valence band with the empty conduction band. If the minimum energy required to transfer an electron from the valence band to the conduction band is below the energy for the longest wavelength of the incident light (~700 nm for infrared), then all visible wavelengths are absorbed and the material appears black. This is the case for carbon black.

Carbon black and dispersions containing them, by nature, strongly absorb all wavelengths of light and therefore are not commonly viewed in transmitted light. However, all carbon blacks are not perfect absorbers. Smaller particle size carbon blacks exhibit higher overall extinction than larger size carbon blacks and therefore develop more jet (blacker) dispersions. The relative absorptivity of carbon blacks is known to increase rapidly from red to blue violet [16]. In other words, the undertone or "color" that a carbon black dispersion would typically exhibit with the wavelength depends on light scattering phenomena.

For very small particle systems, light is scattered rather than reflected. Scattering of light by carbon black dispersions is extremely complex. The small fraction of incident light that is scattered plays an important role in the appearance of carbon black dispersion. When the scattering centers are large relative to the wavelength of light (λ), the scattering has little dependence on wavelength. For somewhat smaller scattering centers, the scattering is proportional to $1/\lambda^2$. In the case of scattering centers, which are small relative to the wavelength of light, the scattering is proportional to $1/\lambda^4$. Therefore, for more blue wavelength (short) radiation is scattered by very small scattering centers than red (long) wavelength.

The undertone of carbon black dispersions depends on the balance between the tendency of the carbon black to absorb versus scatter more than blue wavelengths. If only scattered light is considered, the appearance of systems of small scattering centers would be blue. If only absorbed light is considered, the appearance of carbon black systems would be brown.

9.5.6 Ultraviolet Protection for Polymers

First let us understand the mechanism of polymer degradation. For photochemical degradation to occur, two conditions must be met:

1. The radiation must be absorbed by the polymer.
2. The energy absorbed must be sufficient to cause the degradation to take place.

Plastics containing carbonyl or unsaturated groups readily absorb UV radiations, which is a primary cause of photo degradation. The reaction is autocatalytic in nature, that is, the products that are formed in the early breakdown of the polymer contribute to the further breakdown of the polymer [17].

A major attribute of carbon black is its ability to absorb detrimental UV light and convert it into heat, thereby making plastics, such as polyolefins and polyethylene, more resistant to the UV radiation from sunlight. Fine particle carbon black provides deep jet-black color for plastic products manufactured by injection molding, extrusion, and other processes. In addition to providing a strong UV absorber, which prevents the damaging UV radiation from being absorbed by the polymer, carbon black also acts as an antioxidant. The carbon black surface terminates the oxidative chains and acts as a radical trap and stabilizer.

Important applications requiring long-term UV protection of plastics with carbon black include environmental applications or "weathering" such as pond and landfill liners and barriers to prevent soil erosion. Carbon black-loaded polyethylene cable jacketing is an extremely critical product in terms of UV protection. It is not surprising then that the most recognized study in this area was carried out by Bell Telephone Company during the 1960s [18].

9.5.7 Resin and Film Coloring Agents

Carbon black has high tinting strength and is thermally stable, and therefore it is suitable for coloring resins and films that are heat-formed, when just a small amount is mixed with resins. Resins with carbon black are used in automobile bumpers, wire coverings, and steel pipe linings, which require weather resistance in particular.

Carbon black in coating applications besides the depth of color also provides conductive properties and makes it easy to disperse onto the product. In coatings, oxidized fine particle carbon black is the key to deep jet-black paints. Coarser carbon black is used for tinting and is indispensable for obtaining a desired gray shade or color hue.

Carbon blacks primarily used for tinting are coarse and have broad particle size distributions. Grades with these properties provide good economics coupled with a desired blue tone and best resistance to flooding and flocculation. For use in machine colorant concentrates, coarse carbon blacks, especially the lampblack replacements, are usually specified.

9.5.8 Electric Conductive Agent

Carbon black particles have the graphite-type crystalline structure, providing an excellent electric conductivity. Therefore, carbon black is widely used as conductive filler, being mixed in plastics, elastomer, paints, adhesives, films, and pastes. Fuel caps and fuel-introducing pipes of automobiles, for example, require electric conductivity for preventing static electricity. Therefore, carbon black is used as an excellent antistatic agent.

9.5.9 Electronic Equipment Related Material

Carbon black also provides stable electrical resistance, and therefore is used as electronic equipment related material in various display components, magnetic recording materials, and OA rolls. Furthermore, Mitsubishi Chemical has been developing carbon black with various combined functions for special applications.

9.5.10 Electromagnetic Interference Shielding

Electromagnetic interference (EMI) shielding effectiveness of any material refers to the reflection and/or absorption of electromagnetic radiation by a material, which acts as a shield against the penetration of the radiation through the shield. As an example electromagnetic radiation, particularly that at high frequencies (e.g., radio waves, such as those emanating from cellular phones) tend to interfere with electronics (e.g., computers). In fact, EMI shielding of both electronics and radiation source is gaining increasing importance by governments around the world in view of the high demand of today's society on the reliability of electronics and the rapid growth of radio frequency radiation sources [19].

The primary mechanism of EMI shielding is usually reflection of the radiation by the shield, which must have mobile charge carriers (electrons or holes) that interact with the electromagnetic fields in the radiation. As a result, the shield tends to be electrically conducting, although a high conductivity is not required. A volume resistivity of the order of 1 Ω-cm is typically sufficient. Metals are by far the most common materials for EMI shielding. They function mainly by reflection due to the free electrons in them. However, metal sheets are bulky, so metal coatings made by electroplating or vacuum deposition are commonly used for shielding, though these coatings can sometimes be removed by scratches.

Electromagnetic radiation at high frequencies penetrates only the near surface region of an electrical conductor. This is known as the skin effect. The electric field of a plane wave drops exponentially with increasing penetrating depth into the conductor. The skin depth (δ) for copper is 2.09 mm, whereas it is 0.47 mm for nickel at a frequency of 1 GHz.

Due to the skin effect, a composite material having a conductive filler with a small unit size of the filler is more effective than one having a conductive filler with a large unit size of the filler. Polymer–matrix composites containing conductive fillers are attractive for shielding due to their processability (e.g., moldability), which helps to reduce or eliminate the seams in the housing in the case of metal sheets, which tend to cause leakage of the radiation and diminish the effectiveness of the shield.

For best results the unit size of the filler should be comparable to or less than the skin depth. Therefore, a filler of unit size 1 mm or less is typically preferred. EMI shielding is a rapidly growing application of carbon materials, especially discontinuous short-length carbon fibers (e.g., vapor-grown carbon fiber and carbon nanotubes). For example, a shielding effectiveness of 40 dB at 1 GHz has been attained in a cement–matrix composite containing just 1.5 vol.% discontinuous 0.1 mm diameter carbon filaments [20].

Carbon black, on the other hand, with its nanodimensions is one of the most important and economically viable candidate as a filler material in polymers for EMI shielding, although the dispersion of the filler is more difficult when the filler unit size decreases. In order to compensate small skin depth, a carbon material with high surface area is preferred. The small size of carbon black particles offers high surface area. A seam in a housing that serves as an EMI shield needs to be filled with an EMI gasket (i.e., a resilient EMI shielding material), which is commonly a material based on an elastomer, such as rubber reinforced with nanocarbons. In addition, polymer–matrix composites are attractive for their low density and ease of processability.

9.5.11 Carbon Black as Catalyst Support

In the design of carbon-supported catalyst, three unique virtues of carbon support are exploited:

1. Possibility to tailor pore size distribution and surface area
2. Possibility to modify chemical surface properties
3. Possibility to modify metal/support interaction

One of the reasons for choosing carbon as a catalyst support is the chemical interaction between a carbon support and a catalyst is probably weaker than that found when oxide supports, such as silica and alumina, are used. The strong affinity of most carbons toward oxygen needs more scrutiny. Its accumulation on active sites takes place at room temperature regardless of the nature of the carbon. In high surface-area carbons, which are of greatest interest as catalyst support, surface oxygen may be responsible for as much as 40% of their mass.

The most important current use of carbon as a support is in liquid phase hydrogenation catalyzed by precious metals. In many cases carbon is the only possible choice as a support material; others would dissolve under the often harsh reaction conditions.

9.6 Lampblack

Lampblack is one of the oldest forms of carbon, first produced by the ancient Chinese who made it by collecting the soot from a burning oil lamp. It was used then (and now) as a black pigment for inks and paints.

9.6.1 Production of Lampblack

The lampblack process is the oldest of all processes but is very limited in use. Lampblack is made by burning aromatic oils, such as naphthalene, anthracene, or creosote, in shallow, open pans with limited or controlled air supply. The smoke is drawn off and transported to settling chambers, where the black settles and is predominantly removed. Most current demand of lampblack is related to tint colorant applications for coatings.

It has also traditionally been collected from the inside of oil lamp mantles (the clear glass covering over oil lamps), thus the name. The main criteria for producing it is to burn the material in such a way that combustion is incomplete.

Commercially a preheated feed of petroleum oil is burned in a controlled airflow in a battery of furnaces and the soot generated from the fumes is collected in the colder chamber. During the burning not all of the oil is fully combusted and so some of the hydrocarbons turn into carbon (and also carbon monoxide) instead of carbon dioxide. This is caused by inefficient oxygen reaching the flame. This can be achieved by interrupting the flame with a noncombustible object. The particles are calcined in the flue gas to remove excess residual oil and aromatic compounds and, still entrained by the flue gas, are carried in a large chamber. The sudden expansion of the gases causes the particles to settle by gravity. The feed-to-air ratio in this confined partial combustion is the critical processing factor that controls the characteristics and properties of the resulting lampblack.

Lampblacks tend to be large in particle size (~100–200 nm in diameter) and, to a great degree, control the color and oil adsorption. Its particle shape could be spherical, irregular, needle, or flaky. Lampblack is a soft, flocculent, amorphous material with an apparent density that can vary by a factor of 20, depending on the amount of occluded gases and the shape of the particle. The occluded gases include hydrogen, oxygen, nitrogen, and often sulfur compounds originating from the feed coal tar.

9.6.2 Applications of Lampblack

Lampblack is black pigment and is very simple to turn into an ink, as it does not need further grinding. The simplest ink that can be made from it is Indian ink. This is made by mixing the lampblack with a small amount of alcohol, then adding to this paste a mix of gum Arabic dissolved in water. Lampblack is not at all easily mixed with water. A drop of alcohol helps break the surface tension. The thickness of the ink depends on the amount of gum Arabic used. It produces a very dark black color and should not fade at all, if it is sufficiently absorbed by the paper.

Lampblack can also be used to make oil paint. Lampblack will absorb about 90% to 95% of the incident light and, for most purposes, can be considered a blackbody. Polished metal surfaces, on the other hand, absorb only about 5% of the light. Lampblack's extreme opacity and complete resistance to fading are excellent characteristics for use in the arts. The consumption of lampblack is very extensive in common painting. It helps to modify the brightness of the tones of the other colors. The oil paint applied to iron grates and railing consumes a very large quantity of this black.

9.7 Acetylene Black

Acetylene black is similar to high-grade lampblack but with greater purity, higher liquid-absorption capacity, and higher electrical conductivity because of its unique, three-dimensional structure.

Acetylene black is the specialty grade carbon black and is produced by exothermic decomposition (as by pyrolysis in a retort or by explosion) of acetylene at about 800°C to 1000°C in water-cooled jacketed cylindrical retorts [21]. The reaction produces temperatures at the carbon black surface in excess of 2050°C, which causes partial graphitization of carbon black. Acetylene black is therefore characterized by intermediate particle size, very high structure, and crystallinity by carbon black standards and is mainly used for electric conductive agents. Acetylene black is the purest form of carbon black and must not be confused with the carbon black produced as a by-product during the production of acetylene in the electric arc process.

9.7.1 Applications of Acetylene Black

Because of high graphitic surface, acetylene black exhibits low reactivity and high electrical conductivity. This type of carbon black is therefore used mostly in applications requiring chemical inertness, high electrical, and thermal conductivity.

The principle applications of acetylene black are in the production of dry cells, mainly to improving the depolarizing properties of manganese dioxide. Because of the high electrical conductivity required for reducing the internal resistance of the cell's higher adsorptive capacity for electrolytes and also excellent fluid retention, a required property for electrolytic cells, it is widely used in carbon-zinc cells, dry cells, and batteries. It is also used as a filler in rubber and plastic materials, as an electrically conducting filler in coatings covering conductors and in field boundary layers in high-voltage cables.

Acetylene black is also used as pigment in electrically conducting paints, magnetic tape coatings, conductive concrete, spark plug wires, antistatic elastomerics, rocket propellants, satellite shielding, grease and lubricants, and rocket boosters etc.

9.8 Carbon Nanospheres

Since the discovery of buckminsterfullerene, spherical carbon structures are receiving increased attention from the scientific community [22]. Apart from the filamentous arrangement of the graphene sheets, CNTs, and CNF, carbon can bond in other different ways to create structures with dissimilar properties. The pairing of pentagonal and heptagonal carbon rings can result in the formation of carbon nanospheres (CNSs). This novel nanostructure has only now started to attract significant research activity. In its spherical arrangement, the graphite sheets are not closed shells but rather waving flakes that follow the curvature of the sphere, creating many open edges at the surface. Inagaki classified these spherical nanostructures according to their nanometrical texture, proposing three categories: concentric, radial, or randomly oriented carbon layers [23]. Serp et al. [24] in turn divided these spheres into three groups according to their size, (1) the Cn family and graphitic carbon onions, with diameter in the range 2 to 20 nm; (2) less graphitic carbon spheres with diameters in the range of 50 nm to 1 micrometer; and (3) carbon beads with diameters from one to several microns. Contrary to chemically inert C_{60}, the unclosed graphite flakes provide reactive "dangling bonds" that are proposed to enhance surface reactions, establishing CNS as a good candidate for catalytic and adsorption applications, in composite materials, and energy storage [25].

9.8.1 Growth Model

A carbon sphere is believed to be nucleated from a pentagonal carbon ring followed by a spiral shell growth. This is the growth mechanism first proposed for the formation of large carbon fullerenes. This growth core is believed to be formed in or rather on the solid phase under the action of the oxygen in the lattice of the catalyst, after which the sphere, once formed, separates from the catalyst. The lattice oxygen provided by the catalyst is required to oxidize CH_4. The spiral growth is the key for the nucleation of the sphere, but it is not adequate for growing large spheres because of the limited growth rate along the spiral direction. When the sphere grows larger, graphitic flakes of atom thickness are nucleated on the surface due to the arrival of the pentagonal, hexagonal, and heptagonal carbon rings. A flat graphite layer, on the other hand, is composed of hexagonal carbon rings. The pentagonal, hexagonal, and heptagonal carbon rings are required to modify the graphitic flakes for accommodating the local curvature of the carbon sphere. Geometrical combinations of these carbon rings can make all the structural configurations required to accommodate the surface curvature. Therefore, the spheres are composed of curled graphitic flakes with random twisting because there is almost no correlation among the graphitic flakes.

9.8.2 Synthesis of Carbon Nanospheres

In general, there are three approaches to the preparation of the carbon nanospheres. First, they can be directly synthesized using methods such as chemical vapor deposition, pyrolysis, and hydrothermal treatment. Second, the templating technique using hard templates can be employed. The third method is to first synthesize polymer spheres followed by a thermal treatment in an inert atmosphere. In the third method the size of the CNS could potentially be controlled by tuning the diameter of the polymer nanospheres, which subsequently can be converted into CNS with a controlled particle size after pyrolysis.

9.8.2.1 The CVD Route

The catalytic chemical vapor deposition (CVD) route has been mainly used in the growth of structured carbon filaments, such as CNT and CNF. Knag and Wang grew solid graphitic nanosized spheres (CNS) via the decomposition of methane (in hydrogen) over an iron catalyst at 1100°C and benzene at 1000°C [25–28]. The unexpected shift from the filamentous to spherical carbon was also reported by the decomposition of chlorohydrocarbons such as C_2HCl_3, in the absence of catalyst over the bare silica support [29,30]. This research suggests that the growth of CNS is highly temperature demanding and while the catalyst may play some role in assisting carbon growth and determining some lattice structure, further growth may proceed by a pyrolytic mechanism.

9.8.2.2 Hydrothermal Route

Hydrothermal routes have also been followed for the preparation of hollow carbon nanospheres. Hollow carbonaceous spherules with diameters of a few microns were prepared using glucose as a starting material and sodium dodecyl sulfate as an anionic surfactant [31] before pyrolysis. These could have appreciable surface area ~250 m^2/g.

9.8.2.3 Template Route

A template route has also been commonly used in the synthesis of CNS. In their studies Li et al. prepared mesoporous CNS using melamine-formaldehyde (MF) resin as the carbon precursor and silica as the template. The MF–silica template was carbonized at 800°C–1000°C to obtain the final carbon spheres and the silica template was removed after a hydrofluoric acid (HF) treatment. The spheres with diameter 1.2 microns, with relatively low degree of graphitization and high surface area were obtained [32,33]. Fujikawa et al. used resorcinol-formaldehyde to grow 160–350 nm carbon spheres using the same template method followed by carbonization to 1000°C.

The aforementioned results evidences the huge experimental effort that is underway in order to approach a suitable, economically feasible, and controllable route for CNS production. Despite the large number of preparation routes proposed, highly demanding conditions (temperature and pressure) are commonly required. The presence of catalyst—expected to allow for milder conditions—has been demonstrated to play a minor role.

Only the hydrothermal route stands as a mild, green, fast, and cheap technique; however, only hollow spheres have been obtained by this method.

9.8.3 Microstructure of Carbon Nanospheres

Generally, CNSs are not present as discrete entities but rather as a conglomeration of spherical bodies. Such agglomerates of spheres have been proposed to be the result of extended reaction time as well as consequence of the cooling from synthesis to room temperature. In any case, the coalescence of spheres must be attributed to the presence of reactive dangling bonds on their surface, providing them with a high surface reactivity. Figure 9.11 shows representative scanning electron microscope (SEM) and transmission electron microscope (TEM) micrographs corresponding to carbon nanosized spheres [24].

The detailed analysis of the edges of this material under a high-resolution electron microscope (HREM) revealed (Figure 9.12) that the CNSs consist of a fairly concentric arrangement of carbon layers [24]. From the HRTEM micrographs, it does not appear that these nanometric structures develop any internal porosity.

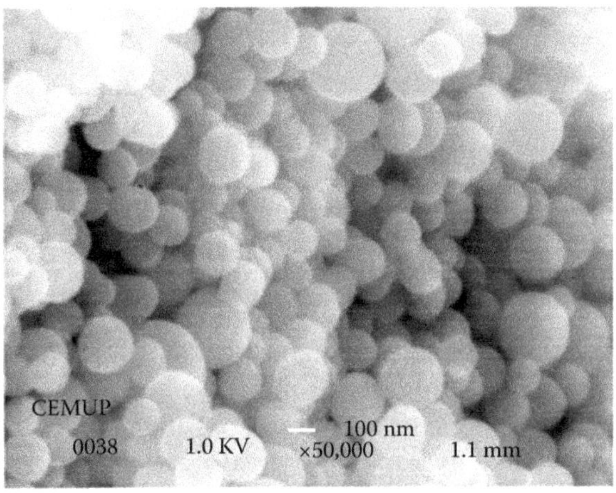

(a)

(b)

FIGURE 9.11
(a) SEM image of nanosphere and (b) TEM low magnification image of nanospheres. (Reprinted from *Carbon* 39, no. 4, P. Serp, R. Feurer, P. Kalck, Y. Kihn, J.L. Faria, and J.L. Figueiredo. A chemical vapour deposition process for the production of carbon nanospheres, 621–626. Copyright 2001, with permission from Elsevier.)

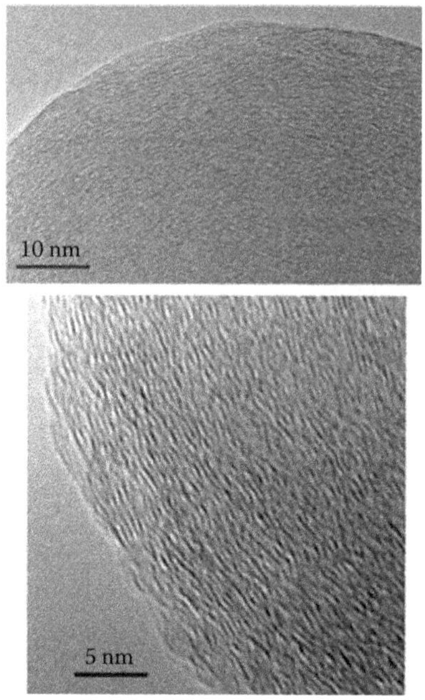

FIGURE 9.12
HRTEM images of the edge of the nanospheres. (Reprinted from *Carbon* 39, no. 4, P. Serp, R. Feurer, P. Kalck, Y. Kihn, J.L. Faria, and J.L. Figueiredo. A chemical vapour deposition process for the production of carbon nanospheres, 621–626. Copyright 2001, with permission from Elsevier.)

Brooks and Taylor proposed a radial arrangement of the graphene layers in these materials, i.e., circular conical surfaces with a common vertex in the center of the spherule.

Regardless of the solid or hollow nature of CNS these present poor graphitization degree as commonly evaluated by x-ray powder diffraction (XRD) and Raman spectroscopy [34]. As shown in Figure 9.13a, XRD patterns associated with CNSs are characterized by a main diffraction peak at $2\theta = 25°$ corresponding to the (002) graphite plane, as well

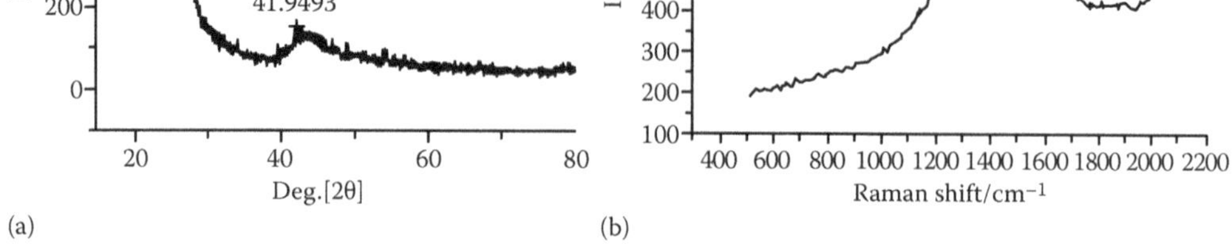

FIGURE 9.13
Representative (a) XRD and (b) Raman spectrum associated with CNS. (Reprinted from *Carbon* 42, no. 4, H.-S. Qian, F.-M. Han, B. Zhang, Y.-C. Guo, J. Yue, and B.-X. Peng. Non-catalytic CVD preparation of carbon spheres with a specific size, 761–766. Copyright 2004, with permission from Elsevier.)

as a secondary peak at ~42° that can be assigned to the (100)/(101) graphite planes. The broadness of these peaks is indicative of a short-range structural order. Interlayer spacing d_{002} is of the order of 3.35–3.60 Å for different CNSs produced by different routes. The average interplane distance between layers has been estimated by x-ray powder diffractometry that indicates a d_{002} = 3.52 Å. This value is intermediate between the d_{002} = 3.42 Å measured for as-grown VGCF and the d_{002} = 3.63 Å for a commercial Hydraffin-activated carbon. These values indicate that the nanospheres consist of a relatively well-organized carbon structure.

In the Raman spectra of CNSs (Figure 9.13b), the D band is associated with the vibrations of carbon atoms with dangling bonds for the in-plane termination of disordered graphite, and the G band corresponds to the E_2g mode of graphite. Subsequently the relative intensities of the D and G bands (I_D/I_G) are commonly used as an index to assess graphitic character. The value (I_D/I_G) for CNSs ranges around 0.8. Regarding the porous structure of CNSs, a wide range of specific surface areas and pore geometries has been recorded in the literature. Carbon nanospheres obtained by high temperature techniques, that is, catalytic chemical vapor decomposition (CCVD), thermal pyrolysis, or pressure carbonization generally show low (5–15 m^2/g) surface area and negligible porosity. However, CNSs obtained by reduction or template route have shown surface areas from a few hundreds to more than a thousand meter square per gram of a solid.

It is found that different production routes produce wide and heterogeneous structure with spherical morphology with relatively poor crystallinity in comparison with other carbon nanostructures, such as CNTs or CNFs. Proper selection of synthesis conditions is therefore required to tailor the CNS for different applications.

9.8.4 Applications of Carbon Nanospheres

Carbon nanospheres can be potentially excellent catalyst support materials. Because of its spherical arrangement, the graphene sheets that form carbon nanospheres are normally unclosed shells with rather waving flakes that follow the curvature of the sphere, creating many open edges at the surface. These create dangling bonds that provide the spheres with high chemical reactivity establishing them as good candidates for catalysis. Kang and Wang [25] dispersed Pt, CdS, and WO_3 nanoparticles, and demonstrated that electron transfer was possible in those systems, indicating their use as catalyst supports. Nitrogen doping of the CNSs strongly influenced the catalytic activity. CNSs have been successfully used as Pt carriers for the electrooxidation of methanol in direct methanol fuel cells (DMFC). The behavior of this catalyst was compared with that of Pt supported on commercial MCMB and Pt/Vulcan XC-72, showing a better catalytic activity, given the enhanced mass transport of methanol in the electrode due to the monodispersion of the spheres.

Li et al. supported Pt on nitrogen-doped nanospheres and reported how proper control of nitrogen doping had a great impact on both Pt dispersion and CO tolerance, outperforming the catalytic behavior of the commercial Pt/Vulcan XC-72 [35] in fuel cells.

Carbon nanospheres have also been tested as electrodes for electrical double layer capacitor (EDLC), as an alternative energy storage device to secondary batteries.

In addition, CNSs have been used as anode for lithium-ion batteries, exhibiting great electrochemical properties for high rate use as well as good stability toward graphite exfoliation. SnO_2–CNS have also shown improved cyclability and high charge/discharge capacity [31].

CNSs have also been used for applications as fillers in nanocomposites as EMI shielding material. Addition to the polymer matrix even in very small concentrations, strongly modifies the low frequency response of the parent polymer as well as improving its thermal stability [36].

Other applications include Ni-loaded, highly porous CNSs for hydrogen storage, drug delivery systems, lubricating properties, and rubber reinforcement. Photoluminescence and multiphoton bioimaging could be some of the other potential applications; however, none have yet been developed on a commercial scale.

9.9 Carbon Nano-Onions

Carbon nano-onions (CNOs), which consist of concentric graphitic shells, represent another new allotropic nanophase of carbon materials. CNOs have already been shown to offer a variety of potential applications such as solid lubrication [37], electromagnetic shielding [38], fuel cells [39], heterogeneous catalysis [40], gas and energy storage [41], and electrooptical devices [42] owing to their outstanding chemical and physical properties. According to a recent study, CNOs can also be used to produce ultrahigh power micrometer-sized super capacitors due to their accessible external surface area for ion adsorption [43].

The closed shell structure of CNOs is expected to have chemical characteristics very different from those with a graphitic flake structure of the spheres. The closed shell structure is likely to be chemically inert because the energy required to break the in-plane covalent bonds of graphite is high, and the carbon atoms located in the graphitic planes have no dangling bonds to combine with other atoms.

Based on their structure these can be considered as concentric, nested hyperfullerenes of carbon, of which C_{60} is the smallest with a concentric arrangement of graphene layers. In the particular case of carbon onions, the large surface-to-volume ratio contributes to the total stability of the structure, eliminating dangling bonds.

9.9.1 Synthesis Route for Carbon Nano-Onions

Since the first observation of the onion-like structure by Iijima in vacuum-deposited amorphous carbon films in 1980 [44], several methods have been reported to prepare CNOs, for example, Ugarte examined the formation of CNOs in a transmission electron microscope (TEM) by irradiating carbon material consisting of different carbon nanostructures (fullerenes, nanotubes, amorphous carbon) [22]. Kuznetsov et al. obtained CNOs by high-temperature annealing of diamond nanoparticles under vacuum [45]. Cabioc'h et al. prepared CNOs by high-dose carbon ion implantation into copper and silver [46]. Sano et al. fabricated CNOs by arc discharge between two graphite rods immersed in water [47]. Recently, Hou et al. reported the high-yield synthesis of CNOs in counterflow diffusion flames [48,49]. In addition, chemical vapor deposition was also considered as a viable method to synthesize CNOs by different groups. Other ways have also been proposed to synthesize CNOs, such as laser irradiation, shock compression, and high-energy ball milling. However, most of the aforementioned methods require high energy input, and CNOs are sometimes of low yield or just as a by-product [50,51]. For applications such as fuel-cell electrodes, large quantities (kilograms) of the material are desired. Nevertheless, CNOs can only be produced in minute quantities by most of the methods discussed earlier. Moreover, it is not practical to use the CNOs obtained by some methods requiring high investment and running costs for ordinary applications such as solid lubrication. Therefore, how to economically synthesize CNOs in large scale is of great importance for their wide applications. A simple, efficient, and economical route for large-scale preparation of CNOs with uniform morphology and size is via the reaction between CaC_2 and $CuCl_2.32H_2O$ at 600°C [52]. The crystal water in $CuCl_2.32H_2O$ plays an important role in the formation of CNOs.

The crystal water in $CuCl_2.32H_2O$ at 110°C immediately reacts with CaC_2 to produce acetylene (C_2H_2). The reaction between water and CaC_2 occurs quite readily even under ambient atmosphere and thus is prior to the reaction between CaC_2 and $CuCl_2$. The overall reaction can be written as

$$CuCl_2.2H_2O + CaC_2 = CaCl_2 + Cu + 2C + 2H_2O \tag{9.3}$$

It is well known that Cu has a very strong effect on graphitization of carbon. Accordingly, the carbon atoms generated will grow into graphitic flakes under the catalysis of Cu. The graphitic flakes obtained are the combination of pentagonal and heptagonal structures together with the planar hexagonal structure of graphene so that they can accommodate the curvature of the CNOs. With continuous heat treatment, the graphitic flakes aggregate into quasi-spherical graphitic shells, giving rise to the formation of CNOs. The multiple shell structure is more energetically favorable than a single or tubular shell. Meanwhile, this process can also result in the elimination of dangling bonds at the edge of the flakes, which can undoubtedly lower the energy of the system.

9.9.2 Classification, Structure, and Physicochemical Properties of Carbon Nano-Onions

The morphology and structure of the obtained products were investigated by field-emission scanning electron microscope (FESEM), high-resolution transmission electron microscopy (HRTEM), x-ray diffraction, Raman spectrometry, and nitrogen adsorption. Large quantities of CNOs consisting of quasi-spherically concentric graphitic shells with high purity and uniform size distribution (about 30 nm) were obtained and are shown in Figure 9.14. The examination by HRTEM (Figure 9.14c) demonstrates that the spheres have an onion-like structure. The interplanar spacing of 0.34 nm corresponds to that of the (002) plane of graphite [52].

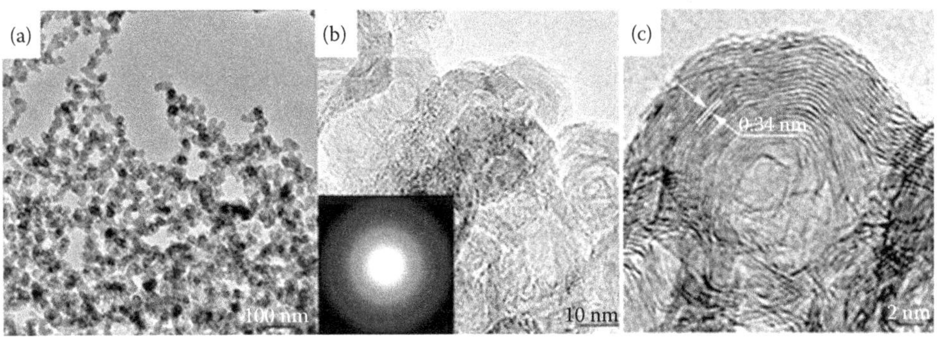

FIGURE 9.14
(a) FESEM image, (b) energy-dispersive x-ray spectroscopy (EDX) spectrum, and (c) HRTEM images of the as-obtained CNOs. The inset in (b) is the corresponding SAED pattern. (Reprinted with permission from F.-D. Han, B. Yao, and Y.-J. Bai. Preparation of carbon nano-onions and their application as anode materials for rechargeable lithium-ion batteries. *The Journal of Physical Chemistry* C 115, no. 18: 8923–8927. Copyright 2011 American Chemical Society.)

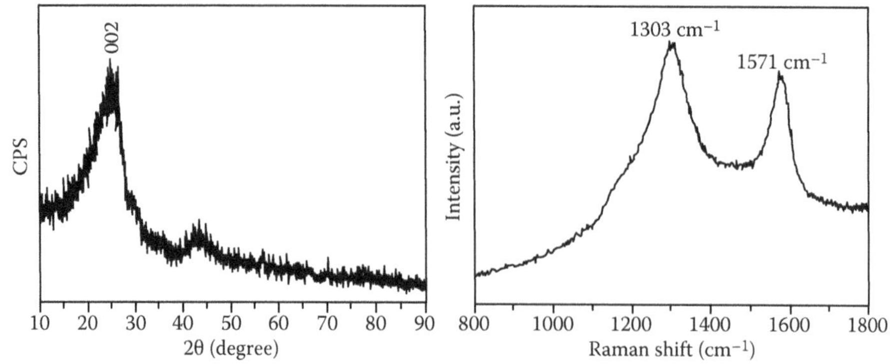

(a) (b)

FIGURE 9.15
(a) XRD pattern and (b) Raman spectrum of the as-obtained CNOs. (Reprinted with permission from F.-D. Han, B. Yao, and Y.-J. Bai. Preparation of carbon nano-onions and their application as anode materials for rechargeable lithium-ion batteries. *The Journal of Physical Chemistry C* 115, no. 18: 8923–8927. Copyright 2011 American Chemical Society.)

The short-range order or turbostratic graphite structure gives rise to concentric graphite layers around a hollow and irregular core about 5 nm in diameter. The corresponding selected area electron diffraction (SAED) pattern (inset of Figure 9.14b) reveals the poor degree of graphitization of the CNOs.

Figure 9.15a shows XRD measurements conducted on CNOs. The main peak at $2\theta = 25.8°$ can be attributed to the (002) diffraction of hexagonal graphite. The broadening nature of the peak is indicative of the long-range disorder structure of the as-obtained CNOs due to their small sizes. Moreover, the disorder structure can be also reflected by the broad band centered at $2\theta = 43.7°$ showing indistinguishability of the (100) and (101) peaks. Raman spectra is used to identify the bonding and structure of the as-obtained CNOs and is shown in Figure 9.15b. The two broad peaks centered at about 1303 and 1571 cm^{-1} are ascribed to the D peak for disordered carbon and the G peak for graphite carbon, respectively. The intensity ratio (I_D/I_G) in Raman spectra is a typical parameter to quantify the disorder degree of carbon materials, that is, a greater value of the ratio means a higher disorder degree for graphite.

9.9.3 Applications

The CNOs as obtained exhibit high capacity and excellent cycling performance as anode materials for lithium-ion batteries, which can deliver a reversible capacity of 391 mAh/g up to 60 cycles [52]. Readers are referred to the literature on various other applications of CNOs [37–43].

References

1. J. Robertson. Amorphous carbon. *Advances in Physics* 35, no. 4 (1986): 317–374.
2. A.G. Gaydon and H.G. Wolfhard. In *Flames*, edited by C.A. Halleds. London, 1960.
3. J.-B. Donnet, R.C. Bansal, and M.-J. Wang. *Carbon Black: Science and Technology*. Marcel Dekker, 1993.
4. H. Marsh, E.A. Heintz, and F. Rodriguez-Reinoso (Eds.). *Introduction to Carbon Technologies*. Secretariado de Publications, 1997.

5. S.J. Harris and A.M. Weiner. Soot particle growth in premixed toluene/ethylene flames. *Combustion Science and Technology* 38, no. 1–2 (1984): 75–87.
6. J. Lahaye. Particulate carbon from the gas phase. *Carbon* 30, no. 3 (1992): 309–314.
7. J. Lahaye and G. Prado. *Chemistry and Physics of Carbon*, 167. Dekker, 1978.
8. K. Homann. Carbon formation in premixed flames. *Combustion and Flame* 11, no. 4 (1967): 265–287.
9. H.F. Calcote. Mechanisms of soot nucleation in flames—A critical review. *Combustion and Flame* 42 (1981): 215–242.
10. B. Warren. X-ray diffraction in random layer lattices. *Physical Review* 59, no. 9 (1941): 693.
11. B. Warren. X-ray diffraction study of carbon black. *The Journal of Chemical Physics* 2, no. 9 (1934): 551–555.
12. T. Ungar, J. Gubicza, G. Ribarik, C. Pantea, and T. Waldek Zerda. Microstructure of carbon blacks determined by x-ray diffraction profile analysis. *Carbon* 40, no. 6 (2002): 929–937.
13. R.D. Heidenreich, W. Hess, and L. Ban. A test object and criteria for high resolution electron microscopy. *Journal of Applied Crystallography* 1, no. 1 (1968): 1–19.
14. A. Schröder, *Thesis*, 2000.
15. P.E. Khizhnyak, A.V. Chechetkin, and A.P. Glybin. Thermal conductivity of carbon black. *Inzhenerno-Fizicheski Zhurnal* 37 (1979): 475–478.
16. J. Donnet and A. Voet. *Carbon Black*, 149–153. Marcel Dekker, 1976.
17. A.W. Birley, B. Haworth, and J. Batchelor. *Physics of Plastics: Processing, Properties, and Materials*. Hanser, 1991.
18. J. Howard and H. Gilroy. Natural and artificial weathering of polyethylene plastics. *Polymer Engineering & Science* 9, no. 4 (1969): 286–294.
19. D. Chung. Electromagnetic interference shielding effectiveness of carbon materials. *Carbon* 39, no. 2 (2001): 279–285.
20. X. Fu and D. Chung. Submicron-diameter-carbon-filament cement-matrix composites. *Carbon* 36, no. 4 (1998): 459–462.
21. V. Schwob. In *Chemistry and Physics of Carbon*, vol. 15, edited by P.L. Walker and P.A. Thrower, 109. Marcel Dekker, 1982.
22. D. Ugarte. Curling and closure of graphitic networks under electron-beam irradiation. *Nature* 359, no. 6397 (1992): 707–709.
23. M. Inagaki. Discussion of the formation of nanometric texture in spherical carbon bodies. *Carbon* 35, no. 5 (1997): 711–713.
24. P. Serp, R. Feurer, P. Kalck, Y. Kihn, J.L. Faria, and J.L. Figueiredo. A chemical vapour deposition process for the production of carbon nanospheres. *Carbon* 39, no. 4 (2001): 621–626.
25. Z. Kang and Z. Wang. Chemical activities of graphitic carbon spheres. *Journal of Molecular Catalysis A: Chemical* 118, no. 2 (1997): 215–222.
26. P. Wang, J. Wei, B. Huang, X. Qin, S. Yao, Q. Zhang, Z. Wang, G. Xu, and X. Jing. Synthesis and characterization of carbon spheres prepared by chemical vapour deposition. *Materials Letters* 61, no. 26 (2007): 4854–4856.
27. Z. Wang and Z. Kang. Graphitic structure and surface chemical activity of nanosize carbon spheres. *Carbon* 35, no. 3 (1997): 419–426.
28. Z. Wang and Z. Kang. Pairing of pentagonal and heptagonal carbon rings in the growth of nanosize carbon spheres synthesized by a mixed-valent oxide-catalytic carbonization process. *The Journal of Physical Chemistry* 100, no. 45 (1996): 17725–17731.
29. A. Nieto-Márquez, J.L. Valverde, and M.A. Keane. Catalytic growth of structured carbon from chloro-hydrocarbons. *Applied Catalysis A: General* 332, no. 2 (2007): 237–246.
30. A. Nieto-Márquez, J.L. Valverde, and M.A. Keane. Selective low temperature synthesis of carbon nanospheres via the catalytic decomposition of trichloroethylene. *Applied Catalysis A: General* 352, no. 1 (2009): 159–170.
31. X.W. Lou, C.M. Li, and L.A. Archer. Designed synthesis of coaxial SnO2@ carbon hollow nanospheres for highly reversible lithium storage. *Advanced Materials* 21, no. 24 (2009): 2536–2539.

32. W. Li, D. Chen, Z. Li, Y. Shi, Y. Wan, J. Huang, J. Yang, D. Zhao, and Z. Jiang. Nitrogen enriched mesoporous carbon spheres obtained by a facile method and its application for electrochemical capacitor. *Electrochemistry Communications* 9, no. 4 (2007): 569–573.
33. W. Li, D. Chen, Z. Li, Y. Shi, Y. Wan, G. Wang, Z. Jiang, and D. Zhao. Nitrogen-containing carbon spheres with very large uniform mesopores: The superior electrode materials for EDLC in organic electrolyte. *Carbon* 45, no. 9 (2007): 1757–1763.
34. H.-S. Qian, F.-M. Han, B. Zhang, Y.-C. Guo, J. Yue, and B.-X. Peng. Non-catalytic CVD preparation of carbon spheres with a specific size. *Carbon* 42, no. 4 (2004): 761–766.
35. Z. Lei, M. Zhao, L. Dang, L. An, M. Lu, A.-Y. Lo, N. Yu, and S.-B. Liu. Structural evolution and electrocatalytic application of nitrogen-doped carbon shells synthesized by pyrolysis of near-monodisperse polyaniline nanospheres. *Journal of Materials Chemistry* 19, no. 33 (2009): 5985–5995.
36. J. Macutkevic, P. Kuzhir, D. Seliuta, G. Valusis, J. Banys, A. Paddubskaya et al. Dielectric properties of a novel high absorbing onion-like-carbon based polymer composite. *Diamond and Related Materials* 19, no. 1 (2010): 91–99.
37. A. Hirata, M. Igarashi, and T. Kaito. Study on solid lubricant properties of carbon onions produced by heat treatment of diamond clusters or particles. *Tribology International* 37, no. 11 (2004): 899–905.
38. O. Shenderova, V. Grishko, G. Cunningham, S. Moseenkov, G. McGuire, and V. Kunetsov. Onion-like carbon for terahertz electromagnetic shielding. *Diamond and Related Materials* 17, no. 4 (2008): 462–466.
39. B. Xu, X. Yang, X. Wang, J, Guo, and X. Liu. A novel catalyst support for DMFC: Onion-like fullerenes. *Journal of Power Sources* 162, no. 1 (2006): 160–164.
40. N. Keller, N.I. Maksimova, V.V. Roddatis, M. Schur, G. Mestl, Y.V. Butenko, V.L. Kuznetsov, and R. Schlogl. The catalytic use of onion-like carbon materials for styrene synthesis by oxidative dehydrogenation of ethylbenzene. *Angewandte Chemie International Edition* 41, no. 11 (2002): 1885–1888.
41. N. Sano, H. Wang, I. Alexandrou, M. Chhwolla, K.B.K. Teo, G.A.J. Amaratunga, and K. Iimura. Properties of carbon onions produced by an arc discharge in water. *Journal of Applied Physics* 92, no. 5 (2002): 2783–2788.
42. E. Koudoumas, O. Kokkinaki, M. Konstantaki, S. Couris, S. Korovin, P. Detkov, V. Kuznetsov, S. Pimenov, and V. Pustovoi. Onion-like carbon and diamond nanoparticles for optical limiting. *Chemical Physics Letters* 357, no. 5 (2002): 336–340.
43. D. Pech, M. Brunet, H. Durou, P. Huang, V. Mochalin, Y. Gogotsi, P.-L. Taberna, and P. Simon. Ultrahigh-power micrometre-sized supercapacitors based on onion-like carbon. *Nature Nanotechnology* 5, no. 9 (2010): 651–654.
44. S. Iijima. Direct observation of the tetrahedral bonding in graphitized carbon black by high resolution electron microscopy. *Journal of Crystal Growth* 50, no. 3 (1980): 675–683.
45. V.L. Kuznetsov, A.L. Chuvilin, Y.V. Butenko, I.Y. Mal'kov, and V.M. Titov. Onion-like carbon from ultra-disperse diamond. *Chemical Physics Letters* 222, no. 4 (1994): 343–348.
46. T. Cabioc'h, M. Jaouen, E. Thune, P. Guerin, C. Fayox, and M.F. Denanot. Carbon onions formation by high-dose carbon ion implantation into copper and silver. *Surface and Coatings Technology* 128 (2000): 43–50.
47. N. Sano, H. Wang, M. Chhowalla, I. Alexandrou, and G.A.J. Amaratunga. Nanotechnology: Synthesis of carbon "onions" in water. *Nature* 414, no. 6863 (2001): 506–507.
48. S.-S. Hou, D.-H. Chung, and T.-H. Lin. High-yield synthesis of carbon nano-onions in counterflow diffusion flames. *Carbon* 47, no. 4 (2009): 938–947.
49. D.-H. Chung, T.-H. Lin, and S.-S. Hou. Flame synthesis of carbon nano-onions enhanced by acoustic modulation. *Nanotechnology* 21, no. 43 (2010): 435604.
50. A.G. Nasibulin, A. Moisala, D.P. Brown, and E.I. Kauppinen. Carbon nanotubes and onions from carbon monoxide using $Ni(acac)_2$ and $Cu(acac)_2$ as catalyst precursors. *Carbon* 41, no. 14 (2003): 2711–2724.

51. J. Kang, J. Li, X. Du, C. Shi, N. Zhao, and P. Nash. Synthesis of carbon nanotubes and carbon onions by CVD using a Ni/Y catalyst supported on copper. *Materials Science and Engineering: A* 475, no. 1 (2008): 136–140.
52. F.-D. Han, B. Yao, and Y.-J. Bai. Preparation of carbon nano-onions and their application as anode materials for rechargeable lithium-ion batteries. *The Journal of Physical Chemistry C* 115, no. 18 (2011): 8923–8927.

Further Reading

Donnet, J.-B. Fifty years of research and progress on carbon-black. *Carbon* 32 (1994): 1305–1310.

Donnet, J.-B., R.C. Bansal, and M.-J. Wang. *Carbon Black*, 2nd ed. Marcel Dekker, 1993.

Harris, P.J.F. New perspectives on the structure of graphitic carbons. *Critical Reviews in Solid State and Materials Sciences* 30 (2005): 235–253.

Harris, S.J. and A.M. Weiner. Chemical kinetics of soot particle growth. *Annual Review of Physical Chemistry* 36 (1985): 31–52.

Palmer, H.P. and C.F. Cullis. The formation of carbon from gases. In *Chemistry and Physics of Carbon, vol. 1,* edited by P.L. Walker, 265. Marcel Dekker, 1965.

Index

D

E

F

G

H

www.ingramcontent.com/pod-product-compliance
Lightning Source LLC
Chambersburg PA
CBHW081140120126
38069CB00011B/3259
* 9 7 8 1 4 9 8 7 0 2 1 0 2 *